SOILS AND QUATERNARY
LANDSCAPE EVOLUTION

SOILS AND QUATERNARY LANDSCAPE EVOLUTION

Edited by
JOHN BOARDMAN
Humanities Department, Brighton Polytechnic

Sponsored by the
Quaternary Research Association

A Wiley–Interscience Publication

JOHN WILEY & SONS

Chichester · New York · Brisbane · Toronto · Singapore

Library of Congress Cataloging in Publication Data:
Main entry under title:

Soils and Quaternary landscape evolution.

'A Wiley–Interscience publication.'
'Based on papers presented at the annual discussion meeting of the Quaternary Research Association at Brighton Polytechnic (6–7 January 1984)' — Preface.
    Includes index.
    1. Paleopedology—Congresses. 2. Geology, Stratigraphic— Quaternary—Congresses. I. Boardman, John.
II. Quaternary Research Association (Great Britain)
QE473.S65   1985   631.4   84-20994
ISBN 0 471 90528 3

British Library Cataloguing in Publication Data:

Soils and Quaternary landscape evolution.
    1. Geology, Stratigraphic—Quaternary 2. Soils
    I. Boardman, John   II. Quaternary Research Association
    551.7'9   QE696

    ISBN 0 471 90528 3

Phototypeset by Dobbie Typesetting Service, Plymouth, Devon
Printed by Page Bros., (Norwich) Ltd

List of Contributors

P. Allen	Geography Section, City of London Polytechnic, Old Castle Street, London 1E 7NT, UK
B. W. Avery	New Farmhouse, Annables Lane, Kinsbourne Green, Harpenden, Herts. AL5 3PR, UK
A. Billard	L. A. 141, CNRS, 1 Place Aristide-Briand, 92195 Meudon, Cedex, France
P. W. Birkeland	Department of Geological Sciences, University of Colorado, Boulder, CO 80309, USA
J. Boardman	Department of Humanities, Brighton Polytechnic, Falmer, Brighton BN1 9PH, UK
P. Bullock	Soil Survey of England and Wales, Rothamsted Experimental Station, Harpenden, Herts. AL5 2JQ, UK
J. A. Catt	Soils and Plant Nutrition Department, Rothamsted Experimental Station, Harpenden, Herts. AL5 2JQ, UK
C. J. Caseldine	Department of Geography, University of Exeter, Exeter EX4 4RJ, UK
E. Derbyshire	Soils Research Laboratory, University of Keele, Staffs. ST5 5BG, UK
M. J. Edge	Soils Research Laboratory, University of Keele, Staffs. ST5 5BG, UK
I. M. Fenwick	Department of Geography, University of Reading, Whiteknights, Reading RG6 2AB, UK
M. Hayward	Department of Geography, University of Reading, Whiteknights, Reading RG6 2AB, UK
V. T. Holliday	Departments of Geography and Anthropology, Texas A&M University, College Station, Texas 77843, USA
D. A. Jenkins	Department of Biochemistry and Soil Science, University College of North Wales, Bangor LL57 2UW, UK
R. A. Kemp	Department of Geography, Birkbeck College, London University, 7–15 Gresse Street, London W1P 1PA, UK
R. Langohr	Department of General Pedology, State University of Ghent, Belgium
M. A. Love	Soils Research Laboratory, University of Keele, Staffs. ST5 5BG, UK

J. A. Matthews	Sub-Department of Geography, Department of Geology, University College Cardiff, Cardiff CF1 1XL, UK
N. Owen	Mapping and Charting Establishment, Elmwood Avenue, Feltham, Middlesex, UK
J. Rose	Department of Geography, Birkbeck College, London University, 7–15 Gresse Street, London W1P 1PA, UK
J. Sanders	Institute for Encouragement of Scientific Research in Industry and Agriculture, Belgium
B. Van Vliet-Lanoë	Centre de Géomorphologie du CNRS, rue des Tilleuls, 14000 Caen, France
C. A. Whiteman	Department of Geography, Birkbeck College, London University, 7–15 Gresse Street, London W1P 1PA, UK

Contents

APPLICATIONS: OTHER AREAS

Preface

This book is based on papers presented at the Annual Discussion Meeting of the Quaternary Research Association held at Brighton Polytechnic (6–7 January, 1984). At the suggestion of the publishers, three papers have been added in order to broaden the scope of the book.

The chosen theme of the meeting, linking soil studies and landscape change, was a new departure for the QRA Discussion Meeting series and not since 1973 had soils featured prominently at what are traditionally multi-disciplinary events. The character of the audience, the aims of the QRA and the often expressed need for soil scientists to be able to explain their ideas to the wider scientific community is, I hope, reflected in the style of the contributions.

The papers have been organized into three sections. In the first section, major soil-forming processes and techniques of examination are discussed. This section is by no means comprehensive and emphasis is given to those processes (e.g. illuviation) and those techniques (e.g. micromorphology) which have proved of special value in studies of soil development through time. Section 2 and 3 include studies at various geographical scales of the relationship of soils and landscape evolution during the Quaternary.

The Quaternary Research Association is based in Great Britain and has a preponderance of British members so that meetings tend to be concerned with what are perceived as local issues and problems. This volume reflects that emphasis but the issues will undoubtedly be of interest to a non-British readership since the problems are encountered in Quaternary studies throughout the world; problems of soil dating and the significance of soil colour are examples. Any tendency to adopt insular approaches is challenged by contributions from Belgium, France and the Alpine area which describe soil and landscape relationships which in large part have not been recognized in Britain. In particular, comments made in discussion at the QRA meeting highlighted the fact that British workers have overlooked the effects of cold-climate processes on soils. The papers of Van Vliet-Lanoë, Langohr and Sanders describe these features and indicate their potential in reconstruction of past phases of periglacial conditions. That such features exist over wide areas of Britain, at least in sediments of middle Pleistocene age, is made clear by the review of the Barham Soil by Rose *et al.* In effect, these papers constitute a plea for equality of treatment for cold-climate pedogenesis. We tend to be impressed by 'interglacial' soils and to downgrade or disregard other features, but that is to risk neglecting a major portion of the Quaternary record.

Similarly, the use of soils as relative dating tools has rarely been attempted in Britain. Their great value in often poorly dated Quaternary sequences in the western United States is discussed by Birkeland. A second American contribution, that of Holliday, demonstrates the use of soil studies at archaeological sites. Though there have been several British publications in this area, many archaeologists still regard soils with some degree of distrust. Perhaps this is because of the difficulty in many situations of distinguishing between a soil and a sediment, a problem referred to by Fenwick in the first chapter. If this book is able to make more Quaternary scientists aware of the existence and potential value of soils as repositaries of information about landscape change, it will have served a useful purpose.

It will be clear to readers of the book that there is in Britain no general agreement as to the spelling of 'paleosol' (palaeosol) though firm views are frequently expressed as to the correctness of the alternatives. Because of the strength of feeling on this matter the choice has been left to individual contributors. Likewise, some contributors have preferred to use 'Holocene' and others the British stage name 'Flandrian' when referring to the last ten-thousand years.

Finally, I would like to thank the many anonymous referees who, working to impossible deadlines, completed their tasks with good humour. Many of the figures were drawn by Stephen Frampton (Brighton Polytechnic) and Dr H. Ainsley, Head of the Department of Humanities is thanked for support. Roger and Ann Smith were most helpful during the Discussion Meeting. Lastly, I am particularly grateful for advice prior to the meeting and during preparation of the book from John Catt, Jim Rose and Peter Bullock.

John Boardman

PROCESS AND TECHNIQUES

Soils and Quaternary Landscape Evolution
Edited by J. Boardman
© 1985 John Wiley & Sons Ltd.

1

Paleosols: Problems of Recognition and Interpretation

IAN FENWICK

ABSTRACT

Although the term 'paleosol' has been employed for several decades the criteria by which such a unit is identified are not clearly defined. Central to an identification of paleosols is an understanding of soil formation and thus of what constitutes a soil. Either we can look at the internal characteristics of the material or at the relationships between similar materials in the landscape—in the form of geographical associations. In recent years the latter approach has been advocated but unfortunately laterally extensive paleosols are rarely observed.

If internal characteristics are to be used, not only do we need to have a clear specification of what combinations are necessary for the recognition of a paleosol, but also we must be sure of the processes responsible for the creation of the various morphological features. This points to the need for work on modern analogues of our fossil soils. Consideration must also be given to the problem of post-burial modification of soils. Work on the decay of amino acids in soil organic matter offers the prospect of progress not only in this direction but may also provide an addition to the range of dating techniques. However, dating of paleosols remains one of the most intractable problems. Interpretation of ^{14}C soil dates is not straightforward and the limitation to <40 ka is a disadvantage when dealing with many paleosols. Accordingly, work is proceeding into the use of U series and thermo-luminescence dating of fossil soil materials.

The long-sought goal whereby paleosols could be used like plant and animal fossils to infer former environmental conditions must remain for some time yet an 'El Dorado'. Only when we have solved the many problems of relating process to morphology and chemistry will the value of the treasure be revealed.

INTRODUCTION

Soil scientists and stratigraphers have employed the term 'paleosol' for several decades but the criteria by which such a unit is identified are still not generally agreed. Indeed, the most quoted, that due to Ruhe (1956), defines a paleosol as 'a soil formed in a landscape of the past'. Many objections have been raised, even to this broad specification, and there is no doubt that it raises real difficulties. For instance, is a 5-year-old dune soil, perhaps represented by a minor accumulation of organic matter, buried this century to be treated as a paleosol? On the other hand, are we to consider a soil developed over the past 120 000 years, but with no unambiguous relict features attributable to climatic changes, as not qualifying as a paleosol? After all, it has undoubtedly formed in 'a landscape of the past'. As Catt (1979) has pointed out, perhaps the majority of soils contain elements which formed as a result of environmental conditions which have now altered. In some soils these elements will be dominant, in others quite minor, but very rarely is it appropriate to refer to the whole soil as being fossilized, i.e. a paleosol. Only when a soil has been isolated from modern processes by burial could it be truly considered paleosolic. It is suggested, therefore, that the term paleosol be confined to soils isolated from present pedogenic processes by reason of burial. 'Paleosolic elements', identified on the basis of criteria to be discussed below, could then be described from soils widely classified as 'relict' at present, or from currently forming soils in which there are relict elements produced under conditions different from the present.

As for the possible criteria to be employed in the recognition of paleosolic elements, the report of the Working Group on the Origin and Nature of Paleosols (1971) indicated a wide spread of attributes—clay and carbonate distribution, structure, colour, organic matter, micromorphology, mineralogy and granulometry. But while some contributors assigned great importance to one or other of these, others cast doubts on their reliability. Colour illustrates this conflict well. Perhaps of all field characteristics colour has been used as the primary means of differentiating between buried soils and sediments. However, sediments can display variations in colour that are due only to the geological parent material. In this case one has to separate parent material colour from pedogenic colour, and in most cases, this can be done.

The Working Group (1971), confronted with the ambivalence of many of the internal criteria for the recognition of paleosols, placed much stress on the importance of their context within the landscape. This was reinforced by Valentine and Dalrymple (1975) who, in studying extensive quarry and pipeline sections in eastern England, considered the identification of a 'paleocatena' as 'the one ubiquitous soil characteristic that sedimentation and diagenesis will not produce'. However, this approach too, contains serious drawbacks. Rarely are we fortunate enough to encounter extensive sections—or even a series of isolated exposures—which reveal the true extent of any catenary relationship. More often, restricted and isolated exposures present themselves for interpretation. Thus we are forced to reconsider which internal characteristics of the material are a response to pedological as opposed to sedimentary processes.

POSSIBLE CRITERIA FOR RECOGNITION

As we have already seen there is no dearth of criteria which are potentially of value. But, are we able to identify positively those features which are specifically pedogenic? Let us consider some of the more obvious attributes and see how reliable they are as pedogenic indicators.

One of the most obvious indicators of a buried soil is a surface horizon enriched in organic matter. Such horizons are not, however, always easily preserved or identified. Being the surface unit necessarily makes them vulnerable to erosion prior to burial. Furthermore, organic matter decomposes by oxidation so that often only vestiges of the original amount may remain. For instance, Valentine and Dalrymple (1976) found that an A/Bwb horizon contained only 1.7 per cent organic matter — 70 per cent more than the Bw/C horizon above and less than 20 per cent more than the Bwb below. In an equivalent modern A horizon we might expect there to be about 10 per cent organic matter. Attempts to establish the existence of pedogenically produced organic matter by resort to micromorphology have often been unproductive because the characteristic organic fabrics are vulnerable to decomposition and compaction. Where burial is recent and/or shallow, micromorphological identification may be conclusive. Perhaps one of the most promising developments in the past decade is that of identifying amino acids originally produced in organic-rich horizons. In allophane-rich materials in New Zealand, Limmer and Wilson (1980) have demonstrated that several amino acids will survive for at least 40 ka, thus providing a method for locating former A horizons which have now lost their distinctive field morphology.

Colour is the most obvious characteristic which attracts our attention to possible paleosols or modern soils with paleosolic elements. It has become commonplace to point to reddened materials, especially in the presumed B horizon position, with the implication that here we have a soil in which the reddening is attributable to pedogenic haematite. Conceivably, this may have been produced during past warmer conditions (Schwertmann *et al.*, 1982). Though this may be the case, the colour may simply arise through *vertical differences* in the parent material. Initial suspicions of this nature surround the Valley Farm paleosol (Rose *et al.*, 1976; Kemp, 1985) where in places (e.g. Stebbing), the colour change seems to be coincident with a sharp boundary between overlying coarse and underlying very fine sand (Figure 1.1). This textural change could have caused a hydrological barrier and thus the production of different iron oxides. Elsewhere at this site, however, the reddened horizon can be seen to transgress obvious lithological boundaries — a key indication of pedogenic influence. Apart from such transgressions, pedogenically coloured horizons can be identified by their tendency to display diffuse lower boundaries and a more intense colour towards the top.

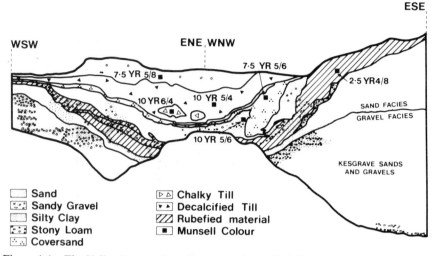

Figure 1.1   The Valley Farm paleosol in the section at Stebbing, Essex (after Whiteman *et al.*, 1983)

Granulometry is an additional means of identifying pedological units. However, on its own it is unlikely to enable us to differentiate between sediment and soil. One might expect that an ideal soil produced by weathering would become progressively finer toward the surface. However, for various reasons, soils rarely display such trends. This may be because weathering in the A horizon may yield soluble components which reform as clay minerals in the B horizon. Again, the soil environment may be most conducive to weathering in the B horizon. Thirdly, the translocation of clay in suspension may contribute to a peak in the clay content in the B horizon. Accordingly, a more normal pedogenic profile may appear, as in Figure 1.2 (Brewer, 1968). Evidence of translocation of clays may be revealed by several means but of relevance here is the ratio of fine ($< 1 \mu$m) to coarse ($> 1 \mu$m) clays. In that it has been shown that fine clays are preferentially translocated (Khalifa and Buol, 1968) an increase in this ratio with depth would be expected in the many environments where lessivage has been prevalent. Confirmation may be provided in the field by coats of oriented clay visible with a hand lens.

Recognition of pedogenic textural gradients may, however, be masked by small atmospheric additions of fines (e.g. loess) or by heterogeneity within the parent material.

Perhaps a better indication of the operation of prolonged pedological weathering can be provided by an analysis of mineral assemblages. Thus Sturdy *et al.* (1979) were able to demonstrate a marked decline in weatherable minerals toward the surface. Alternatively, indices such as the (Zircon and Tourmaline)/(Amphiboles and Pyroxenes) ratio (Ruhe, 1956), are often a more reliable means

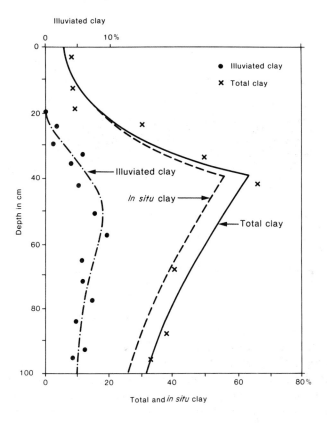

Figure 1.2   Distribution of clay in a profile subject to lessivage

of establishing the pattern of weathering. However, the imposed assumption of a mineralogically uniform sediment is not always met when working with Quaternary sediments (Jenkins, 1985). This problem has been overcome by Locke (1979) who used the depth of etching of hornblende grains as an index of weathering in Arctic soils. He was able to demonstrate that the depth of etching decreased logarithmically with depth in the profile. Indeed, this technique also has potential for dating since etching depth was also shown to be a function of time.

Although the crudest soils may only show minimal disturbance of the original sedimentary structures—as, for instance, in some polar desert soils—it is generally assumed that most soils display a substantial reorganization of the parent material. Several agencies may be responsible for this.

On the one hand, disruption may be due to organisms such as earthworms and nematodes decomposing and mixing organic material with the mineral grains. This in turn will lead to the development of a truly pedogenic structure which will ultimately destroy any geological imprint.

8   *Soils and Quaternary landscape evolution*

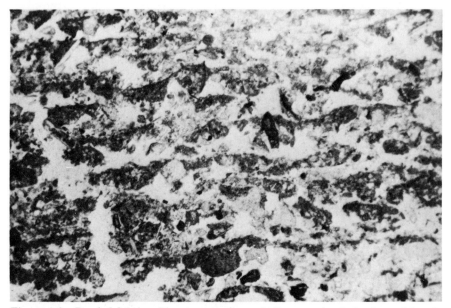

Figure 1.3   Banded fabric with concentrations of silt occurring in horizontal bands

Redistribution of coarse fragments is largely confined to the action of frost or of secondary sedimentary processes such as solifluction. It can be argued that even this should be considered as soil formation since the original sediment is being reworked by processes within the regolith.

Undoubtedly, one of the surest signs of pedological reorganization is provided by the movement of material by water either vertically or laterally. Whereas sand grains are too large to be moved through most voids, the silt fraction may well be mobilized (e.g. Kwaad and Mücher, 1979). In very coarse materials with large voids such translocation is achieved by simple downwashing provided aggregation is relatively weak. However, most examples of translocated silt are drawn from areas of present or former periglacial climates. Work done by Dumanski (1964) (reported by Mermut and St Arnaud, 1981) has demonstrated the effectiveness of freeze/thaw in developing banded fabrics (Figure 1.3) in which the concentrations of silt are supported on a skeleton of sand grains. Much of the movement of silt particles appears to take place during the melt-out of small ice lenses when the soils are very fluid. In like manner silt grains may also accumulate as cappings on coarser material.

Apart from the reorganization of the coarser fractions, soil formation has been shown to involve the removal of clay-size particles from the matrix in the

Figure 1.4 Void linings of strongly oriented clay (crossed polars). Note the extinction band crossing the clay accumulation on the right of the photograph

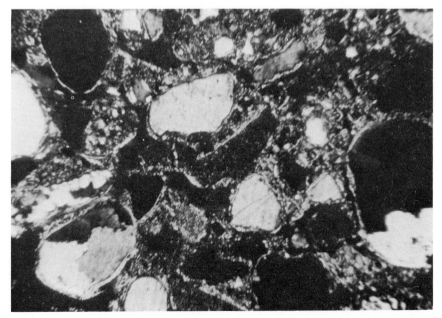

Figure 1.5 Free grain argillans appearing as light aureoles around the grains (crossed polars)

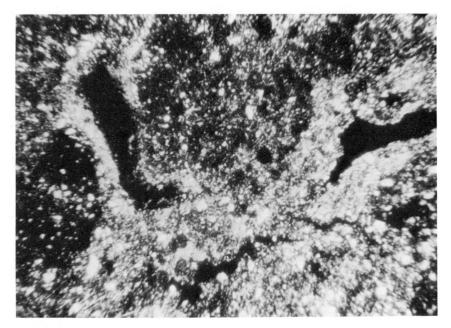

Figure 1.6   Vosepic fabric in which the clay domains show a preferred orientation

E horizon and their translocation to lower (Bt) horizons to form discrete and separable films (Kane and Mott, 1983) that line voids (Figure 1.4) and mantle sand grains. While in most cases grain coatings (Figure 1.5) are attributable to pedogenic illuvial clay, instances have been described where they are not pedogenic at all, but form in fluvial environments during quiescent periods (Bullock and Mackney, 1970; Walker *et al.*, 1978). However, in general, one of the most convincing indicators of soil formation is provided by the build-up of clay as coatings and, in particular, as void linings.

Fines can also show a pedological redistribution in response to stresses— often induced by wetting and drying. Such stresses result in the formation of concentrations of clay which display both an internally preferred orientation and an orientation with respect to some feature (e.g. a void) or direction (Figure 1.6). In one important respect they differ from the illuvial coatings discussed above. Whereas the latter display a sharp boundary with the adjacent material, 'stress concentrations' possess diffuse margins. Although stress seems to be the most likely cause of clay reorganization (e.g. McCormack and Wilding, 1974) there is still a dearth of experimental support for this assertion and the different patterns encountered are not fully understood.

However, it is in the recognition of fossil soils which owe their origins solely to weathering (i.e. those with cambic horizons) that the micromorphological evidence is most difficult to interpret. Examples of micromorphological identification of paleo-cambic horizons are rare but Boardman (1983) felt able to assert that a thin section from the Troutbeck Paleosol showed such features. Of particular importance were the presence of stress cutans and weathering mudstone particles. These properties confirmed the field observation of a deeply weathered soil in which the clasts become more weathered towards the top of the profile.

It is apparent that the mere presence of any one attribute will not be conclusive in identifying a buried paleosol. What is important is whether the totality of attributes — granulometry, colour, micromorphology, boundaries — points in the direction of soil formation. Moreover, it is not always the presence/absence of features, such as argillans, which point to the operation of pedogenesis, but rather it is the depth function which is critical (Brewer, 1972). Above all, it is vital that paleopedologists should be familiar with modern surface soils and the variation of their properties with depth. Only then can they hope to identify convincingly their buried counterparts.

In addition, there is no doubt that Valentine and Dalrymple's (1975) criterion of catenary relationships will help to confirm the existence of a geographically extensive paleosolic unit. However, units which show evidence of pedogenesis may have formed on a variety of parent materials and, in correlating paleosolic exposures, one is not, therefore, looking for textural and lithological similarities between these units but rather for soil-topographical relationships as encountered in modern, surface soils.

MODIFICATIONS TO PALEOSOLS

In seeking evidence of former soil formation we are not always concerned with a distinct soil-stratigraphic unit which has been buried by later deposits. Indeed, some of the most interesting evidence arises from relict elements in surface-soil profiles.

Paleosolic features — for instance, disrupted cutans (Figure 1.7) — may be revealed in a horizon which appears to be dominated by recent modifications. If we can infer with confidence the environmental conditions and the processes which led to the formation of these relict elements then it may be that pedologists can contribute to the story of environmental change, even in the absence of buried paleosols. The possibilities of this approach have been demonstrated by Chartres (1980) and by Bullock and Murphy (1979). Avery (1985) looks specifically at the interpretation of paleoargillic horizons.

However, certain forms of evidence will be preserved more strongly and will be more persistent than others. In a valuable review of the persistence of antecedent pedogenic features Yaalon (1971) points to the fact that the products of self-terminating, irreversible reactions such as calcareous or siliceous

Figure 1.7   Disrupted cutan, Savernake Forest, Wiltshire (plain polarized light)

Table 1.1   Soil diagnostic features and horizons grouped according to their mode of origin and relative persistence. (*Source:* Yaalon, 1971.) Reproduced by permission of the International Society of Soil Science

Altered easily, generally $< 10^3$ years to reach steady state	Relatively persistent slowly adjusting, generally $> 10^3$ years to reach steady state	Persistent
Properties acquired by reversible, largely self-regulating processes.	Mostly steady-state, near-equilibrium features or metastable state.	Features produced by essentially irreversible, self-terminating processes
Mollic horizon (f)	Cambic horizon (f)	Oxic horizon (f)
Slickensides (c)	Umbric horizon (f)	Placic horizon (f)
Salic horizon (c)	Spodic horizon (f)	Plinthite (f)
Gypsic horizon (o)	Fragipan (f)	Durinodes (f)
Mottles (c)	Mottles (o)	Petrocalcic horizon (f)
Gilgai features (c)	Argillic horizon (c)	Gypsic crust (f)
Cambic horizon (o)	Natric horizon (o)	Argillic horizon (c)
Spodic horizon (o)	Calcic horizon (f)	Natric horizon (c)
	Gypsic horizon (c)	Albic horizon (o)
	Histic horizon (c)	Fragipan (o)
		Histic horizon (o)

Letters in parenthesis are an attempt to evaluate semi-quantitatively the frequence or persistence of the feature within the group: f, frequently; c, commonly; o, occasionally; r, rarely.

incrustrations, are among the most permanent and best indicators of paleo-environmental conditions (Table 1.1). On the other hand mottling and spodic (i.e. podzolic Bs) horizons are easily modified by later processes.

However, if only remnants of former soils are available for examination then, clearly, some of the criteria suggested above for identification become difficult to use. In particular, depth functions must be ruled out and we are thrown back once again on relating specific features to pedogenesis.

In the case of buried paleosols, our lack of knowledge on modifications which take place after burial is a major concern. That such changes are occurring is indisputable as shown by the compaction of former surface horizons when buried by alluvial sediment. Perhaps it is advisable to start with the simplest situation— that of soils most recently buried—and it is in work such as that described by Hayward (1985) that we shall gain a first insight into the modifications which occur after soil formation has been terminated.

Yaalon (1971) has pointed out that features such as organic matter which respond rapidly to changed conditions are especially vulnerable to modification after burial. Thus Gardiner and Walsh (1966) have shown that a Neolithic buried soil contains only 25 per cent of the organic carbon and nitrogen of its modern counterpart.

The modification of soil properties is likely to be most substantial where the covering sediment is insufficiently thick to isolate the soil from the effects of modern processes. In these circumstances, Ruhe and Olson (1980) have argued that chemical properties are particularly unreliable characteristics of the soil pre-burial. On the other hand the granulometry and the clay mineralogy may provide a more reliable indication of the true nature of the buried soil. Detailed consideration of changes affecting both chemical, physical and morphological properties will be found in papers by Bullock (1985) and Jenkins (1985).

AGE OF PALEOSOLS

In referring to the age of a paleosol two quite separate concepts need to be considered. On the one hand, we have the date since which the soil has been fossilized by imposition of a sedimentary cover, i.e. *the age of burial*. On the other hand any buried soil possesses a *period of formation*, i.e. the elapsed period during which pedogenesis took place. Likewise paleosolic elements in surface soils must have a period of formation. It is only possible to establish this by application of stratigraphic principles to micromorphological features (Bullock, 1985).

It may be thought that radio-carbon dating of the organic horizon should provide an acceptable estimate of the age of a modern soil in terms of the initiation of the organic accumulation. However, serious problems arise on account of the decomposition and turnover of organic matter (Scharpenseel, 1971). In that organic residues are continuously decomposing it follows that much of the material which comprises soil organic matter will be of recent date with a lesser proportion of older material. ^{14}C activity, therefore, will be an amalgam of these fractions (see Fenwick, 1980, for review) and even modern A horizons give 'dates' of several hundred years BP.

If applied to a buried organic horizon ^{14}C 'age' on a bulk sample from an A or O horizon can only indicate the maximum age of burial. However, Matthews and Dresser (1983) have taken advantage of the fact that in less biotically active soils, such as podzols (Spodosols), organic matter tends to accumulate at the surface with little disturbance to the stratification by bioturbation. Hitherto, the age of burial of such soils has been estimated by radiocarbon determinations on bulk samples from the O horizon with a correction applied for mean residence time, e.g. Griffey (1976). By dating a vertical sequence of samples throughout a buried O horizon Matthews has obtained ^{14}C ages ranging linearly from about 3600 a BP at 15 cm depth (with respect to the buried surface) to about 1000 a BP at 2 cm depth. Where the O horizon has remained complete, inference from the established linear depth/age relationship therefore provides a greatly improved estimate of the age of burial. Of course, the overriding limitation with the radio-carbon technique is that many paleosols are too old to fall within its compass.

Attempts to extend the amino acid racemization technique (especially of L-isoleucine to D-alloisoleucine) to soils have not so far proved successful. Nevertheless, Limmer and Wilson (1980) have shown (see above) that for New Zealand allophanic paleosols buried by ash deposits there is a progressive disappearance of amino acids with age. In the course of 41 000 ^{14}C years, eight of the amino acids found in recent soils became undetectable. It should be possible to date such paleosols merely by the presence or absence of certain amino acids but, before there can be wider application, extensive calibration work is necessary.

In determining the period of formation, a starting point is to date the deposition of the parent material. In this respect, clastic sediments provide the most exciting possibilities at present with thermoluminescence techniques (Wintle and Huntley, 1982) proving increasingly consistent, although often with large error terms. However, through the partial exposure to light of some soil materials the dates obtained must be regarded as 'minimum dates'. Where igneous materials such as ash deposits provide the substrate for soil development, correlation with K-Ar-dated materials in the source area offers the advantages of a well-proven dating mechanism. Such is the case in the Rhine terraces in the Eifel region of Germany and in similar geomorphic situations in the western USA. However, in most circumstances TL is restricted in its application to about 10^2 ka (under very favourable conditions up to about 10^3 ka) while volcanic material is rarely found as soil parent material.

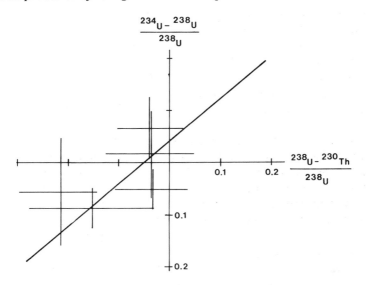

Figure 1.8   Uranium-trend isochron. Error bars are shown as two standard deviations. (*Source:* Atkins, unpublished.)

   The open-system model described by Rosholt (1982) for dating sediments by U-series disequilibria has provoked interest and is currently being tested (Atkins, forthcoming). Whereas U-series dating has been applied successfully to closed systems such as speleothems, much greater problems surround its application to open systems such as highly permeable soil material. In the Rosholt model (see below) the term λ_0 is the decay constant of a flux factor, $F(0)$, thought to cope with the movement of uranium/thorium through the system. However, the precise meaning of $F(0)$ is not clear and its inclusion is one of the weakest links in the procedure. To determine the age of a sample it is necessary to establish a uranium-trend isochron represented by the slope of the line $(^{234}U - {}^{238}U)/(^{238}U - {}^{230}Th)$ (Figure 1.8). By reference to a calibration line it is also possible to derive a value for λ_0 if the concentrations of ^{230}Th, ^{232}Th and ^{238}U have been established. It is then possible to solve for t using the relationship

$$\text{Slope of isochron} = \frac{(^{234}U - {}^{238}U)/{}^{238}U}{(^{238}U - {}^{230}Th)/{}^{238}U} = f\{\lambda_0, \lambda_{230}, \lambda_{234}, t\}$$

where $\lambda_{230} = 0.992 \times 10^{-5}\,\text{yr}^{-1}$ (decay constant of ^{230}Th)
       $\lambda_{234} = 0.28 \times 10^{-5}\,\text{yr}^{-1}$ (decay constant of ^{234}U)
(detail in Rosholt, 1982)

Preliminary results (Atkins, unpublished) indicate that, when applied to paleosols developed in loess on the Rhine terraces it is very difficult to obtain a reliable isochron on account of the large error bars resulting from the $(^{238}U - ^{230}Th)/^{238}U$ and $(^{234}U - ^{238}U)/^{238}U$ determinations. Accordingly if this approach is to bear fruit it seems that means will have to be found of establishing more reliable isochrons and of understanding fully the meaning of $F(0)$.

To date, the possibilities of establishing the actual period of formation by radiometric means are very limited. But, conventional stratigraphic approaches allied with an appreciation of the progression of soil development have been supported by Birkeland (1974) and offer some indication of the intensity and duration of soil formation.

INFERENCE OF ENVIRONMENTAL CONDITIONS

As with the interpretation of the microfossil record so it would be of great benefit if it were possible to infer environmental conditions from the soil record. This is particularly important because soil development is at its maximum precisely when the landscape is most stable, sedimentation rates are slowest and thus the fossil record most compressed. Some progress has been made. For instance, the relationship between colour, hematite content and soil climate is becoming clearer. Schwertmann *et al.* (1982) and Torrent *et al.* (1980) demonstrated a linear correlation between Hurst's (1977) redness rating and hematite content. In these examples, therefore, there seems to be a strong indication that hematite is the cause of the redness. However, the reasons for rubification and hematite formation are less clear.

Thus, Birkeland (1968), while recognizing the effects of climate, suggested that soil redness could also be correlated with age. By illustrations from the western USA he showed that redness increases with age, rapidly at first, but then taking several hundred thousands of years to attain maximum redness. Similarly, Avery (1985) has been cautious in saying only that paleoargillic horizons (i.e. loamy or clayey soils with matrix chroma more than 4 with hues redder than 10 YR) have formed on materials of pre-Devensian age. Whether these paleoargillic horizons formed as a result of a lengthy period of soil formation (in most cases they have formed in materials of uncertain pre-Devensian age) or because of climatic conditions different from the present remains an unanswered question.

On the other hand, continental workers have been more willing to consider the possibility of rubification being associated with distinctive climate regimes. Thus, Boulaine (in Fedoroff, 1966) ventured that red (5 YR 5/8) soils, developed on permeable substrates, in the vicinity of Orléans, probably required a mean annual temperature of 16 °C, precipitation of 800–900 mm and a marked dry and warm season. However, Fedoroff (1966) was of the opinion that these soils had most likely formed when the mean annual temperature was about 12 °C, i.e. only 1 or 2 °C above that of today. This opinion would conform with the evidence advanced by Schwertmann *et al.* (1982) from the Alpine foreland of Bavaria. Here they found that soils of Holocene age developed on coarse, highly permeable materials displayed distinct reddening (7.5 YR 4/4 and redder) and hematite formation. Although some would argue that there might have been a somewhat warmer Boreal episode, Schwertmann *et al.* suggest that it may well be that mean annual temperatures no higher than those of the present (about 7.5 °C) are adequate for some reddening to take place. Circumstantial confirmation of the effects of climate was provided by the increasing redness as the mean annual temperatures along their transect increased from 7.5 °C to 10.3 °C. These data take on an especial significance when it is realized that these variations in redness are in soils of similar ages. Climate is, therefore, the most likely determinant of colour in this case.

Generally the experience in Western Europe with fine textured soils has been that any reddening is confined to much older materials. Fedoroff (1966) assigned rubification in clay soils in the south-west of the Paris Basin to a period prior to the Riss-Würm interglacial and to a temperature 'clearly in excess of the present day' (p. 104). In contrast, Bresson (1974), working in the Jura reports rubification (5 YR 5/8) in some silty clay soils of late Würm age. The dilemma is that it is extraordinarily difficult to disentangle the effects of age and climate since climatic conditions have varied so much over time. In bringing together evidence from a wide range of sources in Europe and North America, Boardman (1984) has argued that the evidence for interglacial climates warmer than the Holocene is often non-existent or is conflicting. Again, he suggests that the evidence for relatively short, warm stages remains largely unproven. Moreover, many soils which show rubification may have been subjected to pedogenesis through more than one warm stage. In short, Boardman is pointing once again to the possibility that time may be critical in determining the degree of reddening.

If we can distil this often conflicting evidence, reddening seems to be dependent not only on climatic conditions but also on soil texture and conceivably on the duration of soil formation. Inferences of warmer climate on the basis of reddened soils is quite clearly a gross oversimplification. What would seem to be indicated is that a warmer and seasonally drier *soil* climate accelerates rubification but that the same end product may be achieved over a longer period in fine textured soils or in a cooler climate. Therefore there can be no unique interpretation of reddened soils in environmental terms.

In interpreting paleosolic elements an understanding of micromorphological features offers one of the best hopes of progress. For instance, papules (concentration of clay minerals with sharp external boundaries and set in the matrix of skeleton grains and clay particles) have been attributed, via cryoturbation or solifluction, to cold climate conditions (e.g. Bullock, 1974; Chartres, 1980). Although this may be a valid inference in many cases, we must be aware that bioturbation or reworking of soil material by streams may produce similar papular forms. Purely pedogenic processes, such as the concentration of expanding lattice clays in one horizon, may also result in the disruption of clay domains as a consequence of shrinkage and swelling. Therefore, although we may not be able to infer the precise climatic conditions which held sway during the formation of clay papules we can say that these features indicate a period of pedological instability and, in most cases, of landscape instability. Similar difficulties are met in interpreting other features, first in terms of processes and then in terms of prevailing environmental conditions. Only the most general of assertions can be made at present and then only with caution.

CONCLUSIONS

The investigation of paleosolic features has reached the stage where there is a proliferation of characteristics which can be used to establish the identity of a pedogenic horizon. Unfortunately, the possession of no one attribute is both necessary and sufficient for this purpose. Most importantly, for a unit to be identified as paleosolic, it must be seen to possess several of the features capable of being attributed to soil formation. If we are to rely on 'internal properties' for identification then, wherever possible, the depth functions of the various features must be compared with those of modern soils.

Nevertheless, most frequently we are confronted by the occasional relict feature rather than by units which contain a diversity of possible paleosolic elements. Confirmatory judgements based on other pedological features or on depth functions are not then possible. The paramount need is to be able to interpret features with certainty and, if progress is to be made in this area, laboratory-based simulations and analogue studies must be given priority. In particular, field examination of modern soils from different climatic areas will enable sounder interpretations to be made of paleosols and relict features. Perhaps the most intractable problem remains that of assessing the effect of time on soil formation and hence of interpreting paleosols in environmental terms. For this, the most valid approach would seem to be space/time substitution in the field situation.

ACKNOWLEDGEMENTS

Grateful thanks are due to Professor Peter Birkeland for his most helpful comments on an earlier draft of this paper.

REFERENCES

Avery, B. W. (1985). 'Argillic horizons and their significance in England and Wales', in *Soils and Quaternary Landscape Evolution* (Ed. J. Boardman), Wiley, Chichester (in press).

Birkeland, P. W. (1968). 'Correlation of Quaternary stratigraphy of the Sierra Nevada with that of the Lake Lahontan area', in *Means of Correlation of Quaternary Successions* (Eds R. B. Morrison and H. E. Wright), pp. 469–500, International Association of Quaternary Research, VII Cong., Proc. v. 8.

Birkeland, P. W. (1974). *Pedology, Weathering and Geomorphological research*, Oxford Univ. Press, New York.

Boardman, J. (1983). 'The role of micromorphological analysis in an investigation of the Troutbeck paleosol, Cumbria, England', in *Soil Micromorphology* (Eds P. Bullock and C. P. Murphy), pp. 281–288, AB Academic Publishers, Berkhamsted.

Boardman, J. (1984). 'Comparison of soils in Midwestern United States and Western Europe with the Interglacial record', *Quaternary Research* (in press).

Bresson, L. M. (1974). 'A study of integrated microscopy; rubefaction under wet temperate climate in comparison with Mediterranean rubefaction', in *Soil Microscopy* (Ed. G. K. Rutherford), pp. 526–541, Limestone Press, Kingston.

Brewer, R. (1968). 'Clay illuviation as a factor in particle-size differentiation in soil profiles', *Trans. 9th Int. Cong. Soil Sci., Adelaide, IV*, pp. 489–499.

Brewer, R. (1972). 'Use of macro- and micromorphological data in soil stratigraphy to elucidate surficial geology and soil genesis', *J. Geol. Soc. Australia*, **19**, 331–334.

Bullock, P. (1974). 'The use of micromorphology in the new system of soil classification for England and Wales', in *Soil Microscopy* (Ed. G. K. Rutherford), pp. 607–631, Limestone Press, Kingston.

Bullock, P. (1985). 'The role of micromorphology in the study of Quaternary soil processes', in *Soils and Quaternary Landscape Evolution* (Ed. J. Boardman), Wiley, Chichester (in press).

Bullock, P. and Mackney, D. (1970). 'Micromorphology of strata in the Boyn Hill Terrace Deposits, Buckinghamshire', in *Micromorphological Techniques and Applications*, Tech. Monog. No. 2, (Eds D. A. Osmond and P. Bullock), pp. 97–105, Soil Survey, Harpenden.

Bullock, P. and Murphy, C. (1979). 'Evolution of a paleoargillic brown earth (Paleudalf) from Oxfordshire, England', *Geoderma*, **22**, 225–252.

Catt, J. A. (1979). 'Soils and Quaternary geology in Britain', *J. Soil Sci.* **30**, 607–642.

Chartres, C. J. (1980). 'A Quaternary soil sequence in the Kennet Valley, central southern England', *Geoderma*, **23**, 125–146.

Dumanski, J. (1964). *Micropedological Study of Eluviated Horizons*, Unpublished M.Sc. Thesis, Univ. of Saskatchewan, Saskatoon, Canada.

Fedoroff (1966). 'Contribution à la connaissance de la pedogenèse Quaternaire dans le S-W du Bassin Parisien', *Bull. Ass. francaise Etude Quat.* **2**, 94–105.

Fenwick, I. M. (1980). 'Palaeosols', in *Geomorphological Techniques* (Ed. A. Goudie), pp. 342–345, Allen & Unwin, London.

Gardiner, M. J. and Walsh, T. (1966). 'Comparison of soil material buried since Neolithic times with those of the present day', *Proceedings Royal Irish Acad.* **65C**, 29–35.

Griffey, N. J. (1976). 'Stratigraphical evidence for an early Neoglacial glacier maximum of Steikvassbreen, Okstindan, north Norway', *Norsk Geologisk Tidsskrift*, **56**, 187–194.

Hayward, M. (1985). 'Soil development in Flandrian floodplains: River Severn case-study', in *Soils and Quaternary Landscape Evolution* (Ed. J. Boardman), Wiley, Chichester (in press).

Hurst, V. J. (1977). Visual estimation of iron in saprolite', *Geol. Soc. Amer. Bull.* **88**, 174–176.

Jenkins, D. A. (1985). 'Chemical and mineralogical composition in the identification of palaeosols', in *Soils and Quaternary Landscape Evolution* (Ed. J. Boardman), Wiley, Chichester (in press).

Kane, P. and Mott, C. J. B. (1983). 'Non-destructive separation of discrete pedogenic features from soils', *J. Soil Sci.* **34**, 205–212.

Kemp, R. (1985). 'The Valley Farm Soil in southern East Anglia', in *Soils and Quaternary Landscape Evolution* (Ed. J. Boardman), Wiley, Chichester (in press).

Khalifa, E. M. and Buol, S.W. (1968). 'Studies of clay skins in a Cecil (Typic Hapludult) soil: I. Composition and genesis', *Soil Sci. Soc. Amer. Proc.* **32**, 857–861.

Kwaad, F. J. P. M. and Mücher, H. J. (1979). 'The formation and evolution of colluvium on arable land in northern Luxembourg', *Geoderma*, **22**, 123–192.

Limmer, A. W. and Wilson, A. T. (1980). 'Amino acids in buried paleosols', *J. Soil Sci.*, **31**, 147–153.

Locke, W. W. (1979). 'Etching of hornblende grains in Arctic soils: an indicator of relative age and paleoclimate', *Quaternary Research*, **11**, 197–212.

McCormack, D. E. and Wilding, L. P. (1974). 'Proposed origin of lattisepic fabric', in *Soil Microscopy* (Ed. G. K. Rutherford), pp. 761–771, Limestone Press, Kingston.

Matthews, J. A. and Dresser, P. Q. (1983). 'Intensive ^{14}C dating of a buried palaeosol horizon', *Geologiska Föreningens i Stockholm Forhandlingar*, **105**, 59–63.

Mermut, A. R. and St Arnaud, R. J. (1981). 'Microband fabric in seasonally frozen soils', *Soil Sci. Soc. Amer. Journal*, **45**, 578–586.

Rose, J., Allen, P. and Hey, R. W. (1976). 'Middle Pleistocene stratigraphy in Southern East Anglia', *Nature*, **263**, 492–494.

Rosholt, J. N. (1982). 'Mobilization and weathering', in *Uranium Series Disequilibrium: Applications to Environmental Problems* (Eds M. Ivanovitch and R. S. Harmon), pp. 167–170, Oxford University Press.

Ruhe, R. V. (1956). 'Geomorphic surfaces and the nature of soils', *Soil Sci.*, **82**, 441–455.

Ruhe, R. V. and Olson, C. G. (1980). 'Soil Welding', *Soil Sci.*, **130**, 132–139.

Scharpenseel, H. W. (1971). 'Radiocarbon dating of soils—problems, troubles, hopes', in *Paleopedology: Origin, Nature and Dating of Paleosols* (Ed. D. H. Yaalon), pp. 77–88, Israel Universities Press, Jerusalem.

Schwertmann, U., Murad, E. and Schulze, D. G. (1982). 'Is there Holocene reddening (hematite formation) in soils of axeric temperate areas?' *Geoderma*, **27**, 209–223.

Sturdy, R. G., Allen, R. H., Bullock, P., Catt, J. A. and Greenfield, S. (1979). 'Paleosols developed on chalky boulder clay in Essex', *J. Soil Science*, **30**, 117–137.

Torrent, J., Schwertmann, U. and Schulze, D. G. (1980). 'Iron oxide mineralogy of some soils of two river terrace sequences in Spain', *Geoderma*, **23**, 191–208.

Valentine, K. W. G. and Dalrymple, J. B. (1975). 'The identification, lateral variation, and chronology of two buried paleocatenas at Woodhall Spa and West Runton, England', *Quaternary Research*, **5**, 551–590.

Valentine, K. W. G. and Dalrymple, J. B. (1976). The identification of a buried paleosol developed in place at Pitstone, Buckinghamshire, *J. Soil Science*, **27**, 541–553.

Walker, T. R., Waugh, B. and Crone, A. J. (1978). 'Diagenesis in first-cycle desert alluvium of Cenozic age, southwestern United States and northwestern Mexico', *Geol. Soc. Amer. Bull.*, **89**, 19–32.

Whiteman, C. A., Kemp, R. and Wilson, D. (1983). 'Stebbing', in *Quaternary Research Association Field Guide: Diversion of the Thames* (Ed. J. Rose), pp. 149–161, Birkbeck College, London.

Wintle, A. G. and Huntley, D. J. (1982). 'Thermoluminescence dating of sediments', *Quat. Sci. Reviews*, **1**, 31–53.

Working Group on Origin and Nature of Paleosols (1971). 'Criteria for the recognition and classification of paleosols', in *Paleopedology: Origin, Nature and Dating of Paleosols* (Ed. D. H. Yaalon), pp. 153–158, Israel Universities Press, Jerusalem.

Yaalon, D. H. (1971). 'Soil-forming processes in time and space', in *Paleopedology: Origin, Nature and Dating of Paleosols* (Ed. D. H. Yaalon), pp. 29–39, Israel Universities Press, Jerusalem.

Soils and Quaternary Landscape Evolution
Edited by J. Boardman
© 1985 John Wiley & Sons Ltd.

2

Chemical and Mineralogical Composition in the Identification of Palaeosols

DAVID A. JENKINS

ABSTRACT

Because palaeosols may preserve the imprint of past environments, chemical and mineralogical analyses are of value both in their recognition and in the extraction of information for the Quaternary geologist and pedologist. Ease of recognition and interpretation increases with divergence between palaeosol and its contemporary equivalent. Although gross major element changes may not be manifested, examples are given of changes in trace element, carbonate, extractable Al/Fe & P, and organic matter contents, all of which reflect pedogenic history and the influence of the biosphere (and man) in particular. The biosphere may be preserved directly in the form of fossils — pollen, spores, beetles, snails, diatoms, phytoliths, etc. — some of which have yet to be studied in detail. Soil mineralogy records events both through weathering of inherited material, quantifiable by microscopy in terms of mineral ratios or degree of etching, and also through genesis of minerals characteristic of the changing soil environment. Subtle changes in clay mineralogy may be difficult to detect and interpret by XRDA, but in temperate regions such as the UK the presence of kaolinite, gibbsite or haematite is often cited as proof of a palaeosol. However, their presence might also arise through inheritance, contamination or hydrothermal agencies, and the conditions of their formation are not necessarily exclusively linked with extreme weathering: more detailed studies on crystallinity, impurities, isotopic composition, etc., may resolve such problems. Co-ordination of information is advocated from as many different forms of analysis and sources as possible, including archaeology and micromorphology.

INTRODUCTION

Palaeosols are of value as a special case. From them the pedologist may deduce the influence of variations in climate, vegetation and, in particular, time on pedogenesis. To the Quaternary geologist, they are valuable as a potential source of information about the past environmental systems of which they were a component. Invoking uniformitarianism, it is possible to interpret fossil features in terms of contemporary processes, although account must be taken of post-burial modification (Valentine and Dalrymple, 1976). For both the pedologist and Quaternary geologist there are the problems of recognition and validation. A palaeosol may be obvious if it is buried and retains its morphology but recognition becomes increasingly difficult with subsequent disturbance. This is especially so when a palaeosol occupies the present land surface, having been re-exposed by erosion or perhaps never buried, and bears the superimposed imprint of contemporary soil processes. Conversely, considering the changes known to have occurred in climate, vegetation and usage during the current phase of post-glacial pedogenesis, it is probably safer to assume that most contemporary profiles embody relict features to a greater or lesser extent. It is then a matter of definition as to when one chooses to recognize palaeosols (INQUA Com. Palaeoped., 1971).

The pedologist and Quaternary geologist therefore turn to the intrinsic properties of soil material in the search for criteria which would identify a palaeosol. Any palaeosol was a product of processes acting on a parent material, controlled by such environmental factors as climate, vegetation and topography, and operating over a specific period of time. The palaeosol will therefore be most easily distinguished where one of these factors differed so markedly from those involved in the genesis of contemporary post-glacial soils as to result in contrasting compositional features. Priority must therefore be given to establishing diagnostic features and, if necessary, developing the analytical techniques required to identify them. These features include the composition of the soil material, whether in chemical terms, or in terms of discrete components such as minerals, rocks, organic tissue or fossils. Within this diversity of material there is a wide range of chemical and mineralogical analytical techniques that can be applied, and these form the subject of this paper.

This review will not attempt to be comprehensive but will concentrate on the more useful areas and those of recent or potential development. It will also limit itself mainly to examples relevant to studies in the UK, that is to areas which have experienced glaciations during the Quaternary. Certain aspects will be excluded. For example, at a higher level of organization, the components mentioned above are rearranged during pedogenesis to produce a distinctive microfabric. This is perhaps one of the most fruitful areas in which to seek diagnostic criteria of palaeosols, modifications being achieved relatively rapidly in materials which may otherwise show no change of bulk or component composition. However, this subject has been reviewed by Mücher and Morozova (1983) and is specifically dealt with by Bullock (1985). Although it will not be considered further here, it must be remembered that many of the components described below often achieve their fullest significance only in the context of the microfabric in which they occur: this chapter should therefore be considered in conjunction with that on micromorphology. Similarly, important dating techniques constitute a specialized technology which is discussed by Fenwick (1985). However, it should be noted that the interests of the pedologist and Quaternary geologist are paralleled by those of the archaeologist who seeks similar information in the interpretation of stratigraphic sequences, which may include palaeosols, and in the reconstruction of palaeoenvironments. Valuable information regarding techniques and interpretation is therefore presented in the archaeological literature (e.g. Davidson and Shackley, 1976; Evans, 1975).

CHEMICAL ANALYSIS

Pedogenesis is marked by changes during which components are lost, or less commonly gained, such that the chemical composition diverges progressively from that of the parent material. These changes are by now well established (e.g. Loughnan, 1970), but are usually only manifested in a useful form at an advanced stage of soil development. Bulk chemical composition tends to be relatively insensitive at earlier stages, especially when the difficulty of establishing a soil's homogeneity and detailed parentage is taken into account. In view of the expertise and expense required by silicate analysis it is therefore not surprising that indiscriminate total analysis is rarely resorted to in the recognition of palaeosols. There are, however, several situations where specific chemical analysis can be helpful.

Trace elements

Whilst total analysis for major elements is not often justified in the study of palaeosols, that for trace elements may be. This is mainly because arc spectography provides a relatively cheap and rapid method of analysis for some 20 or so trace elements simultaneously, requiring samples of the order of 10 mg only. That such analysis may be only semi-quantitative (e.g. ± 50 per cent) is compensated for by a trace element's concentration often varying by a factor of 10^3 between soils, unlike those for major elements. As a specific example, the differentiation of post and pre-industrial (i.e. 'fossil') autochthonous organic topsoils may be cited. In such cases atmospheric input assumes significance in relation to material inherited from parent materials (Jenkins and Bower, 1974), so that contemporaneous peaty horizons are significantly enriched in the atmospherically dispersed products of industrial activity (Sn, Pb, Zn, Cu $\pm$ Ag, Mo, Ge, Be, etc.), compared to buried horizons (Figure 2.1). Conversely, of course, a fossil soil surface in an archaeological excavation which marked a past phase of metal working may be detectably contaminated by Sn, Cu, Pb, Ag or Au. The sensitivity of the archaeological test can be increased by analysis of charcoal which appears to selectively absorb and concentrate trace elements

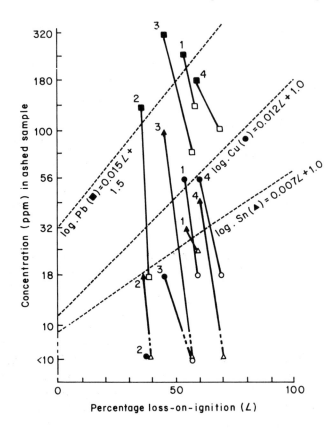

Figure 2.1   Enrichment of Sn, Cu and Pb in four contemporary (●) as compared to buried (○) organic top-soils from Snowdonia: the effect of general enrichment with increasing organic matter content ('loss-on-ignition') is indicated by the appropriate regression lines for Snowdonian soils generally

from percolating soil solutions and thus preserve a record of past soil chemistry: this may also be true for 'non-archaeological' palaeosols but has yet to be tested.

Phosphorus

Apart from carbon, hydrogen and nitrogen, phosphorus is probably the most characteristic element of the biosphere. It is accumulated in living tissue, deposited in bones and concentrated in excreta: moreover it often has the advantage of relative immobility within subsequent soil environments. Again, this can be most obvious in an archaeological context where up to a ten-fold enhancement of phosphorus may be detectable in the ('fossil') soils of occupation

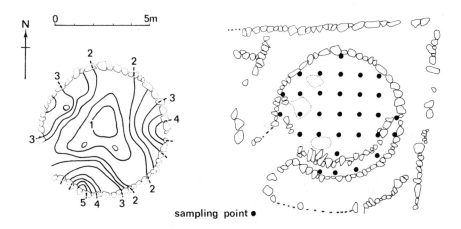

Figure 2.2 Total phosphorus concentration contours (ppm) in a Romano-British hut excavated at Graeanog, Gwynedd, Wales. Reproduced with permission from Conway, 1983, Journal of Archaeological Science vol.10 p.122. Copyright 1983 by Academic Press Inc. (London) Limited

surfaces (Proudfoot, 1976). The patterns of concentration can be most revealing, especially when subjected to Trend Surface Analysis (Conway, 1983). For example, in a circular Iron Age hut at Graeanog, Gwynedd, the concentration contours even suggest the routine of the inhabitants' activity (Figure 2.2). Differentiation of the form of the phosphorus provides further details; total values are enhanced in hearths, where phosphorus is presumably fixed on ignition, whilst drains are more clearly seen in the contours of the more ephemeral phosphorus extractable by weak acids. Even the cryptic form of bodies long since vanished may still be detectable in phosphorus levels (J. S. Conway, pers. comm.). Whilst all these examples derive from special archaeological situations, they illustrate the recovery of information by the detailed spatial chemical analyses of deposits such as palaeosols where the localized influence of the biosphere may be preserved.

Carbonates

The situation for carbonates is atypical in that chemical analysis proves to be a convenient assessment of a mineralogical component. Calcite, the dominant soil carbonate, is a mineral which is progressively depleted from leached soils of calcareous parentage, but which accumulates in soils of more arid regimes (Gile *et al.*, 1966). Together with the less abundant aragonite and other carbonates, it is readily estimated chemically by titration or CO_2 evolution. Its abundance can therefore be used as an indicator of past climates and, since the processes involved operate on a scale of 10^3 years, the final pattern of calcite distribution records the integrated effect of time. For example, depth of decalcification of calcareous tills — with or without secondary enrichment in a subjacent Ck horizon — is of the order of 1 m in post-glacial British soils, the actual depth varying with permeability, drainage regime, rainfall and related factors. There is therefore the possibility of relating depth to time and using it as a criterion of soil age (Catt, 1979).

An early attempt at such analysis was reported by Boulton and Worsley (1965), in which a bimodal distribution in the depth of decalcification, and consequently age of soil, was suggested for tills on the Cheshire Plain: although these conclusions were subsequently challenged on geomorphological grounds (Poole, 1966) the study illustrated the possibilities offered by carbonate analysis. These possibilities also exist in the development — as well as destruction — of calcic horizons, and include the application of ^{14}C dating, although here many problems of interpretation are clearly involved (Bowler and Polach, 1971).

Extractable Fe & Al

Although total analysis for major elements may be an insensitive approach, chemical differentiation of the forms in which elements are present may be more productive. An obvious example would be the various pedogenic forms of Fe and Al which characterize the bulk of our brown earths and podzolic soils and which, indeed, are used in their classification. Because these forms tend to be poorly crystalline, chemical analysis is again more convenient than standard mineralogical techniques, such as XRDA. Thus organic complexes only are extractable by pyrophosphate, whilst acid oxalate also extracts amorphous and poorly crystalline forms (e.g. ferrihydrite) and citrate-buffered dithionite further extracts more aged crystalline forms such as goethite, lepidocrocite etc. (McKeague *et al.*, 1971). Few examples of this approach to the identification and classification of palaeosols have yet been reported, but that by Dormaar and Lutwick (1983) identified cryptic 'palaeogley' features and highlighted the factor of post-burial modification that must be taken into account in the study of buried soils, the original pyrophosphate extractable fraction apparently having decreased with time since burial.

Organic matter

The organic fraction is usually the most characteristic feature of temperate soil profiles and the indications are that a steady-state condition is achieved within a period of 10^2–10^3 years, more rapidly than for other soil properties. The organic fraction is especially obvious where there is a surface accumulation of humus and indeed many buried profiles are first recognized by the darker pigmentation of their original organic-rich surface horizons. Organic content can be quantified by crude (loss-on-ignition) or by more precise (organic carbon by wet oxidation) analysis. Under suitable conditions, these components also offer the possibility of ^{14}C dating, the limitations to which have been discussed by Scharpenseel (1971).

The organic fraction contains further information in its detailed chemical composition which reflects the nature of the plant input and the manner and degree of its breakdown, as expressed in such parameters as the C/N ratio. Whilst there is now a considerable body of information accumulated about the detailed organic chemistry of soils (Gieseking, 1975), this does not yet appear to have been applied in many studies of palaeosols, although Stephenson (1969) noted the progressive diagenetic polymerization that occurred in organic matter after burial. One example, however, is that reported by Dormaar and Lutwick (1969) who distinguished between former grass and forest vegetation from infrared analyses of the humic acids extracted from buried Ah horizons and also from their fatty acid suites. A different example of a time-related process involving organic material is the progressive conversion of the L-enantiomer of the amino acids in living proteins to an equilibrium mixture (approximately 1:1) of the D and L forms—i.e. 'racemization'. This has enabled molluscan and mammalian samples to be dated back over 10^6 years, but temperature and biological origin are also variables and so far its application has been confined to marine or cave deposits. Amino acids have been cited as indices of biological activity of value in identifying palaeosols (Thompson *et al.*, 1981) and it has also been suggested that their differential persistence could provide a basis for identifying and dating palaeosols up to 105 years old (Limmer and Wilson, 1980).

Isotopic analysis

Chemical analysis is being continuously developed and refined as new techniques become available, so adding to the range of information that can be extracted from palaeosols. Isotopic analysis by mass spectrography is one such new development which shows considerable promise. It is already being utilized specifically in the measurement of ^{14}C, increasing the sensitivity and thus range of this important dating method (Hedges, 1981). It is also applicable to nonradioactive isotopes which, it has been shown, may be detectably fractionated where processes such as evaporation are significantly influenced by the small differences in mass between isotopes. This has been extensively developed and applied in $^{18}O/^{16}O$ studies of carbonates deposited in the marine environment where the isotopic ratio has been correlated with temperature to provide a definitive stratigraphy for the Quaternary (Shackleton, 1977a and b). Oxygen and hydrogen isotopic analysis is also applicable to clay minerals and other pedogenic minerals (Lawrence and Taylor, 1972), where it can again be correlated with environmental conditions. Specific applications will be considered in the relevant sections that follow, as will those of other physico-chemical techniques such as Mossbauer spectroscopy.

MINERALOGICAL ANALYSIS

The inorganic portions of soils comprise an assemblage of chemical compounds (minerals) variously at equilibrium with their environment. Those inherited from the parent material tend to be existing metastably outside their stability fields and they therefore 'weather' progressively. By contrast the pedogenic minerals are characteristic of the particular soil environment in which they formed. Analysis of these two fractions is more conveniently dealt with separately, especially since they are generally concentrated in different size fractions and so require contrasting analytical techniques.

Coarse inherited fractions

Weathering of the mineral components of a rock may lead to either disaggregation and 'rotting' *in situ*, or to the development of a distinct 'rind' or 'crust'. The former arise more commonly with coarse-grained igneous rocks, and, when these are present as clasts in a till, it is strong evidence for intense post-depositional weathering which may imply more than just post-glacial pedogenesis (e.g. Mosedale; Boardman, 1984). In the latter case, the 'rind' is made up of a white surface zone, formed by selective dissolution of minerals leaving a porous framework of the more stable minerals, with or without deposition of brown ferric hydrous oxides in a subsurface zone. It is more commonly seen in fine-grained silicic rocks (e.g. Figure 2.3) wherein trends can be followed chemically (XRF, EPMA) or mineralogically (XRDA, thin-section

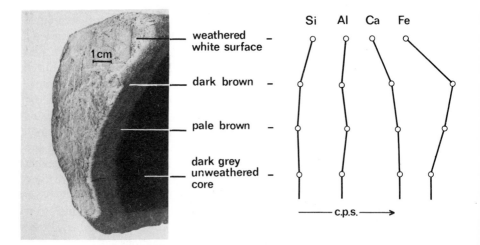

Figure 2.3   Weathering rind developed in a rhyolite pebble buried in Post-glacial peat, Cwm Idwal, Gwynedd: selected electron microprobe analyses (cps)

microscopy). The depth of weathering is a function of time and, under suitable circumstances, it can be used for relative dating (Birkeland, 1964) and thus for the recognition of relict features or palaeosols.

Fine sand mineralogy

Although very rewarding, the opportunities for studying palaeosols directly overlying their parental rock outcrops are rare in most parts of Britain due to

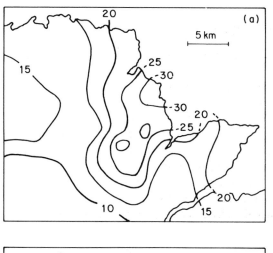

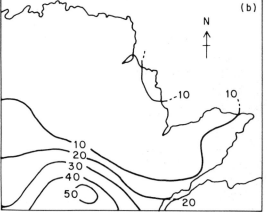

Figure 2.4 Percentage heavy mineral contours in the non-opaque heavy mineral (SG > 2.95) fine sand (63–200 μm) fractions of soils from north east Anglesey, North Wales (from Younis, 1983). (a) Total zircon. (b) Total clinopyroxenes

removal by glacial erosion or to deep burial by later glacial, fluvial or aeolian deposits. Contemporary soils inherit the latter deposits as their parent materials and possible criteria for recognizing a palaeosol within a section are therefore either a lithologic discontinuity, arising from a change of parent material, or contrasting mineralogies arising from different degrees of weathering. Both these possibilities can be investigated by mineralogical analysis using established optical microscopic techniques. These are conveniently applied to the fine sand fraction (e.g. 200–60 μm), particularly its 'heavy minerals' (SG >2.95): finer fractions pose increasing difficulties of microscopic technique which hands over to XRDA, whilst the particles of coarser fractions tend to be polymineralic and too thick for normal transmitted light microscopy. However the 2.0–0.6 mm fraction can be resin-impregnated and thin-sectioned and its petrography established: this is particularly useful in estimating rock types such as shale which may be poorly represented in both the fine sand mineralogy and coarser stone fractions.

Pedogenesis generally involves the progressive depletion through weathering of less-stable mineral species such as the ferromagnesians (pyroxenes, amphiboles) as against relatively stable species such as zircon, tourmaline, etc. Differential weathering effects are manifested in bulk mineral composition over periods of 10^4 or more years. Where the time factor for a palaeosol differed on this scale it would therefore be detectable by quantitative mineral analysis, although depth within the profile, permeability, and similar factors must also be taken into account. For example, Ruhe (1956) used these mineral ratios to differentiate erosion surfaces in tills dating from different periods back to >10^5 years in North America. In British soils, however, heavy mineral studies of fine sand/silt fractions have revealed a variable involvement of loess in the parentage (Catt, 1977). For example, studies in Anglesey, North Wales (Figure 2.4) have shown that an already complicated pattern of solid and drift geology is further confused by aeolian dispersion of loess (63–20 μm) and coversands (200–360 μm). This indicates the need for detailed knowledge of both the mechanical composition and mineralogical background before attempting to use heavy mineral data in the recognition of palaeosols.

Grain morphology

On a shorter time scale (10^3 years), weathering is visibly evident in the progressive etching of individual grains before it is seen in total assemblage composition. This is most obvious in amphiboles (e.g. Figure 2.5a) and pyroxenes, the pattern of dissolution being controlled by the chain structure of the silicates. The degree of etching, quantified as the mean depth of etching on 100 grains, has been shown by Locke (1979), in his studies of glacial deposits on Baffin Island, Canada, to be related to depth within the profile, age and subsequent climate and thus to be of value for relative dating. It is interesting to note that soils in Snowdonia, North Wales, contain a very mixed assemblage

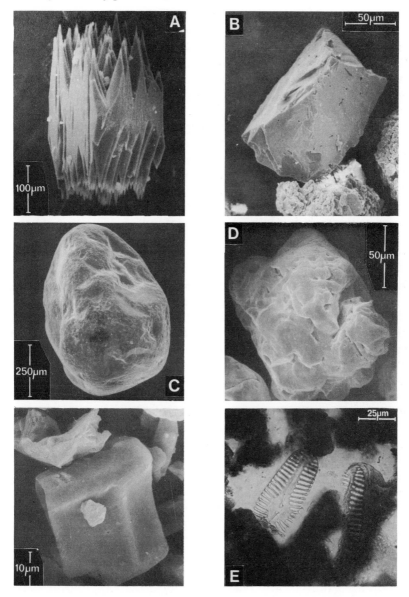

Figure 2.5 SEM (a–e) and optical (f) micrographs of soil grains. A. Grain of hornblende deeply etched IIc in an ultramafic sapropel from Ife, Nigeria. B. Quartz grain with freshly fractured surfaces in a late Devensian till from Llyn Lydaw, Gwynedd. C. Rounded quartz grain from desert sand, Hofuf, Saudi Arabia. D. Corroded quartz grain from an ultisol, Ife, Nigeria. E. Diatom test (*Pinnularia* sp.) from the buried Oh horizon of a stagnogley profile from Clwyd, North Wales. F. Distinctive silicified cell (phytolith) of *Molinia caerulea* from a buried organic horizon, Gwynedd

of clinopyroxenes with regard to etching (Jenkins, 1964), raising the possibility of the incorporation of otherwise undetectable relict material from earlier interglacial/preglacial phases of pedogenesis into contemporary soils: however, it would be necessary to establish that other variables (e.g. compositional : Ti-content) were not involved before this hypothesis could be accepted.

Apart from etching, the shape of mineral grains as seen under the SEM in particular can provide useful information about their sedimentary history. It is suggested (e.g. Krinsley and Doornkamp, 1973) that glacial, fluvial, aeolian and chemical agencies all impose their own recognizably characteristic fracture and etch pattern on grain surfaces (e.g. Figure 2.5b, c and d). This can allow the relative age (Douglas and Platt, 1977) and provenance of a soil's parent material to be deciphered, although the latter may prove to be a difficult procedure as many different phases of a grain's history may be superimposed prior to its incorporation into a soil or palaeosol.

Pedogenic minerals of the finer fractions

In contrast to inherited minerals, pedogenic minerals are by definition stable within the particular soil environment in which they are formed. But the soil environment will itself change over time with a consequent further modification of the mineral assemblage, this being particularly true for the clay minerals. If preserved unchanged, pedogenic minerals within a palaeosol may therefore

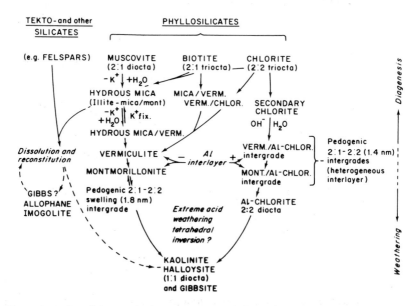

Figure 2.6   Possible weathering sequences of clay minerals (after Jackson, 1964)

be useful indicators of the environment in which they formed. Overall a general sequence of development of the clay minerals with pedogenesis/time has been established (Figure 2.6), although this will vary with parent material, leaching intensity etc. This sequence tends to be an empirical summary and only in a few cases has theoretical knowledge advanced to the point where the stability fields of individual minerals can be adequately defined in terms of soil parameters

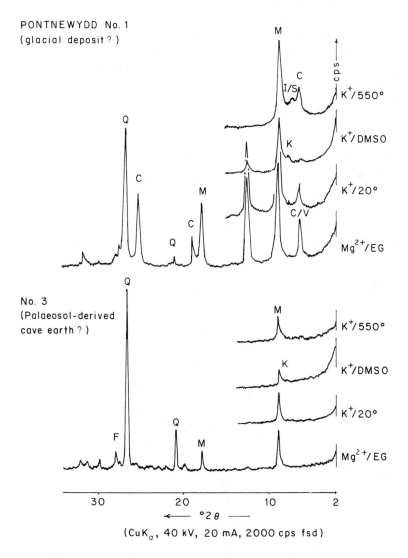

Figure 2.7 XRDA traces of the clay fraction of selected sediments from Pontnewydd Cave, Clwyd, North Wales

(e.g. Palygorskite—Singer and Norrish, 1974). Usually the time factor in pedogenesis is only a fraction of that required for the total clay mineral sequence so that only limited sections are involved, often resulting in subtle changes (e.g. illite/vermiculite/smectite) and therefore requiring careful detailed analysis to detect them. Since pedogenic minerals are commonly concentrated in the fine clay fractions, identification relies mainly on XRDA and this poses its own specialist problems many of which have yet to be solved (Brindley and Brown, 1980). Nevertheless, there is potential in the use of clay mineral analysis in interpreting the influence of climatic and time factors as recorded in palaeosols (Singer, 1980). For example, from their studies of palaeosols on tills in north-eastern, USA, Novak *et al.* (1971), concluded that the weathering sequence observed (mica vermiculite kaolinite) was indeed time-controlled, whilst Bronger (1971) was able to recognize warmer climatic episodes in the clay mineralogy (illite/smectite) of palaeosols on loess from Yugoslavia.

Sometimes distinctive changes in clay mineralogy mark out a palaeosol or its derivative. For example, a sequence of sediments from a cave at Bontnewydd in North Wales, is being studied in detail because of its Palaeolithic hominid remains (Green, 1984). Within this sequence one stratum stands out by its orange-brown colour which contrasts with the grey-brown strata above and below. Whereas the heavy mineralogy suggests a common parentage, the clay mineralogy of the orange-brown horizon (Figure 2.7) differs in the absence of chlorite (C), vermiculite (V) and inter-stratified material (I/S), the relative enhancement of quartz (Q) and felspars (F), and the persistence of kaolinite (K) and hydrous mica (M). Such a mineralogy is consistent with the adjacent horizons (e.g. No. 1) being derived with little change in mineralogy from the likely source rocks, probably via glacial processes, and the orange-brown horizon (No. 3) from soils which had reached an advanced stage of weathering, resulting in the removal of the 1.4 nm minerals. This conclusion is reinforced by the highly corroded nature of the few chlorite grains which survive in the fine sand fraction.

Kaolinite, gibbsite and haematite

In Britain, it is the occurrence of one of the minerals at the end of the weathering sequence which is most often cited as evidence for a palaeosol. Kaolinite and gibbsite are not obvious products of contemporary pedogenesis in this country and these minerals, together with haematite, tend to be interpreted as products of intense weathering such as characterizes the older soils of hotter climates.

Kaolinite is a distinctive 0.7 nm 1:1 clay mineral which is readily detected by XRDA, DTA and IRA when abundant. However, it can be difficult to detect when present in only minor amounts, particularly when dominated by Fe-chlorite whose 0.7 nm (002) peak also decreases on heating to 500 °C, causing misidentification of the kaolinite: unfortunately this association is not uncommon in British soils. A more specific, albeit protracted, identification is therefore needed for kaolinite such as that involving DMSO (Lim *et al.*, 1981). Even if the presence of kaolinite is confirmed the problem of origin remains. Since it is widely dispersed as an original constituent of rocks and derived sediments, and is also a product of hydrothermal processes, there is always the possibility of kaolinite being inherited or a glacial/aeolian 'contaminant' rather than authigenic in the soil. Further relevant information may be gained here by 'typing' the kaolinite (Range *et al.*, 1969), and determining its 'crystallinity' which apparently increases to a maximum at the midddle stages of development of tropical soils (Hughes, 1980).

Unlike kaolinite, gibbsite is not common in North European rocks and sediments and therefore less likely to be present due to 'contamination'. It is consequently a more probable indicator of a palaeosol (e.g. Roaldset *et al.*, 1982). It also poses fewer problems in identification by XRDA and DTA and, in addition, can usually be picked up microscopically through its autofluorescence under UV. Unfortunately, although common in some oxisols and occupying a terminal position in the weathering sequence, its presence is not necessarily conclusive proof of such extreme weathering conditions. For example, recent studies have revealed its presence in alpine and humid temperate soils and suggest its ephemeral development at an early stage of pedogenesis prior to its reappearance at the final stages (Macias Vasquez, 1981).

Haematite is perhaps the most commonly cited indicator of palaeosols by virtue of the visible reddening which is ascribed to its presence in the soil. Indeed, the colour (7.5 YR or redder, or 5 YR or redder in textures coarser than sandy silt loam) is used as a criterion for the recognition of the 'Paleo-argillic Soil Group' in England and Wales (Avery, 1980). Certainly haematite is the distinctive pigmenting component of red tropical soils and the cause of 'rubification' in many Mediterranean soils. Red colours are therefore a useful prompt as to the possibility of palaeosols (e.g. Bullock *et al.*, 1973), but several difficulties arise in confirming the identification. First, it is again possible for soils to inherit haematite from a number of parent rock types in Britain (e.g. Lawson, 1983) and from the red continental sandstones/mudstones (Devonian/Triassic) in particular: this possibility can sometimes be eliminated by careful heavy mineral analysis and study of grain morphology. Patchy red colours are also a common feature of pyrite oxidation in shales, some of which may be hydrothermal in nature. Secondly, the climatic control of pedogenic haematite is not fully known. Schwertmann *et al.* (1982), for example, have shown from detailed studies in Germany that haematite had formed under temperate axeric conditions during the Holocene (boreal?) period, especially in gravelly (and calcareous) materials: they therefore question the necessary correlation of reddening with earlier warmer interglacial conditions (e.g. Chartres, 1980). Thirdly, the relationship between red colours and haematite is not necessarily exclusive since the physical state of other ferric minerals (lepidocrocite, maghemite, ferrihydrite, ferroxyhite, etc.) may result in similar effects, especially in gley mottles. And finally, the identification of haematite at low concentrations by XRDA is not always easy, especially—as is often the case—in the presence of goethite. To confirm its presence, it is often necessary to refine the XRDA technique in order to resolve overlapping peaks and to increase poor sensitivity (e.g. by Differential XRDA, Schulze, 1981) or to have recourse to more sensitive but less available techniques (e.g. Mössbauer spectroscopy; Schwertmann *et al.*, 1982), and it can also be helpful to pre-concentrate the iron oxide minerals from the clay fraction magnetically (Sanmugadas, 1971). Other properties which may prove relevant in the study of palaeosols are the Al-content and crystallinity of goethite which have been correlated with age and palaeo-environment (Fitzpatrick and Schwertmann, 1982).

Similar difficulties therefore beset the use of all three minerals as proof of the existence of palaeosol. Difficulties can arise in detection and identification; the limits to the pedogenic formation in terms of climate are uncertain; and the possibility of hydrothermal, inherited or contaminant material must be eliminated. Nevertheless, these minerals are valuable in alerting the investigator to the possible presence of a palaeosol and, with due precautions, confirming its identification.

'Deep weathering'

A special case of considerable interest and relevance to the study of palaeosols is 'deep weathering', where the depth and degree of alteration is on a scale incompatible with post-glacial weathering and more reminiscent of tropical sapropels. This has led investigators of 'deep weathering' in Britain to invoke more intense and/or prolonged periods of weathering such as might have occurred in pre-glacial (Tertiary) or possibly inter-glacial times. Where substantiated by detailed stratigraphical, mineralogical and geochemical data (e.g. Basham, 1974), these sites provide valuable, if truncated, evidence for palaeosols. However, a problem exists in that the igneous rocks which most commonly show these features to best effect are also subject to hydrothermal alteration by late-stage solutions permeating joint systems in the solidified rock; these alter the original minerals with effects which may prove indistinguishable from those of 'surface' weathering. Here there is a pressing need to establish and apply criteria to distinguish 'subaerial' from hydrothermal processes.

One possibility is provided where hydrothermal activity enhances the rock selectively with characteristic trace elements (B, F, rare earths, etc.). For example, 'deep weathering' was suggested for a site at 800 m OD in Snowdonia, Wales, where gibbsite occurs within a zone of rotted granite (Ball, 1964). Subsequent mineralogical analyses revealed the presence of a niobium-rich rutile (struverite) and of unusual varieties of brookite and anatase, all of which strongly suggests hydrothermal alteration. However, the problem is not resolved in that deep 'subaerial' weathering is not precluded: it may still have occurred subsequently, penetrating deeper into the zone already weakened by hydrothermal activity, and producing a palaeosol with pedogenic gibbsite, protected relics of which have survived.

Another possible criterion for distinguishing subaerial and hydrothermal products is isotopic analysis, particularly for $^{18}O/^{16}O$ and D/H ratios. Sheppard (1977) utilized this technique in a study of the important Cornish kaolin deposits and concluded, contrary to previous general opinion, that they are predominantly of subaerial rather than of hydrothermal origin. When substantiated by further studies, isotopic analysis should prove a very valuable technique in the validation of palaeosols.

Biological components

Microscopic examination of soils, both disaggregated and in thin section, usually reveals the presence of many distinctive components of biological origin such as spores, charcoal or skeletal material. In palaeosols these are effectively 'fossils' and can fulfil a correlative role similar to that in classical stratigraphy. There are species which have become extinct during the course of the Quaternary, but the time-span is generally too brief and only rarely is it possible to use the fossil remains of an extinct species as unambiguous proof for a palaeosol: fossils in palaeosols tend to be of more use as indicators of changing environment. Some of these biological components in soil have been studied in great detail whilst others remain unidentified. The study of plant pollen evolved into palynology, now an indispensable, though somewhat separate, branch of Quaternary science (Birks and Birks, 1980). The earlier unique position of palynology as a stratigraphic aid has been supplemented by the study of other resistant fossils such as beetle elytra (Coleoptera) which, its protagonists would claim, are particularly responsive to climatic change (Coope, 1977). In calcareous soils the shells of land snails (Mollusca) have proved valuable as indicators of microenvironment and for stratigraphic correlation (Kerney, 1977).

Other obvious terrestrial microscopic fossils include the opaline silica tests of diatoms (Figure 2.5e) and sponge spicules. Due to their relatively low density, these are readily concentrated from the fine sand fraction using heavy liquids (SG >2.2). They have a distinctive morphology and can provide useful palaeoenvironmental information. More ubiquitous, though less distinctive, components of the same low-density fraction are the silicified plant cells (phytoliths) which survive in acid soils (Figure 2.5f). There are a few distinctive forms whose origins can be identified in terms of plant species, but the systematic study of phytoliths has not yet advanced to the point where they can be used generally in the study of palaeosols (Rovner, 1983). Relative abundance has, however, been used by Dormaar and Lutwick (1969) to distinguish forest from grass vegetation that colonized palaeosols. There is obviously considerable scope for future development in the use of phytoliths and other microfossils in the recognition and interpretation of palaeosols as techniques of recovery and identification are developed and information accumulated.

CONCLUSIONS

A palaeosol is a valuable source of information about a past system in which environmental conditions were imprinted with varying clarity and permanence. Much of this information is of a chemical or mineralogical nature and its successful extraction is dependent on the availability of techniques which are sufficiently sensitive to detect often subtle distinguishing features. These features often relate to a difference in the time factor, expressed in degree of weathering and the pedogenic mineralogy, or in the climatic or biotic factors. The latter include the direct and indirect effects of man and there is therefore a valuable overlap with archaeological research in terms of both technique and data.

Established soil analytical techniques can often provide positive evidence for the existence of a palaeosol. For this, the properties must be clearly differentiated from, and incompatible with, those of contemporary soils. Examples of the successful use of analysis for trace elements, carbonates, extractable iron, phosphorus, or of mineralogical observations by optical and scanning electron microscopy or XRDA can be cited. But often the evidence is ambiguous because it tests knowledge of the rates and nature of pedogenic processes, and of the conditions of formation and persistence of minerals, beyond its present limits: indeed, it is often by the study of palaeosols that these limits can be extended. This is illustrated by the difficulties of using the presence of even such obvious minerals as kaolinite, gibbsite and haematite as criteria for the presence of palaeosols in the UK. Here the future availability of more refined analyses for isotopic composition, structural details, and other physical properties holds much promise. Nevertheless, if all sources of chemical, mineralogical and micromorphological information are co-ordinated with due care, they can provide an adequate basis for the recognition and interpretation of palaeosols.

REFERENCES

Avery, B. W. (1980). 'Soil classification for England and Wales: Higher categories', *Soil Survey Techn. Monogr.* No. 14, Harpenden, UK.

Ball, D. F. (1964). 'Gibbsite in altered rock in North Wales', *Nature*, **204**, 673–674.

Basham, I. R. (1974). 'Mineralogical changes associated with deep weathering of gabbro in Aberdeenshire', *Clay Mineralogy*, **10**, 189–202.

Birkeland, J. (1964). 'Pleistocene glaciation of the northern Sierra Nevada north of Lake Tahoe, California', *Journal of Geology*, **72**, 810–825.

Birks, H. J. B. and Birks, H. H. (1980). *Quaternary Palaeoecology*, Edward Arnold, London.

Boardman, J. (1984). 'The Troutbeck paleosol, Cumbria, England', in *Soils and Quaternary Landscape Evolution* (Ed. J. Boardman), Wiley, Chichester (in press).

Boulton, G. S. and Worsley, P. (1965). 'Late Weichselian glaciation in the Cheshire-Shropshire basin', *Nature*, **207**, 704–706.

Bowler, J. M. and Polach, H. A. (1971). 'Radiocarbon analysis of soil carbonates', in *Palaeopedology* (Ed. D. H. Yaalon), pp. 97–108, Israel Universities Press, Jerusalem.

Brindley, G. W. and Brown, G. (1980). *Crystal Structures of Clay Minerals and their X-ray Identification*, Mineralogical Society, London.

Bronger, A. (1971). 'Zur Genese und Verwitterungs intensität fossiler Loessboden in Jugoslavien', in *Palaeopedology* (Ed. D. H. Yaalon), pp. 271–281, Israel Universities Press, Jerusalem.

Bullock, P. (1985). 'The role of micromorphology in soil studies', in *Soils and Quaternary Landscape Evolution* (Ed. J. Boardman), Wiley, Chichester (in press).

Bullock, P., Carroll, D. M. and Jarvis, R. A. (1973). 'Palaeosol features in Northern England', *Nature Phy. Sci.*, **242**, 53–54.

Catt, J. A. (1977). 'Loess and coversands', in *British Quaternary Studies: Recent Advances* (Ed. F. W. Shotton), pp. 221–231, Clarendon Press, Oxford.

Catt, J. A. (1979). 'Soils and Quaternary geology in Britain', *Journal of Soil Science*, **30**, 507–542.

Chartres, C. J. (1980). 'A Quaternary Soil sequence in the Kennet Valley, central southern England', *Geoderma*, **23**, 125–146.

Conway, J. S. (1983). 'An investigation of soil phosphorus distributions within occupation deposits from a Romano-British Hut Group', *Journal of Archaeological Science*, **10**, 117–128.

Coope, G. R. (1977). 'Quaternary coleoptera as aids in the interpretation of environmental history', in *British Quaternary Studies: Recent Advances* (Ed. F. W. Shotton), pp. 55–68, Clarendon Press, Oxford.

Davidson, D. A. & Shackley, M. L. (1976). *Geoarchaeology: Earth Science and the Past*, Duckworth, London.

Dormaar, J. F. and Lutwick, L. E. (1969). 'Infra-red spectra of humic acids and opal phytoliths as indicators of palaeosols', *Canadian Journal of Soil Science*, **49**, 29–37.

Dormaar, J. F. and Lutwick, L. E. (1983). 'Extractable Fe and Al as an indicator for buried soil horizons', *Catena*, **10**, 167–173.

Douglas, L. A. and Platt, D. W. (1977). 'Surface morphology of quartz and age of soils', *Soil Science Society of America, Journal*, **41**, 641–645.

Evans, J. G. (1975). *The Environment of Early Man in the British Isles*, Elek, London.

Fenwick, I. (1985). 'Paleosols: problems of recognition and interpretation', in *Soils and Quaternary Landscape Evolution* (Ed. J. Boardman), Wiley, Chichester (in press).

Fitzpatrick, R. W. and Schwertman, U. (1982). 'Al-substituted goethite — an indicator of pedogenic and other weathering environments in South Africa', *Geoderma*, **27**(4), 335–347.

Gieseking, J. E. (1975). *Soil Components, Vol. 1 Organic Components*, Springer, Berlin.

Gile, L. H., Peterson, F. F. and Grossman, R. B. (1966). 'Morphological and genetic sequences of carbonate accumulation in desert soils', *Soil Science*, **101**, 347–360.

Green, H. S. (1984). *Pontnewydd Cave, a Lower Palaeolithic hominid site in Wales: The First Report*, National Museum of Wales, Cardiff.

Hedges, R. E. M. (1981). 'Radiocarbon dating with an accelerator — review and preview', *Archaeometry*, **23**, 3–18.

Hughes, J. C. (1980). 'Crystallinity of kaolin minerals and their weathering sequence in some soils from Nigeria, Brazil and Colombia', *Geoderma*, **24**, 317–325.

INQUA: Working group on the origin and nature of palaeosols (1971). 'Criteria for the recognition and classification of palaeosols', in *Palaeopedology* (Ed. D. H. Yaalon), pp. 753–758, Israel Universities Press, Jerusalem.

Jackson, M. L. (1964). 'Chemical composition of the soil', in *Chemistry of the Soil* (Ed. F. E. Bear), pp. 71–141, Reinhold Publishing Co., New York.

Jenkins, D. A. (1964). *Trace Element Studies on Some Snowdonian Rocks, Minerals and Related Soils*, Unpublished PhD Thesis, University of Wales.

Jenkins, D. A. and Bower, R. P. (1974). 'The atmospheric contribution to the trace element content of soils', *Transactions 10th International Soil Science Conference, Moscow, VI(II)*, 466–473.

Kerney, M. P. (1977). 'British Quaternary non-marine mollusca — a brief review', in *British Quaternary Studies: Recent Advances* (Ed. F. W. Shotton), pp. 31–42, Clarendon Press, Oxford.

Krinsley, D. A. and Doornkamp, J. C. (1973). *'Atlas of Quartz Sand Surface Texture'*, Cambridge University Press, London.

Lawrence, J. R. and Taylor, H. P. (1972). 'Hydrogen and oxygen isotope systematics in weathering profiles', *Geochemica et Cosmochimica Acta*, **36**, 1377–93.

Lawson, T. J. (1983). 'A note on the significance of a red soil on dolomite in North West Scotland', *Quaternary Newsletter*, **40**, 10–12.

Lim, C. H., Jackson, M. L. and Higashi, T. (1981). 'Intercalation of soil clays with dimethylsulfoxide', *Soil Science Society of America, Journal*, **45**, 433–436.

Limmer, A. W. and Wilson, A. T. (1980). 'Amino acids in buried paleosols', *Journal of Soil Science*, **31**, 147–153.

Locke, W. W. (1979). 'Etching of hornblende grains in arctic soils: an indicator of relative age and palaeoclimate', *Quaternary Research*, **11**, 197–212.

Loughnan, F. C. (1970). *Chemical Weathering of Silicate Minerals*, Elsevier, New York.

McKeague, J. A., Brydon, J. E. and Miles, N. M. (1971). 'Differentiation of forms of extractable iron and aluminium in soils', *Soil Science Society of America Proceedings*, **35**, 33–38.

Macias Vasquez, F. (1981). 'Formation of gibbsite in soils and saprolites of temperate humid zones', *Clay Minerals*, **16**, 43–52.

Mücher, H. J. and Morozova, T. D. (1983). 'The application of soil micromorphology in Quaternary geology and geomorphology', in *Soil Micromorphology* (Eds P. Bullock and C. P. Murphy), pp. 151–194, A.B. Academic Publishers, Berkhamsted.

Novak, R. J., Motto, H. C. and Douglas, L. A. (1971). 'The effect of time and particle size on mineral alteration in several Quaternary soils in New Jersey and Pennsylvania U.S.A.', in *Palaeopedology* (Ed. D. H. Yaalon), pp. 211–224, Israel Universities Press, Jerusalem.

Poole, E. G. (1966). 'Late Weichselian glaciation in the Cheshire-Shropshire basin', *Nature*, **208**, 507.

Proudfoot, B. (1976). 'The analysis and interpretation of soil phosphorus in archaeological contexts', in *Geoarchaeology* (Eds. D. A. Davidson and M. L. Shackley), pp. 93–114, Duckworth, London.

Range, J. K., Range, A. and Weiss, A. (1969). 'Fire-clay type kaolinite or fire-clay mineral?' *Proceedings International Clay Conference 1969 Tokyo*, **1**, pp. 3–13. Israel Universities Press, Jerusalem.

Roaldset, E., Petterson, E., Longva, O. and Mangerud, J. (1982). 'Remnants of Preglacial Weathering in Western Norway', *Norsk Geologisk Tidsskrift*, **62(3)**, 169–178.

Rovner, I. (1983). 'Plant opal phytolith analysis: major advances in archaeobotanical research', in *Advances in Archaeological Method and Theory*, Vol. 6 (Ed. M. Schiffer), pp. 225–266, Academic Press, New York.

Ruhe, R. V. (1956). 'Geomorphic surfaces and the nature of soils', *Soil Science*, **82**, 441–455.

Sanmugadas, K. (1971). *Trace Elements Associated with the Iron Oxide Fraction of Soils*, Unpublished PhD Thesis, University of Wales.

Scharpenseel, H. W. (1971). 'Radiocarbon dating of soils—problems, troubles, hopes', in *Palaeopedology* (Ed. D. H. Yaalon), pp. 77–88, Israel Universities Press, Jerusalem.

Schulze, D. G. (1981). 'Identification of soil iron oxide minerals by differential X-ray diffraction', *Soil Science Society of America, Journal*, **45**, 437–440.

Schwertmann, U., Murad, E. and Schulze, D. G. (1982). 'Is there Holocene reddening (haematite formation) in soils of axeric temperate areas?' *Geoderma*, **27**, 209–223.

Shackleton, N. J. (1977a). 'Oxygen isotope stratigraphy of the middle Pleistocene', in *British Quaternary Studies: Recent Advances* (Ed. F. W. Shotton), pp. 1–16, Clarendon Press, Oxford.

Shackleton, N. J. (1977b). 'The oxygen isotope record of the Late Pleistocene', *Philosophical Transactions of the Royal Society*, **B280**, 169–182.

Sheppard, S. M. F. (1977). 'The Cornubian batholith, S. W. England: D/H and $^{18}O/^{16}O$ studies of kaolinite and other alteration minerals', *Journal Geological Society, London*, **133**, 573–591.

Singer, A. (1980). 'The palaeoclimatic interpretation of clay minerals in soils and weathering profiles', *Earth Science Reviews*, **15**, 303–326.

Singer, A. and Norrish, K. (1974). 'Pedogenic palygorskite occurrences in Australia', *American Mineralogist*, **59**, 508–517.

Stevenson, F. J. (1969). 'Pedohumus: accumulation and diagenesis during the Quaternary', *Soil Science*, **107**, 470–479.

Thompson, M. L., Smeck, N. E. and Bigham, J. M. (1981). 'Parent materials and paleosols in the Teays River Valley, Ohio', *Soil Science Society America, Journal*, **45**, 918–925.

Valentine, K. W. G. and Dalrymple, J. B. (1976). 'Quaternary buried palaeosols: a critical review', *Quaternary Research*, **6**, 209–222.

Younis, M. G. A. (1983). *Mineralogical Studies on Soils in North-West Wales*, Unpublished PhD Thesis, University of Wales.

Soils and Quaternary Landscape Evolution
Edited by J. Boardman
© 1985 John Wiley & Sons Ltd.

3

The Role of Micromorphology in the Study of Quaternary Soil Processes

P. BULLOCK

ABSTRACT

Micromorphology is the branch of soil science concerned with studying soils in an undisturbed state at magnifications from 2 to 50 000 times using stereo, petrological and scanning electron microscopes. Most studies to date have concentrated on examination of thin sections under the petrological microscope. The paper demonstrates some of the potential of micromorphology for identifying the nature and properties of the main soil minerals, and particularly the degree of weathering which they have undergone. An example of the increasing possibilities for quantification of components in thin sections is given.

The paper emphasizes the application of micromorphology to the study of Quaternary soil-forming processes. A distinction is made between processes associated with stable periods of soil formation, *viz*. weathering and neoformation, decalcification, podzolization, clay translocation, gleying, rubification and laterization, and those associated with unstable periods, e.g. cryoturbation, erosion.

Micromorphology has been used to order chronologically soil-forming processes during the Quaternary and three case studies are discussed.

INTRODUCTION

Micromorphology is a technique for studying soils in their natural state, i.e. with fabric and structure undisturbed, at a level of observation beyond that obtainable with the naked eye. Three main levels of observation are within the realm of micromorphology.

1 Stereomicroscopic observations. These are usually made within a magnification range of 1–30 times. They involve the use of a binocular stereomicroscope on samples in the field or laboratory. Samples can be

45

examined in their natural state, without prior drying or impregnation, and can be dissected during observation.

2   Examination of thin sections. The use of this technique necessitates the collection of samples from the field, removal of the moisture from them by air-drying, freeze-drying or acetone exchange, followed by their impregnation with a polyester or epoxy resin. Once the impregnated samples have gelled, they are sliced with a diamond saw, and slices mounted on a microscope slide before grinding down to about 30 μm. The size of samples that can be treated in this way has increased in recent years in an effort to obtain more representativity and thin sections up to 20×13 cm are now produced. Thin sections can be examined at a range of magnification from about 6 to 1000 times to provide information on soil components, their spatial distribution and features associated with various soil-forming processes (Bascomb and Bullock, 1976; Brewer, 1976; Bullock *et al.*, 1984).

3   Scanning electron microscope (SEM) observations. Using this technique, small samples are mounted on aluminium stubs, coated with gold, carbon, palladium or a combination of these to enhance conductivity of the specimen, and examined under the scanning electron microscope at magnifications of 100–50 000 times. The technique has the advantage of a large depth of field and a large magnification (Eswaran and Shoba, 1983).

Of these three approaches, the examination of thin sections has been the dominant one in soil micromorphology and it is this approach which is emphasized in this chapter. Micromorphology is now used in several branches of soil science including mineralogy, physics, chemistry, biology and pedology, and in a number of different sciences, e.g. agriculture, archaeology, ecology and soil mechanics. Historically, it has been used most widely in pedology to study soil processes, soil formation and to aid the development of criteria in systems of soil classification.

Rather than attempting to cover the whole spectrum of the uses to which micromorphology has been applied, this paper concentrates on those related to soil processes and soil formation. It attempts to show how micromorphology can be valuable in answering the following questions which Quaternary scientists might ask.

1   Is a particular material a soil or a sediment?
2   If it is a soil, what have been the main processes responsible for its formation?
3   Can soil-forming processes associated with particular periods of the Quaternary be identified?
4   To what extent is it possible to recognize sequences of soil-forming events within Quaternary soils?

MICROMORPHOLOGICAL DESCRIPTION

Description is fundamental to the use of micromorphology for interpreting soil processes and understanding soil formation. It is important for detailing the characteristics of soil whether at the simple level of a quartz grain or more organized levels, for example structure and fabric. It provides a basis for comparing one soil with another, for classifying soils and for developing models to explain soil genesis. It also acts to support other analyses and forms a basis for selecting further analyses.

A number of systems for thin-section description, many dealing with specific aspects such as organic matter and structure, have been produced. The recently published *Handbook for Soil Thin Section Description* (Bullock *et al.*, 1984), prepared under the auspices of the International Society of Soil Science, represents an attempt to provide a comprehensive international system.

Most of the important components of soils, with the exception of gases and liquids, can be seen in thin section. In describing thin sections, the following three groups of constituents are generally recognized: basic components, microstructure including aggregates and pores, and pedofeatures, i.e. features formed from the various soil-forming processes. These are usually treated separately for descriptive purposes.

Basic components

This term is used to describe the simplest units, such as sand grains, clay particles, organic fragments, in a thin section. These components constitute the basic building blocks of the groundmass, structure, fabric or pedofeatures. They are generally inherited from the parent material (mineral or organic), but some, particularly in the finer fractions, may be formed by alteration of primary particles. In describing a thin section, a distinction is usually made between coarse and fine basic components. The division is somewhat flexible but is usually between 2 and 10 μm, dividing on the one hand particles capable of being readily resolved under the microscope from fine particles mostly below the resolution of the microscope. Coarse mineral basic components include discrete mineral grains, quartz being by far the most common in most soils, rock fragments and, more rarely, inorganic residues of biological origin, e.g. diatoms, and artefacts such as pieces of pot, brick, etc. Description of these components should include the following attributes: nature, size, shape, colour, abundance and degree of weathering.

The mineral fine material is particularly important because it influences many soil properties. For the most part, it consists of clay-size particles. Because the particles cannot be resolved individually, the optical properties of fine material as a whole are described including nature, colour, birefringence and limpidity. The degree of preferred orientation of the fine material expressed by its birefringence is an important property indicating much about the soil environment (Brewer, 1976).

The coarse organic fraction mainly consists of tissue or organic fragments. These are dominantly floral rather than faunal. They can be described according to size, shape, colour, opacity, internal structure and degree of preservation. The origin of the fragments can be inferred sometimes from the description provided the fragments are reasonably fresh.

Fine organic material includes cell residues and amorphous organic material in which the original plant structure is no longer discernible. It is described mainly according to colour and opacity.

In many topsoils it may not be possible to distinguish between organic and mineral fine material since the two are intimately associated. The fine material is termed organo-mineral in such cases and described according to colour, limpidity and birefringence where present.

Microstructure

Soil microstructure is concerned with the size, shape and arrangement of the basic components and voids in both aggregated and non-aggregated material and the size, shape and arrangement of any aggregates present. The term 'fabric' is used in the sense of the arrangement of constituents. In many soils, the basic components are welded together into aggregates. Often these aggregates are recognizable in the field and thin sections are used to confirm the grade of structure and to aid identification of the smaller ones. The main types of aggregates in soils are crumbs, granules, subangular and angular blocks, plates and prisms. The type present together with size and degree of development should be described.

In some soils, no aggregates are present and in such cases the microstructure is defined through the types of voids present. Brewer (1964) identified the following types of voids in thin sections: simple packing, compound packing, vughs, channels, vesicles, chambers and planes.

A large number of different types of microstructure are recognizable in thin section as a result of different combinations of basic components, voids and aggregates. Some twenty-six are listed and illustrated in Bullock *et al.* (1984).

Pedofeatures

Soil-forming processes leave their imprint in soils in the form of particular features, termed pedofeatures. Six different groups of pedofeatures have been described in the *Handbook for Soil Thin Section Description*.

1   *Textural.* These are pedofeatures due to mechanical translocation and subsequent deposition of particles in the soil. They are formed mainly of mineral particles, the most common being clay coatings.

2 *Depletion*. These form by the chemical depletion of certain components. They occur most commonly at the edges of pores and aggregates and usually involve the loss of iron, manganese and/or calcium carbonate with respect to amounts present in the soil matrix.

3 *Crystalline*. These are features formed by the recrystallization of compounds from solution. The most common example in the Quaternary soils of Britain involves calcite. Forms include coatings on voids, infillings of voids and nodules within the matrix.

4 *Amorphous and crypto-crystalline*. These features are formed of substances without clearly defined discrete crystals. They are commonly of manganese, iron or calcium carbonate taking the form of coatings, infillings and nodules. Mottles in poorly drained soils fall into this category.

5 *Fabric*. These features result from modification of the fabric more or less *in situ*, for example where faunal activity or roots cause compaction of one part of the matrix compared with another. Such pedofeatures are rather rare and sometimes difficult to recognize.

6 *Excrements*. Faunal excrements are common in many soils, particularly in upper horizons. Where fresh, they can be readily identified and often assigned to specific soil animals.

The descriptive system for pedofeatures varies with the particular pedofeature involved. Some or all of the following properties are usually described: nature; size; external morphology, e.g. shape, rugosity, sharpness of boundaries; internal fabric; contrast with adjacent material; abundance; variability; and distribution and orientation.

QUANTITATIVE MICROMORPHOLOGY

In the past micromorphology has been essentially a descriptive and interpretational branch of soil science. This is no longer the case as dramatic developments in new technology now enable rapid accurate measurements to be made of components, structures and features in thin sections, polished impregnated blocks and pieces of unimpregnated soil or rock.

There are now a number of instruments that can be used to make a microchemical analysis of components and features in thin section. These include electron probe micro-analyser, SEM coupled with an energy dispersive X-ray analyser, auger electron spectroscopy, electron spectroscopy for chemical analysis, ion microscopy, ion microprobe analyser and laser microprobe mass analyser. With some of the latter instruments all the elements present in a mineral or a pedofeature can be analysed and local concentrations as small as 10 mg/kg can be measured.

Image analysing computers are now being used for accurate measurement of pore space and structural units in soils. Images can be derived from thin

sections, by photography in UV light of flat faces of impregnated blocks, or
by using back-scattering electron images derived from the SEM. It is now
possible to measure pores of 1 μm diameter and larger.

Using ion-beam thinning, it is possible to produce sections sufficiently thin
for direct use on a transmission electron microscope enabling the fine structure
and fabric of soils to be investigated.

The combination of micromorphological study of thin sections and these
submicroscopic techniques represents an important new era in soil studies. An
example of a study involving thin sections and electron probe microanalysis
is given in Figure 3.1 (by kind permission of Dr D. A. Jenkins). Useful
summaries of recent advances in the use of submicroscopic techniques are given
in Bisdom (1981) and Bisdom and Ducloux (1983).

**In the B₂C of a ferrallitic
soil from E. Africa.
(Buganda Series)**

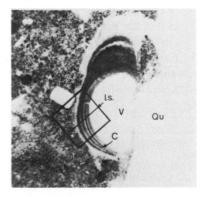

Fe x-ray image   **x-ray line scans**

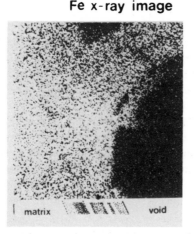

Figure 3.1   X-ray image and line scan across a laminated iron-enriched clay coating

MICROMORPHOLOGICAL STUDIES OF
QUATERNARY SOIL PROCESSES

Micromorphology has been used extensively to develop and improve models of soil formation in the UK. To this end it has been used in two ways. At its simplest, it has been used to extend the field description of soils. By combining macro- and micromorphological descriptions it is possible to relate features seen in thin sections to soil profiles and to landforms. Over the past 15 years some 12 000 thin sections of samples from some 3000 soil profiles have been made in the Soil Survey of England and Wales representing most, if not all, of the soil types in Britain. At a more detailed level, micromorphological studies have been integrated with various mineralogical, chemical, physical and biological studies to examine certain soil processes and their effects on soil formation in-depth. These two approaches have been fundamental to our current knowledge of soil formation and to the development of soil classification systems. The concept of diagnostic soil horizons has been introduced into several classification systems. These are horizons which express the dominance of particular soil-forming processes over others, e.g. the argillic horizon formed by significant translocation and deposition of clay (see Avery, 1985). It is now possible to characterize the main diagnostic horizons according to their micromorphology.

There are a number of soil-forming processes associated with soil development in Britain during the Quaternary, some of which have given rise to diagnostic horizons, others which have been significant in other aspects of soil formation. In effect, most of the major soil-forming processes known throughout the world appear to have been active at some stage during the Quaternary. These include, (a) weathering of minerals and rock fragments and neoformation of minerals, (b) alteration of organic material, (c) solution and precipitation, (d) podzol-ization, (e) translocation of clay and other particulate matter, (f) gleying, (g) rubification and laterization. These processes are characteristic of periods of stability, e.g. interglacial periods, but the Quaternary has also been marked by periods of instability. During these periods of instability, permafrost, solifluction, glacial erosion and deposition have affected soil development but compared with soil-forming processes during stable periods, much less is known of their effect on soil formation.

The main strength of micromorphology has been in the identification of the effects of the various soil processes. It is generally not possible to measure the processes themselves but, with varying degrees of reliability, it is possible to relate many macro- and microscopic soil features to particular processes. Soils in the landscape are rarely the result of a simple process but rather a number of interacting processes, and most soils have a complex history of soil formation. Micromorphology is a useful technique with which to unravel this history. Some examples of the way in which micromorphology can be used to model the main soil-forming processes in the Quaternary are given.

Weathering and neoformation of mineral components

Although most Quaternary soils are dominated by quartz in the sand and silt fractions there is a variety of other minerals and sometimes also rock fragments present. These can be used to determine the source of the soil material, its relationship to underlying rocks and the extent and type of weathering that may have taken place.

Desirable microscopic data includes: identification of primary mineral species and rock fragments and their abundance, changes in the optical properties with weathering, patterns of distribution of weathering products within mineral grains and rock fragments and identification and amounts of the secondary minerals.

Identification and counting of primary and some secondary minerals is best achieved by examination of loose grains under a petrological microscope (Jenkins, 1985). Thin sections are most valuable for identifying the stages in the weathering of minerals and rock fragments, the pattern of distribution of the weathering products and the composition and other properties of the neoformed minerals. The International Society of Soil Science (ISSS) Sub-Commission on Soil Micromorphology has recently produced guidelines for the description of mineral alterations in thin sections (Stoops *et al.*, 1979). The guidelines include the following important characteristics of weathering mineral grains, (a) the secondary porosity, including the arrangement of pores and secondary products, their shape and classification, (b) the pattern of mineral alteration and the location of the alteration within the mineral grain, (c) the degree of alteration, and (d) the nature of the secondary products. They also include a terminology for different weathering patterns and a scale of five classes to describe the degree of mineral alteration.

Soil minerals vary in their stability and a number of stability/weatherability sequences have been proposed (e.g. Goldich, 1938; Brewer, 1976). Furthermore, a particular mineral may weather to different secondary (neoformed) minerals depending on the prevailing physico-chemical conditions. There have been a number of useful weathering studies using separated grains, thin sections and submicroscopic techniques. Meyer and Kalk (1964) established a weathering sequence in Holocene soils in Central Europe for the following minerals: biotite, muscovite, chlorite, feldspars, augite, epidote and hornblende. Bisdom (1967) and Meunier (1977) demonstrate the value of an integrated approach to mineral weathering involving a number of techniques including micromorphology. The ISSS Sub-Commission on Soil Micromorphology has recently published detailed studies of the weathering of olivine (Delvigne *et al.*, 1979) and biotite (Bisdom *et al.*, 1982). Both show convincingly how a particular mineral can alter to a variety of new minerals. For example, biotite may weather to vermiculite, smectite, kaolinite and iron oxyhydrates, or gibbsite and iron oxyhydrates.

The extent to which pedogenic neoformation of minerals has occurred during the Quaternary is not well established. Neoformed minerals most commonly occurring in thin sections of Quaternary soils are, anhydrite, barite, calcite, celestite, goethite, gypsum, haematite, jarosite, kaolinite, pyrite, siderite, smectite, vermiculite, vivianite. However, such minerals although occurring in the soil may not always be due to pedogenesis. Many may be associated with neoformation in older materials now re-worked to form a new parent material, or may be formed during deposition of the parent material.

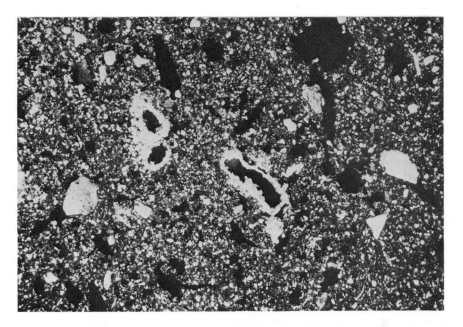

Figure 3.2   Calcite coatings (white) around voids in a brown calcareous earth. Frame length: 5 mm. Cross polarized light

There are some instances in which neoformed minerals can be associated specifically with pedogenesis. The following are some examples from thin sections of Quaternary soils.

1   Calcite coating voids in some B horizons below A horizons leached of carbonate (Figure 3.2).
2   Vivianite present in some alluvial soils and in soils on archaeological sites.
3   Gypsum formed by oxidation of pyrite in soils developed in alluvial deposits, i.e. in some soils of the Fens.
4   Goethite present as coatings to voids in some soils subject to periodic waterlogging.
5   Jarosite forming coatings to voids in some alluvial soils.

Many more studies of weathering and neoformation in soils of the Quaternary are needed. In undertaking such studies, the following points should be borne in mind.

1   Ideally the study should involve petrographic analysis of loose mineral grains, thin sections and submicroscopic techniques.
2   Parent materials of soils may already contain weathered and/or neoformed minerals. Hence, it is important to determine the nature of the parent material also.
3   Wherever possible, whole profiles should be chosen in order to establish weathering trends and to relate them to particular environmental conditions.

Alteration of Organic Material

Micromorphology has aided morphological studies of organic matter. Its principal use has been to define the humus profile, i.e. the profile that develops at the soil surface as a result of the gradual breakdown of plant litter. A number of humus profiles, referred to as humus forms, have been recognized and defined micromorphologically; for example, Kubiena (1953) defined thirty-one different forms. In some humus forms (e.g. mor, moder), there is differentiation of the humus profile into a number of layers: litter (L), fermentation (F) and humification (H). In other humus forms (e.g. mull), such layers are absent because breakdown of the plant material is rapid and humified organic material combines with clay to form clay–humus complexes. Micromorphology has proved a useful technique for describing in each layer and for each form the main morphological forms of plant material, the stages in decomposition of the plant material under different ecological conditions and the activity of soil fauna (Figures 3.3–3.6). A valuable review of the micromorphology of the main humus forms is given by Babel (1975).

In addition to defining the main humus forms, micromorphology can be used to identify the influence of soil fauna on the breakdown of plant material and the formation of structural aggregates. Many soil animals produce distinctive excrements from which their activity in particular layers of the soil can be inferred (Figures 3.5 and 3.6). Some of the main micromorphological forms of excrements have been described by Bal (1982) and the *Handbook for Soil Thin Section Description* (Bullock *et al.*, 1984) contains a chapter on their recognition.

Whereas the pathways and stages by which mineral grains undergo alterations are now fairly well documented, the same is not true for organic particles. Although micromorphology has been used effectively to determine the extent of breakdown of plant material in particular layers there have been few detailed studies of the stages in the breakdown. This deficiency may well be remedied

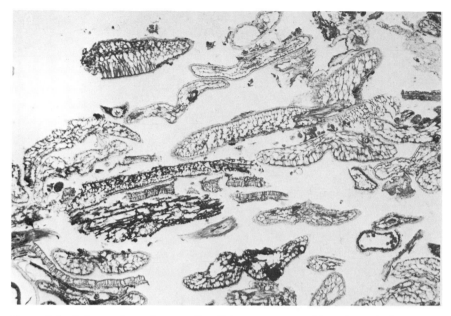

Figure 3.3   L layer of mor humus. Parallel arrangement of loosely packed, more or less fresh, pine needles. Humo-ferric podzol. Frame length: 5 mm. Plane polarized light

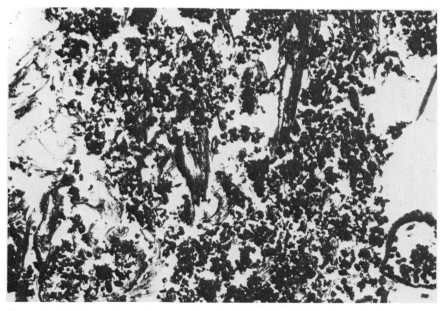

Figure 3.4   F layer of mor humus. Abundant blackened faunal excrements, partially coalesced, together with occasional severely decomposed plant fragments. Humo-ferric podzol. Frame length: 5 mm. Plane polarized light

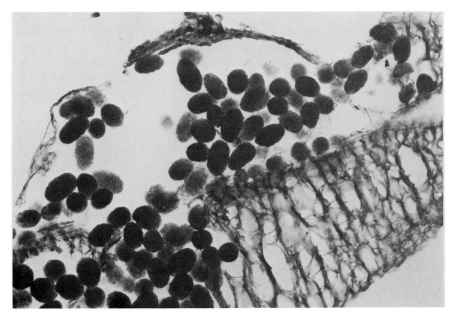

Figure 3.5   Oribatid mite excrements in pine needles. Humo-ferric podzol. Frame length:
1.3 mm. Plane polarized light

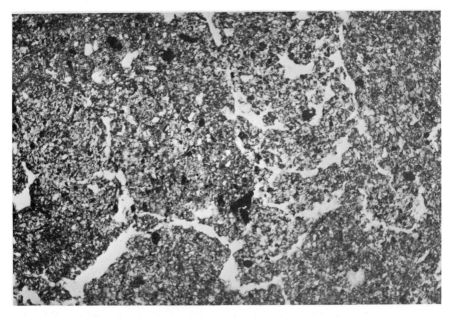

Figure 3.6   Mull humus. Note earthworm casting partially filling channel right of centre
and lack of plant remains compared with Figures 3 and 4. Typical brown earth. Frame
length: 5 mm. Plane polarized light

to some extent in a forthcoming paper by U. Babel (pers. comm.) in which a number of stages have been distinguished and described. Babel (1984) has also produced a simple micromorphological classification of organic matter in soils and suggested a descriptive system for the properties of each class.

Solution and re-precipitation

Some Quaternary deposits (e.g. Chalky Boulder Clay), which form extensive soil parent materials, are calcareous. The calcium carbonate occurs as fragments of chalk or limestone, tiny ($<20\,\mu$m) crystallites of calcite disseminated homogeneously throughout the matrix, and, in the case of loess, sometimes as nodules.

The onset of soil formation on these parent materials involves leaching of the calcium carbonate. Field evidence for this lies in the absence or lower content of $CaCO_3$ in upper soil horizons than in the parent material itself. Examination of thin sections of such soils provides added evidence for leaching of carbonate during soil formation in the form of two features.

1 Depletion pedofeatures in partially leached upper horizons, the edges of peds, channels and planes in contact with percolating water having lost $CaCO_3$ compared with adjacent ped interiors which still contain crystallites of calcite.
2 Horizons beneath the presumed leached horizons contain evidence for re-precipitation of carbonate in the form of calcitic void coatings and infillings (Figure 3.2).

From micromorphological studies Brewer and Sleeman (1969) postulated a possible relationship between the form and degree of concentration of carbonates with age. In soils with low amounts of carbonate, they suggested a progression with age from crystallites in the matrix, through diffuse irregular nodules to dense, usually fine-grained, discrete nodules. Crystallaria and void coatings of calcite were most common in horizons containing rather large proportions of carbonate. Different successions were suggested for gravelly and fine-textured materials. The sequences proposed are based on Australian experience (see Birkeland, 1985). No similar relationships have been described in Britain perhaps because few detailed studies of our calcareous soils have been made.

Podzolization

Podzolization is the process whereby carbon, iron and aluminium released in upper soil horizons are translocated downwards and redeposited to form B horizons enriched in these elements. Horizons diagnostic of the podzolization process are termed spodic (Soil Survey Staff, 1975) or podzolic (Avery, 1980).

With respect to Quaternary soils it is important to distinguish between lowland podzols on permeable parent materials in which the B horizons can be quite thick and upland stagnopodzols in which deposition of the iron is mainly in the form of a thin iron pan.

Micromorphology has contributed to the study of podzols. Most activity has centred on the microstructural properties of the B horizons. Four main types of microstructure of such horizons have been described by De Coninck and Righi (1983).

1 *Cemented.* This is the most common form of microstructure in lowland podzols. It is formed by iron and aluminium, possibly in combination with organic carbon, coating individual mineral grains, welding them into a more or less cemented entity (Figure 3.7). The degree of cementation depends on the extent of coatings and particularly the degree of bridging between grains.

2 *Friable.* This type of microstructure has also been termed 'pellety' because it consists of a loose arrangement of small, 20–200 μm pellet-like aggregates (Figure 3.8). They occur mainly in the upper parts of B horizons and particularly where the clay content exceeds about 10 per cent. There is considerable controversy over their origin, since some researchers attribute the microstructure to biological activity, and others to precipitation of organic matter — Fe/Al complexes.

3 *Placic (thin iron pan).* In thin section, it can be seen that the pan is composed of a number of layers. The upper layer is the darkest, most isotropic and most impervious. Layers below are generally less dense and more ochreous or reddish in colour and the pores in them appear to have been fossilized by deposition of iron and/or manganese. Globular masses of iron oxyhydrate are characteristic of several layers of the pan.

4 *Nodular.* This is characteristic of horizons in which iron-rich patches alternate with iron-poor areas. The iron-rich patches are nodular, the nodules being reddish brown and dense. The iron-poor areas are loose.

The recognition of these four different types of microstructure has led to much discussion about their origin. More information on their origin, particularly their connection with the podzolization process, is needed before they can all be used to aid field identification of podzols and as criteria for defining podzol B horizons. At present only the cemented and placic microstructures are accepted as being directly related to podzolization.

Clay Translocation

Clay translocation is one of the more recent soil-forming processes to be recognized. The process involves the dispersion of clay particles in upper horizons, their translocation downwards in suspension in water, and their deposition in lower horizons. The horizon of deposition, termed the argillic horizon, is an important diagnostic horizon in several classification systems. Clay translocation has been a significant process during stable interglacial periods of the Quaternary.

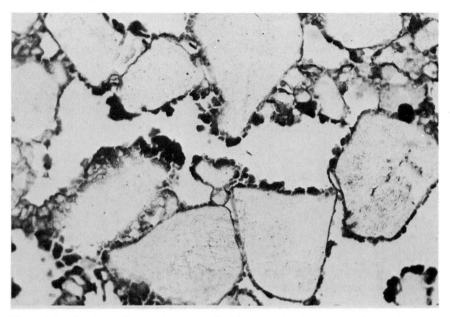

Figure 3.7   Coatings of amorphous sesquioxides on grains in the Bs horizon of a humo-ferric podzol. These coatings form the cementing material for the horizon. Frame length: 1.4 mm. Plane polarized light

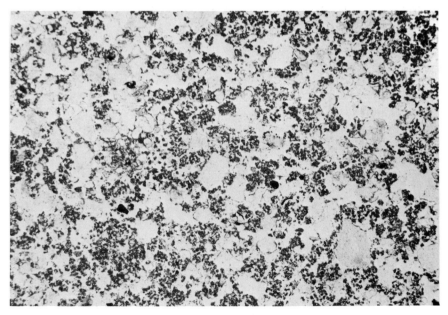

Figure 3.8   Pellety microstructure of Bs horizon of a brown podzolic soil. Frame length: 5 mm. Plane polarized light

Micromorphology has aided the characterization of argillic horizons. Identification of clay movement in the field in the form of clay coatings around voids and peds is sometimes difficult and its ultimate recognition is often only possible in thin sections.

Illuvial clay has a number of pronounced optical properties which aid its distinction from non-illuvial or inherited clay. These properties include, strong preferred orientation, optical continuity, lamination, textural contrast with matrix material, sharp boundary with adjacent material in plane or cross-polarized light.

A distinction between argillic horizons formed in the post-Devensian and those formed in earlier interglacials has been made by Avery (1973, 1985), the latter being referred to as paleo-argillic. There are marked differences between the two microfabrics (Bullock, 1974). By contrast with the argillic horizon, the paleo-argillic horizon has a complex fabric with the following features: most of the illuvial clay is within the matrix rather than as coatings to void or ped surfaces; a large proportion of the illuvial clay occurs as fragments of coatings; any argillans present are egg-yellow or sometimes partially reddened; and there may be some red segregations within the matrix. In argillic horizons formed in Devensian or later deposits, most of the illuvial clay occurs as brown to dark-brown coatings on voids and peds (Figure 3.9), although some may be transferred to the matrix if biological activity is intense. Red segregations are largely absent and the matrix clay generally has a much lower degree of preferred orientation than its paleo-argillic counterpart.

Clay-size particles are not the only ones to move through the soils (Harris, 1981). There is increasing evidence of coatings around pores and peds composed of a range of both mineral and organic particles. Such coatings are associated with the introduction of agriculture under which stability of the structure decreases and more particles become available for transport.

Micromorphology has been used more widely to study the process of clay translocation than any other processes. A number of reviews have been made of its role in the study of clay translocation, e.g. McKeague and co-workers (1981), McKeague (1983) and Bullock and Thompson (1985).

Gleying

Gleying is the process associated with periodic waterlogging of the soil during which the soil becomes starved of oxygen and there is reduction of iron and manganese. The reduced forms of iron and manganese are more mobile than the oxidized forms and migrate within the soil, resulting in enrichment in some areas and impoverishment in others (Figure 3.10). As the soil dries out and becomes oxidized, the enriched parts appear as rust or ochreous mottles which contrast strongly with the grey depleted zones.

A number of macro- and micromorphological features are associated with gleying which Veneman *et al.* (1976) have related to the length of periods of

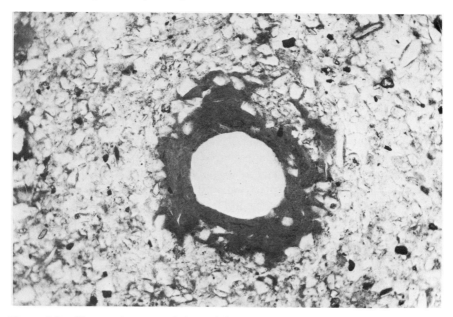

Figure 3.9 Clay coating around channel (in cross section) in an argillic brown earth developed in loess. Frame length: 1.3 mm. Plane polarized light

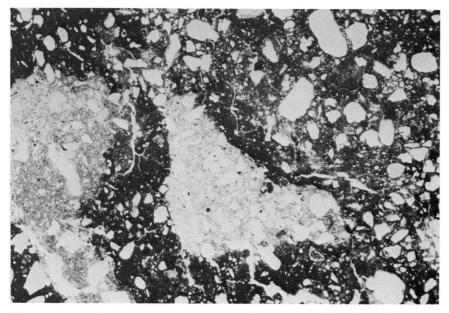

Figure 3.10 Variegated morphology of a gley soil. White areas represent zones of depletion, black areas zones of accumulation of iron oxides. Typical stagnogley. Frame length: 5 mm. Plane polarized light

saturation by water. With short periods of saturation of 1 day or less, ped mangans and Mn nodules formed and, less prominently, some concentrations of iron. Short periods of saturation of a few days combined with high matric potentials for several months produced chromas of two within peds, ped ferrans and neoferrans and a few manganese cutans or nodules. Occurrence of chromas of one within peds, ped and channel neoalbans (depletion pedofeatures) and virtual lack of manganese cutans and nodules were associated with periods of continuous saturation for several months.

The investigations by Veneman *et al.* (1976) provide a basis for interpreting gley morphology in terms of water regime. There appears to be a good relationship between specific macro- and micromorphological features and period of saturation though care in interpretation needs to be exercised because some gley features may be formed in previous water regimes and subsequently fossilized.

Rubification and laterization

The term rubification is used for the development of red colours (5 YR or redder) due to partial crystallization of amorphous iron oxides in a stage towards the formation of haematite. The process has been associated with average annual temperatures of $> 15\,°C$, average annual precipitation of about 800–900 mm and a marked dry season (Fedoroff, 1966). Age also seems to play a part, older soils tending to be redder than younger soils on similar parent material.

At present it is not possible to use micromorphology to relate different degrees of rubification to different proportions of haematite and goethite, even with the support of submicroscopic techniques. Such information has to be obtained from X-ray analyses and chemical techniques (Jenkins, 1985). Micromorphology can be used to observe the extent of reddening and particularly to identify the features with which rubification is associated.

Rubification has been identified in a number of interglacial soils and evidence so far suggests that it is associated with earlier rather than more recent interglacials (Bullock and Murphy, 1979; Chartres, 1980). However, more evidence of its significance in interglacial soils is required particularly in view of the fact that the process has been reported in some Flandrian soils in France and southern Germany. At present, a combination of micromorphology with X-ray and chemical techniques offers the best means of determining its form and distribution. It would be a step forward if this could be done on a range of Quaternary soils of known age.

Laterization has only rarely been described in Quaternary soils in the UK (Bullock *et al.*, 1973). The process is related to a hot, humid climate under which weathering is so intense that with the exception of quartz, virtually all primary minerals have disappeared. In their place are kaolinite, gibbsite and ferric oxide (haematite and goethite).

Lateritic soils in thin sections show the following properties (Stoops, 1983).

1 Few, if any, weatherable minerals in the sand and silt fractions.
2 Very low silt content.
3 Little or no illuvial clay.
4 A microstructure dominated by small aggregates.
5 Little or no preferred orientation to the fine material which tends to be isotropic due to the influence of iron oxides.

COLD PERIOD AND LATERAL TRANSPORT PROCESSES

During interglacial periods, conditions have been sufficiently stable for soils to develop. Subsequent glacial periods have removed, decapitated and/or buried, or at least partially modified the previously formed soils. While there have been several studies of interglacial soils, the effect of glacial stages on soil formation and disruption has received much less treatment. At its most drastic the effect is to remove existing soils from an area entirely either by the ice sheet itself or by solifluction. Occasionally relicts of soil material can be observed in tills and solifluction deposits but they are rather rare presumably because of extensive dilution with other sediments.

In post-Devensian times, wind and water erosion and colluviation, particularly following deforestation, have laterally transported material to new sites on slopes or in valley bottoms. The micromorphology of colluvial soils has been studied extensively by Mücher and co-workers (Mücher, 1974; Mücher and Morozova, 1983). Micromorphological criteria for colluvial soils include, papules, discrete nodules, plant and root fragments parallel to the surface, lithorelicts derived from outside the immediate area, buried mudcrusts and various pedorelicts.

Some interglacial soils survived extensive erosion and have been more or less preserved, either buried beneath later deposits or remaining at or near the surface to undergo further soil development. Such soils often bear the imprint of a cold period. This may be in the form of large-scale features such as zones of cryoturbation, ice wedges (Rose *et al.*, 1985), indurated layers or smaller-scale features such as platy structures, redistribution phenomena, silty cappings on grains, rock fragments and peds (Romans *et al.*, 1966). Thin sections provide a valuable means of identifying and describing the smaller features and combined with field studies can help to document the effects of the larger features on soil structure and fabric. Much valuable work has been done in Belgium and France on micromorphological features associated with cold periods (Van Vliet-Lanoë, 1982, 1985; Langohr and Sanders, 1985) which can form a basis for further investigations of Quaternary soils in the UK.

CHRONOLOGICAL ORDERING OF SOIL PROCESSES

Many Quaternary soils have a complex history of development. Some formed in earlier interglacials were subsequently buried too deeply under thick deposits to be subject to further pedogenic processes; others from the same period escaped burial or erosion, remaining at the surface for over a million years to experience a variety of processes; still others were subsequently buried by thin deposits with subsequent pedogenesis affecting both new deposits and the buried soil.

Attempts have been made to use micromorphology to unravel the complex history of Quaternary soils, of which the following case studies are examples.

Buried paleocatenas from Woodhall Spa and West Runton (Valentine and Dalrymple, 1975)

The authors investigated two paleocatenas.

1   A ferric podzol, humo-ferric podzol, humo-ferric gley podzol, sandy gley sequence beneath marine clay and fen peat (dated 4100 yr BC) at Woodhall Spa, Lincolnshire; and
2   A similar sequence in Beestonian sands and gravels under a layer of Cromerian organic muds at West Runton, Norfolk.

At the Woodhall Spa site, the microfeatures in the sandy paleosols demonstrated differential translocation and deposition of organic matter, silicate clay and iron oxides similar to characteristics of podzol to gley sequences described in lowland England. Translocation of clay appears to have occurred first, to be followed by podzolization. In the lower profiles, gleying was the most significant process, together with changing organization of the clay plasma under repeated wetting and drying cycles. The authors concluded that the marine clay and peat sealed off the paleosols from further soil development on the grounds that there was no evidence for translocated material within the overlying deposits.

At the West Runton site, only in the upper part of the sequence was there possible soil development in the form of reorganized clay plasma, clay cutans and the redistribution of iron-rich compounds but these microfeatures lacked a clear vertical profile pattern. Non-illuviated grain cutans occurred in all layers probably derived from shrinking and swelling and wetting and drying *in situ*. Throughout the lateral sequence, the upper horizons retained some degree of sand-grain orientation due to sedimentary deposition, implying little disturbance by soil-forming processes. The authors concluded that the apparent vertical differentiation in this profile was produced by post-burial subsurface weathering and iron translocation rather than by pedological development at the surface.

The North Leigh paleosol, Oxfordshire (Bullock and Murphy, 1979)

From a detailed macro- and microscopic study of a paleo-argillic brown earth in Plateau Drift on the Oxford Plateau, the authors identified eight kinds of pedofeature and related these to a probable sequence of soil-forming events. Three 'stable' periods of soil formation were recognized, one pre-dating emplacement of the Plateau Drift, the other two associated with one or more interglacial periods. Clay illuviation and reduction and segregation of iron oxides are the main processes recognized in all three periods and rubification is associated with periods before the last glaciation. Although the soil was probably subject to several unstable periods only two are distinguished. The first is associated with erosion, disruption and mixing of soils and sediments, followed by transport and emplacement of the Plateau Drift. The second is related to the Devensian glaciation and is characterized by erosion, cryoturbation and deposition of loess on the eroded surface. Post-Devensian soil formation has involved progressive leaching, clay translocation, reduction, mobilization and segregation of iron oxides and the onset of podzolization.

Soils on terraces of the River Kennet, Berkshire (Chartres, 1980)

The author has compared soil development on four chronologically separate terraces of the River Kennet. The older two terraces are believed to be Anglian or older, the next youngest Wolstonian and the youngest late Devensian.

Using micromorphology as the principal technique Chartres described the pedofeatures in profiles from each of the terraces and used the data to postulate the following sequence of development.

1   A period of strong clay illuviation associated with manganiferous staining and rubification of the illuvial clay (Hoxnian?).
2   Disturbance of the rubified illuvial features (Wolstonian?).
3   A second period of clay translocation giving strongly oriented egg-yellow argillans (Ipswichian?).
4   A period of cryoturbation with disruption of existing soils (Devensian).
5   A further period of clay illuviation with reddish-orange to yellowish-brown clay coatings (Holocene).

Thus, the oldest terraces were characterized by three distinct periods of clay illuviation and two periods of disruption whereas on the youngest terrace only one period of clay illuviation was evident. In the older three terrace soils, the eluvial horizons were disrupted and contained aeolian silts derived from outside the Kennet catchment.

CONCLUSIONS

Micromorphology is a valuable approach to studies of Quaternary soil processes. It has the advantage that in thin sections the structure and fabric are undisturbed and even features associated with incipient soil processes can be identified. Ideally, micromorphology should be used in conjunction with detailed field studies and other analytical techniques. Although in the past it has been mainly a descriptive branch of soil science, accurate measurements of components and features are now possible using a range of submicroscopic techniques. Micromorphology has considerable potential for unravelling the complex history of soil developments and it is anticipated that considerable advances will be made as Quaternary soil parent materials are more precisely dated.

ACKNOWLEDGEMENTS

The author is grateful to Dr D. A. Jenkins, University College of North Wales for permission to use Figure 3.1. Dr J. Pidgeon, ICI Plant Protection Ltd, kindly supplied the soil on which Figure 3.1 is based.

REFERENCES

Avery, B. W. (1973). 'Soil classification in the Soil Survey of England and Wales', *J. Soil Sci.* **24**, 324–338.

Avery, B. W. (1980). 'Soil Classification for England and Wales (Higher Categories)', *Soil Surv. Tech. Monogr.* **14**, Harpenden.

Avery, B. W. (1985). 'Argillic horizons and their significance in England and Wales', in *Soils and Quaternary Landscape Evolution* (Ed. J. Boardman), Wiley, Chichester (in press).

Babel, U. (1975). 'Micromorphology of soil organic matter', in *Soil Components*, Volume 1. *Organic Components* (Ed. J. E. Gieseking), pp. 369–473, Springer-Verlag, New York.

Babel, U. (1984). 'Basic organic components', in *Handbook for Soil Thin Section Description* (P. Bullock, N. Fedoroff, A. Jongerius, G. Stoops and T. Tursina), Waine Research Publications, Wolverhampton.

Bal, L. (1982). *Zoological Ripening of Soils*, Agricultural Research Report 850, Pudoc, The Netherlands.

Bascomb, C. L. and Bullock, P. (1976). 'Sample preparation and stone content', in *Soil Survey Laboratory Methods* (Eds B. W. Avery and C. L. Bascomb), pp. 5–13, Soil Survey Technical Monograph, No. 6, Harpenden.

Birkeland, (1985). 'Quaternary soils of the Western United States', in *Soils and Quaternary Landscape Evolution* (Ed. J. Boardman), Wiley, Chichester (in press).

Bisdom, E. B. A. (1967). 'Micromorphology of a weathered granite near the Ria De Arosa (NW Spain)', *Leidse Geologische Mededelingen*, **37**, 33–67.

Bisdom, E. B. A. (1981). *Submicroscopy of Soils and Weathered Rocks*, Pudoc, Wageningen.

Bisdom, E. B. A. and Ducloux, J. (Eds) (1983). *Submicroscopic studies of soils*, Elsevier, Amsterdam.

Bisdom, E. B. A., Stoops, G., Delvigne, J., Curmi, P. and Altemüller, H. J. (1982). 'Micromorphology of weathering biotite and its secondary products', *Pedologie*, xxxii, 225–252.

Brewer, R. (1976). *Fabric and Mineral Analysis of Soils*, Krieger, New York.

Brewer, R. and Sleeman, J. (1969). 'The arrangement of constituents in Quaternary soils', *Soil Sci.* **107**, 435–441.

Bullock, P. (1974). The use of micromorphology in the new system of soil classification for England and Wales' in *Soil Microscopy* (Ed. G. K. Rutherford), pp. 607–631, The Limestone Press, Kingston, Ontario.

Bullock, P., Carroll, D. M. and Jarvis, R. A. (1973). 'Palaeosol features in northern England', *Nature Phy. Sci.*, **242**, 53–54.

Bullock, P., Fedoroff, N., Jongerius, A., Stoops, G. and Tursina, T. (1984). *Handbook for Soil Thin Section Description*, Waine Research Publications, Wolverhampton.

Bullock, P. and Murphy, C. P. (1979). 'Evolution of a paleo-argillic brown earth (Paleudalf) from Oxfordshire, England', *Geoderma*, **22**, 225–252.

Bullock, P. and Thompson, M. L. (1985). 'Micromorphology of Alfisols', in *Soil Micromorphology and Soil Classification* (Eds L. A. Douglas, J. Bigham, J. A. McKeague and R. Protz), Soil Science Society of America Special Publication.

Chartres, C. J. (1980). 'A Quaternary soil sequence in the Kennet valley, central southern England', *Geoderma*, **23**, 125–146.

De Coninck, F. and Righi, D. (1983). 'Podzolisation and the spodic horizon', in *Soil Micromorphology*, Volume 2, *Soil Genesis* (Eds P. Bullock and C. P. Murphy), pp. 389–417, AB Academic Publishers, Berkhamsted.

Delvigne, J., Bisdom, E. B. A., Sleeman, J. and Stoops, G. (1979). 'Olivines, their pseudomorphs and secondary products', *Pedologie*, XXIX, 247–309.

Eswaran, H. and Shoba, S. (1983). 'Scanning electron microscopy in soil research', in *Soil Micromorphology*, Volume 1, *Techniques and Applications* (Eds P. Bullock and C. P. Murphy), pp. 19–52, AB Academic Publishers, Berkhamsted.

Fedoroff, N. (1966). 'Contribution à la connaissance de la pédogenèse quaternaire dans le S-W du Bassin Parisien', *Bulletin de l'Association française pour l'Etude du Quaternaire*, **1966-2**, 94–105.

Goldich, S. S. (1938). 'A study rock weathering', *J. Geol.* **46**, 17–23.

Harris, C. (1981). 'Microstructures in solifluction sediments from South Wales and North Norway', *Biul. Peryglac.* **28**, 221–226.

Jenkins, D. A. (1985). 'Chemical and mineralogical composition in the identification of palaeosols', in *Soils and Quaternary Landscape Evolution* (Ed. J. Boardman), Wiley, Chichester (in press).

Kubiena, W. L. (1953). *The Soils of Europe*, Thomas Murby and Co., London.

Langohr, R. and Sanders, J. (1985). 'The Belgian Loess Belt in the last 20 000 years. Evolution of soils and relief in the Zonien Forest', in *Soils and Quaternary Landscape Evolution* (Ed. J. Boardman), Wiley, Chichester (in press).

McKeague, J. A. (1983). 'Clay skins and the argillic horizon', in *Soil Micromorphology*, Volume 2, *Soil Genesis* (Eds P. Bullock and C. P. Murphy), pp. 367–388, AB Academic Publishers, Berkhamsted.

McKeague, J. A., Wang, C., Ross, G. J., Acton, C. J., Smith, R. E., Anderson, D. W., Pettapiece, W. W. and Lord, T. M. (1981). 'Evaluation of criteria for argillic horizons (Bt) of soils in Canada', *Geoderma*, **25**, 63–74.

Meunier, A. (1977). 'Les Mécanismes de l'Alteration des Granites et le Rôle des Microsystèmes', Thèse, Université de Poitiers.

Meyer, B. and Kalk, E. (1964). 'Verwitterungs-Mikromorphologie der Mineral-Spezies in mittel-europäischen Holozan Boden aus Pleistozanen und Holozanen Lockersedimenten', in *Soil Micromorphology* (Ed. A. Jongerius), pp. 109–130, Elsevier, Amsterdam.

Mücher, H. J. (1974). 'Micromorphology of slope deposits: the necessity of a classification', in *Soil Microscopy* (Ed. G. K. Rutherford), pp. 553–566, The Limestone Press, Kingston, Ontario.

Mücher, H. J. and Morozova, T. D. (1983). 'The application of soil micromorphology in Quaternary geology and geomorphology', in *Soil Micromorphology*, Volume 1, *Techniques and Applications* (Eds P. Bullock and C. P. Murphy), pp. 151–194, AB Academic Publishers, Berkhamsted.

Romans, J. C. C., Stevens, J. H. and Robertson, L. (1966). 'Alpine soils of northeast Scotland', *J. Soil Sci.* **17**, 184–199.

Rose, J., Allen, P., Kemp, R. A., Whiteman, C. A. and Owen, H. (1985). 'The Early Anglian Barham Soil of eastern England', in *Soils and Quaternary Landscape Evolution* (Ed. J. Boardman), Wiley, Chichester (in press).

Soil Survey Staff (1975). *Soil Taxonomy*, US Department of Agriculture, Soil Conservation Service, Agriculture Handbook 436.

Stoops, G. (1983). 'Micromorphology of the oxic horizon', in *Soil Micromorphology*, Volume 2, *Soil Genesis* (Eds P. Bullock and C. P. Murphy), pp. 419–440, AB Academic Publishers, Berkhamsted.

Stoops, G., Altemüller, H. J., Bisdom, E. B. A., Delvigne, J., Dobrovolsky, V. V., FitzPatrick, E. A., Paneque, G. and Sleeman, J. (1979). 'Guidelines for the description of mineral alterations in soil micromorphology', *Pedologie*, XXIX, 121–135.

Valentine, K. W. G. and Dalrymple, J. D. (1975). 'The identification, lateral variation and chronology of two buried paleocatenas at Woodhall Spa and West Runton, England', *Quaternary Research*, **5**, 551–590.

Van Vliet-Lanoë, B. (1982). 'Structures et microstructures associées à la formation de glace de ségrégation: leurs conséquences', *Proceedings 4th Canadian Permafrost Conference, Calgary 1981*, pp. 116–122.

Van Vliet-Lanoë, B. (1985). 'Frost effects in soils', in *Soils and Quaternary Landscape Evolution* (Ed. J. Boardman), Wiley, Chichester (in press).

Veneman, P. L. M., Vepraskas, M. J. and Bouma, J. (1976). 'The physical significance of soil mottling in a Wisconsin toposequence', *Geoderma*, **15**, 103–118.

Soils and Quaternary Landscape Evolution
Edited by J. Boardman
© 1985 John Wiley & Sons Ltd.

4

Argillic Horizons and their Significance in England and Wales

B. W. AVERY

ABSTRACT

An argillic horizon is defined in the US Soil Taxonomy as a subsurface soil horizon in which silicate clay has accumulated by illuviation, as shown by the vertical distribution of clay-size particles coupled with recognition of translocated clay in the field or in thin sections.

Soils with argillic horizons are widely distributed from boreal regions to the tropics in various parent materials that provide significant amounts of silicate clay, but are mainly restricted to relatively stable ground surfaces of pre-Flandrian age in areas with a seasonal soil–water deficit. Two genetically distinct variants have been distinguished in England and Wales on the basis of detailed soil surveys and associated pedological studies. Those of the first kind, with a brownish or grey and brown mottled 'ordinary argillic B horizon', are typically developed in initially calcareous Devensian sediments but also occur on various older deposits. Those of the second, with a strong brown to red or red-mottled 'paleo-argillic B horizon', are in older drift or residuum from pre-Quaternary rocks, and have complex or polycyclic profiles considered to incorporate the more or less disturbed remains of pre-Devensian interglacial soil formations.

Recognition of argillic horizons is important in Quaternary studies because failure to do so can lead to misinterpretation of textural variations in near-surface deposits, and because these horizons generally mark phases of ground-surface stability representing particular stages in the evolution of the landscapes in which they occur. Although pre-Devensian surfaces cannot as yet be dated with any certainty by intrinsic soil characteristics alone, the recognition and mapping of soils with paleo-argillic B horizons has cast fresh light on the evolution of polycyclic landscapes like those of southern England.

INTRODUCTION

The argillic horizon, conceived as a subsurface soil horizon 'in which layer-lattice silicate clays have accumulated by illuviation to a significant extent', is one of several named 'diagnostic horizons' defined as a basis for differentiating classes in the USA system of soil taxonomy (Soil Survey Staff, 1975). The terms Bt and argillic B (Avery, 1980) denote essentially the same horizon concept.

Although translocation of clay in suspension was recognized as a possible horizon-forming process early in this century (e.g. Glinka, 1914; Robinson, 1932), soils with argillic horizons were not consistently distinguished from other kinds of soil. In the preceding American classification (Baldwin *et al.*, 1938), for example, forested or formerly forested soils with clay-illuvial B horizons were described as podzolic (e.g. grey-brown podzolic; red-yellow podzolic) on the basis that they had light-coloured, eluvial horizons resembling those of podzols (spodosols), whilst in Britain they were grouped, as brown earths or brown forest soils, with brownish soils lacking illuvial B horizons. In western Europe, they were first set apart as a morphologically and genetically distinct class by French pedologists (Aubert and Duchaufour, 1956), who termed them *sols lessivés* and the formative process *lessivage*.

According to current concepts, soils with argillic horizons are widely distributed from boreal regions to the tropics in various parent materials that contain or yield significant amounts of silicate clay. They include both well-drained soils, in which the argillic horizon is normally brown or reddish, and periodically wet or artificially drained soils in which it is mottled with greyish and rusty colours attributable to reduction, mobilization and redeposition of iron (gleying).

IDENTIFICATION AND GENESIS

A subsurface horizon that meets prescribed depth and thickness requirements is identified as argillic by the following criteria.

1   More clay ($<2\,\mu$m) by specified amounts than an overlying horizon (normally E or A) interpreted as eluvial, and normally a higher ratio of fine ($<0.2\,\mu$m) to total clay than in overlying and underlying horizons: these conditions may not be met if the parent material is stratified or if the soil has been truncated.
2   Presence, throughout or in the lower part, of discrete strongly oriented clay bodies recognizable in the field as coats on ped or grain surfaces or lining pores; or in thin section as ferri-argillans or micromorphologically similar clay concentrations embedded in the matrix (Brewer, 1964; Bullock and Murphy, 1979): this requirement may be waived if the horizon contains more than 35 per cent clay-size particles of 2:1 lattice types, there is evidence of swelling, and other evidence (criterion 1) indicates that clay translocation has occurred. (The clay bodies interpreted as illuvial may or may not contain appreciable amounts of iron oxides closely bonded to the layer silicates.)

Recognition that clay translocation has played a major role in the development of these horizons comes from detailed profile studies and laboratory experiments on mobilization and deposition of clay, reviewed in several publications (e.g. Brewer and Sleeman, 1970; Duchaufour, 1982). According to the commonly accepted conceptual model (Soil Survey Staff, 1975), rain water penetrating dry or partially dry soil causes disruption of aggregates and dispersion of some clay, which remains in suspension as the water moves through non-capillary voids. If the lower part of the rooting zone is also dry, as happens during the growing season wherever evapo-transpiration exceeds rainfall and the water table is out of reach, the downward-moving water is withdrawn into finer capillary pores and the suspended clay is deposited on the walls of the non-capillary voids. Deposition may also result from any marked change in pore-size distribution which restricts water flow. The argillans so formed may remain *in situ* or they may be disrupted and incorporated in the matrix as a result of disturbance by soil fauna and flora, frost action, or seasonal shrinking and swelling, and in this state may be identifiable in thin section as sharply bounded clay concentrations unassociated with voids. In materials rich in swelling clay, however, disrupted argillans may become indistinguishable from plasma separations (Brewer, 1964; Holzhey and Yeck, 1974), hence the provision under criterion 2 above.

The extent to which clay translocation takes place and eventually produces an argillic horizon is evidently influenced by several factors (McKeague, 1983). Among those apparently favouring clay movement are a seasonally dry pedoclimate; absence of flocculating or cementing agents such as finely divided carbonates and amorphous sesquioxides; pH (in water) between about 4.5 and 6.5 or high pH associated with exchangeable sodium; and zero point of net charge different from soil pH. Mobilization of clay is inhibited in well-aggregated humus-rich surface horizons, but soluble organic matter may aid dispersion under acid conditions. Leaching of carbonates leaves voids providing favourable sites for clay deposition, so that an argillic horizon commonly occurs immediately above unleached material in originally calcareous deposits such as loess or till. Fine clay is translocated more readily than coarse clay, and certain clay minerals, particularly smectites, may be mobilized preferentially. Finally, since there are generally few indications of clay movement in soils in recent Flandrian deposits, it seems that by contrast with certain other soil processes, such as incorporation of humified organic matter, gleying and podzolization, longer periods of ground-surface stability are ordinarily needed for its cumulative effects to become discernible.

Clay translocation may proceed without giving rise to a B horizon of maximum clay content if there is much pedoturbation; if the parent material is stratified and originally contained most clay in the uppermost layer, as in many alluvial deposits; or if the coarser fractions contain rock (e.g. shale) fragments or silicate minerals that weather to yield clay-size particles in the upper part of the profile, so offsetting concomitant loss by eluviation (Duchaufour, 1982, pp. 271–272). Conversely, a subsurface horizon can be richer in clay than the overlying horizon without being argillic if, as is commonly the case, the overlying horizon is developed in a thin, initially coarser textured deposit of alluvial, colluvial or aeolian origin. Apparent loss of clay from the upper horizons of a soil can also result from differential erosion or lateral eluviation of fine material on slopes

Table 4.1 Particle size and micromorphological data on profiles representing four soil series

Horizon[a]	Depth (cm)	CaCo₃ (%)	Sand 600 μm– 2 mm (%)	200– 600 μm (%)	60– 200 μm (%)	Silt 2–60 μm (%)	Clay <2 μm (%)	Illuvial clay[b]
Charity series in silty head (grid ref. SP 833040)[c]								
Ah	0–8	—	5	9	70	16	—	
Eb	8–28	—	5	8	74	13	—	
Bt1	28–43	—	5	7	67	21	**	
Bt2	43–69	—	3	5	48	44	***	
Bt3	69–84	—	3	4	47	46	***	
Cu	84–99	33.0	4	5	53	38		
Whimple series in loamy head over weathered Triassic mudstone (grid ref. SP 282993)[d]								
Ap	0–24	—	3	12	16	49	20	—
Eb	24–34	—	4	10	14	50	22	*
2Bt(g)	34–68	—	<1	1	4	50	45	***
2BC	68–92	4.7	<1	6	6	59	35	***
Bardsey series in loamy head over weathered Carboniferous clay-shale (grid ref. SP 218980)[d]								
Ap	0–30	—	3	23	21	24	29	—
Eg	30–42	—	5	26	16	30	23	—
2Bg	42–77	—	1	8	5	35	51	*
2BCg1	77–88	—	2	3	5	28	62	*
2BCg2	88–112	—	<1	<1	1	36	63	*
Tendring series in coverloam over Beestonian(?) fluvial deposits (grid ref. TM 110245)[e]								
Ah	0–8	—	3	9	10	66	12	—
Eb(g)1	8–31	—	2	9	9	69	11	—
Eb(g)2	31–45	—	2	6	8	70	14	*
Btg	45–63	—	1	6	8	60	25	**
2Btg	63–82	—	4	19	33	25	19	***
3Bt(g)	82–100	—	3	35	23	9	30	****

[a] After Avery (1980).
[b] Abundance of illuvial clay (argillans and intrapedal concentrations) in representative thin sections: *0.2–2%; **2–4%; ***4–10%; ****>10%.
[c] Avery *et al.* (1959); Bullock (1974).
[d] Whitfield and Beard (1980).
[e] Sturdy and Allen (1981).
Particle size distribution calculated on carbonate-free basis. Data for argillic horizons are italicized.
Data reproduced by permission of Soil Survey of England and Wales.

and possibly from differential decomposition of clay minerals under acid, seasonally wet, conditions (Brewer and Sleeman, 1970; Brinkman, 1979).

Consistent recognition of argillic horizons therefore depends on distinguishing illuvial clay from clay inherited from the parent material or formed *in situ*. This presents fewest problems where the original material is coarse to medium textured (e.g. loess); argillans are then clearly distinguishable from the matrix in the field as well as in thin section, particularly when the soil is dry. Uncertainty usually increases where the horizon is clayey, especially if it is also gleyed, in which case bleached pressure faces may be wrongly identified as illuviation argillans. Problems also arise in long-cultivated soils, subsurface horizons of which commonly contain 'agricutans' (Jongerius, 1970) composed of organic matter, silt and clay washed down cracks and worm channels from the ploughed layer. These can again be confused with argillans proper.

When field observations are supplemented by particle-size analyses and micromorphological studies, however, the horizons can be identified with reasonable confidence in a wide range of soils. Besides evaluating the vertical distribution of total and fine clay, particle-size analyses serve to confirm or detect lithostratigraphic discontinuities. For this purpose emphasis is placed on

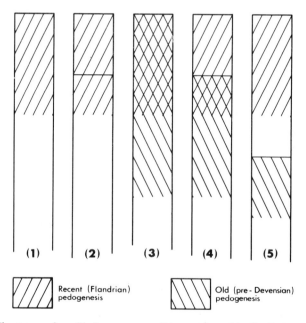

Figure 4.1 Five types of profile in one or two lithostratigraphically distinct layers (partly according to Duchaufour, 1982). (1) Simple (monocyclic) profile. (2) Composite (monocyclic) profile. (3) Polycyclic profile. (4) Complex (composite polycyclic) profile. (5) Simple profile with buried paleosol. Reproduced by permission of George Allen and Unwin

proportions of coarser non-carbonate fractions which, providing they consist largely of resistant minerals, constitute an immobile 'skeleton' unaffected by translocation, residual accumulation or decomposition of clay. The data reproduced in Table 4.1 illustrate four common situations. Those for the Charity soil (Avery *et al.*, 1959) on chalky head indicate that it has a simple profile (Figure 4.1(1)) formed in initially uniform material, with a well-developed argillic B horizon resulting from decalcification by leaching and subsequent clay translocation. Both the Whimple and Bardsey soils, in thin loamy head over pre-Quaternary mudstones, have composite profiles (Figure 4.1(2)) with marked lithostratigraphic discontinuities at 34 and 42 cm respectively (Table 4.1), strongly suggesting that the pronounced clay increases at these depths are mainly inherited. In both profiles, micromorphology indicates some additional clay translocation, but only in the Whimple does the B horizon qualify unequivocally as argillic according to the criteria currently used in England and Wales (Avery, 1980). The distribution of sand and silt fractions in the Tendring profile (Table 4.1) shows that it is also developed in a succession of lithologically distinct layers, again with evidence of superimposed clay translocation, but in this case the upper boundary of the argillic B horizon is within the relatively silty uppermost layer.

ORDINARY AND PALEO-ARGILLIC B HORIZONS

In the hierarchical classification used by the Soil Survey of England and Wales (Avery, 1980), soils with argillic B horizons are differentiated at group or subgroup level as divisions of major soil groups defined by other characteristics. Except among predominantly sandy soils, a further distinction is drawn between soils with ordinary and paleo-argillic B horizons. In terms of the Quaternary stages of Mitchell *et al.* (1973), the former are considered to have simple or composite (monocyclic) profiles (Figures 4.1(1) and 4.1(2)) resulting primarily from Flandrian or Late Devensian pedogenesis, whereas the latter are interpreted as complex or occasionally polycyclic profiles (Figures 4.1(3) and 4.1(4)) which also bear the impress of one or more pre-Devensian interglacial phases of soil formation.

The paleo-argillic B horizons have been distinguished from ordinary argillic B horizons in pre-Ipswichian deposits by colour and micromorphology. First the horizon is required to have Munsell moist soil colours as follows, either throughout or in the lower part, that are not directly inherited from a red or red-mottled pre-Quaternary rock (Avery, 1980).

1   If the particle-size class is silty clay loam or finer, dominant matrix chroma more than 4 (moist or dry) and hue 7.5 YR or redder; if coarser, the hue should be 5 YR or redder.
2   Many coarse mottles with chroma more than 5 and hue 5 YR or redder, or common to many mottles redder than 5 YR.

Secondly, a paleo-argillic B horizon should show micromorphological features that distinguish it from ordinary argillic B horizons in Devensian deposits of similar composition. In particular, the extent and degree of orientation of the clay-size (plasma) fraction and the proportion consisting of plasma concentrations unassociated with voids are normally significantly greater; and ferruginous segregations, where present, are denser and redder in colour.

Argillic horizons with these characteristics were considered by the author to incorporate the more or less truncated or disturbed remains of interglacial soils for several reasons. First, loamy or clayey argillic horizons in Devensian sediments seemed generally to have dominant chromas of 4 or less or mottles no redder than 7.5 YR if gleyed. In contrast, those of buried paleosols developed in similar materials (Figure 4.1(5)), including the widespread Sangamon soil in North America (Ruhe *et al.*, 1974) and stratigraphically comparable paleosols in north-west Europe (Paepe, 1968; Jamagne, 1972), were reported to resemble the paleo-argillic horizons in having stronger chromas and/or redder hues, coupled with thicker sola, more pronounced pedological reorganization of the clay-size material and more alteration of weatherable minerals. Secondly, the horizons with paleo-characteristics are typically associated with plateaux, terraces and other (pediment-like) landscape facets that lie south of the presumed Devensian glacial limits and are truncated in places by younger surfaces attributable to Devensian or earlier phases of valley incision.

Further evidence comes from the common association of the horizons with Devensian 'cover deposits' and cryoturbation effects. Although paleo-argillic B horizons occur in places immediately beneath the topsoil, particularly on eroded slopes in agricultural land, they usually underlie less-strongly coloured subsurface horizons qualifying as E or B. In widely separated parts of England, these have evidently formed wholly or partly in thin lithostratigraphically distinct layers containing Devensian loess or coversand which has been irregularly mixed with pre-existing surface materials mainly by cryoturbation or solifluction (Avery *et al.*, 1959; Pigott, 1962; Perrin *et al.*, 1974; Catt, 1977). The soils therefore have complex profiles as depicted in Figure 4.1. In places the upper boundary of the paleo-argillic horizon is sharp and smooth, suggesting that a former profile was truncated prior to addition of fresh material. Elsewhere it is wavy or irregular, and periglacial disturbance is indicated by occasional involutions, ice-wedge casts or vertically oriented stones. Fragipan features believed to have originated under cold conditions (Van Vliet-Lanoë, 1985) also occur locally in or above the horizon where it is loamy in texture.

That disturbance involved pre-existing soil materials, as well as affecting the subsequent development of pedogenic horizons, is supported by micro-morphological studies (e.g. Avery *et al.*, 1972, 1982; Bullock and Murphy, 1979; Sturdy *et al.*, 1979; Chartres, 1980). These have shown that horizons recognized as paleo-argillic in the field commonly have complex macrofabrics in which the plasma contains seemingly fragmented, bright yellow or red, strongly oriented, clay bodies (papules or more irregular concentrations). Many of these appear to have originated as argillans but are clearly distinguishable from less-strongly coloured argillans lining some voids. In profiles like that in Table 4.1 (Tendring series), a Bt (ordinary argillic) horizon has evidently formed in a loess-rich cover deposit and contains similar void argillans, confirming that substantial clay translocation took place after deposition of the loess and also affected the older soil material below. Sections have also been recorded, for example in the Wallingford Fan Gravels (Horton *et al.*, 1981), in which a paleo-argillic B horizon can be traced laterally into a buried paleosol (Figure 4.1(5)) separated from an ordinary argillic B horizon in the cover deposit by a discontinuous calcareous C horizon.

Some paleo-argillic B horizons, particularly in well-drained soils on pre-Devensian fluvial deposits and limestones, have nearly uniform colours. Those associated with sands and gravels are commonly termed 'hoggin' and often appear to have developed in superincumbent solifluction gravels (Gibbard, 1982; McGregor and Green, 1983). In places, however, the lower part of the paleo-argillic horizon retains sedimentary stratification and resembles generally thinner argillic bands formed in Devensian coversands in that the fine ($<2\,\mu$m) fraction consists almost entirely of argillans. Over calcareous rocks and chalky head deposits, paleo-argillic horizons are represented wholly or partly by 'decalcification clays', including the Clay-with-flints *sensu stricto* of Loveday (1962) and similar deposits lining irregular limestone surfaces. These deposits seem from their composition and micromorphology to consist partly of solution residues and partly of illuvial clay derived from overlying materials. Again there is evidence (Thorez *et al.*, 1971; Chartres and Whalley, 1975) that essentially similar horizons may have formed during or since the Devensian stage, but those considered as paleo-argillic are normally brighter coloured, and thicker except where they have evidently been truncated by Devensian or earlier erosion.

Other paleo-argillic horizons, particularly in less-well-drained soils on pre-Ipswichian Quaternary deposits, are more or less prominently mottled, at least in their lower parts. Certain of the gley effects they display, including strong brown mottles and greyish iron-depleted zones bordering structural faces and old root channels, are closely paralleled in gley soils in Devensian tills, suggesting that these features originated, along with some void argillans, in later Devensian or Flandrian times. The horizons identified as paleo-argillic are normally thicker than those in Devensian deposits, however, and most show additional features, chiefly iron segregations with 5 YR or redder hues, which are evidently rare in British Devensian deposits.

These reddened (rubified) zones, like the red-clay concentrations mentioned above, are normally distinguishable in thin section from reddish matrix materials inherited from pre-Quaternary red or red-mottled beds, and their hues indicate they contain haematite (Torrent *et al.*, 1983). Since identical pedological features are common in argillic horizons of tropical and subtropical regions, and haematite formation in soils is evidently favoured by warm dry conditions (Schwertmann and Taylor, 1977), these features have been ascribed to transformation or segregation of iron oxides under climatic regimes warmer and with more pronounced dry seasons than at present. There is also evidence, however, that well-drained soils on calcareous deposits in south-east France (Bresson, 1976) and southern Germany (Schwertmann *et al.*, 1982) were reddened during the Flandrian, so summer temperatures may not have been very much higher. This conclusion is consistent with paleo-entomological evidence (Coope, 1977) that mean July temperatures in southern England reached 20 °C (about 3° higher than at present) during the Ipswichian Interglacial.

PROBLEMS OF IDENTIFICATION AND INTERPRETATION

During the last 10 years, soil surveys and related investigations have broadly confirmed the postulates on which the paleo-argillic horizon concept was based, but have also posed various problems of identification and interpretation in addition to those attending the recognition of argillic horizons generally.

In the first place argillic horizons retaining the impress of interglacial pedogenesis clearly cannot always be distinguished by colour and microstructural characteristics alone, since both features are affected by variations in parent material and hydrologic conditions as well as age and climatic factors (Catt, 1979). The proposed criteria therefore have to be treated as guidelines in some cases, and reference must be made to other evidence such as solum thickness, geomorphological and stratigraphic relationships, and comparison with geographically associated or lithologically similar soils whose age can be confidently inferred. Prominent among such marginal cases are horizons that are uniformly strong brown (7.5 YR 5/6–8) or mottled with strong brown and grey, and show no unexceptionable relict features in thin section.

A second common problem is distinguishing old soil horizons somewhat disturbed *in situ* (e.g. by cryoturbation) from layers of soliflucted soil materials, and Avery (1980) recognized accordingly that some horizons qualifying as paleo-argillic by internal characteristics may have been emplaced by Devensian solifluction. Although this can often be discounted geomorphologically or by the presence of a Devensian cover deposit, it is often difficult to establish the extent to which micromorphological features originated as a result of pedogenesis *in situ* rather than as pedorelicts (Brewer, 1964) derived from older soils elsewhere. Thus among the features recognized by Jamagne (1972), Bullock (1974) and Bullock and Murphy (1979), more or less reddened (haematitic) argillans, irregular or linear intrapedal clay concentrations, and reddened iron segregations with irregular or diffuse outlines, are unlikely to have survived transport except as inclusions in larger bodies, but strongly sepic plasmic fabrics, rounded papules and embedded grain argillans may well have done so.

Where horizons with paleo-argillic characteristics are overlain by less-weathered loess, and have compositions consistent with derivation from an underlying little-altered pre-Ipswichian deposit, it is reasonable to suppose that they result from pre-Devensian weathering *in situ*. This situation is exemplified by the deeply leached Hornbeam and Oak soils over Anglian chalky till in Essex investigated by Sturdy *et al.* (1979). However, the presence of coarser textured, apparently reworked layers in closely associated soils, and the occurrence in the Oak profile studied of papules embedded in reddened ferruginous segregations, suggested to these authors that the materials in some horizons may have been transported, perhaps by Wolstonian solifluction, from nearby sites where weathering and clay illuviation first took place. That the reddening originated during the Ipswichian is supported first by entomological evidence (Coope, 1977) suggesting that summer temperatures during the Hoxnian climatic optimum were close to those of today, and secondly by recent surveys showing that similar deeply weathered soils with red mottled paleo-argillic B horizons occur in tills dated as Wolstonian near Moreton-in-March (W. A. D. Whitfield, pers. comm.) and in other widely separated localities in the Midlands (Jones, 1983; Beard, 1984).

Distinguishing the effects of pedogenesis *in situ* presents greater problems in heterogeneous high-level deposits of uncertain age, including the Chiltern Plateau Drift (Loveday, 1962) and the Northern (Plateau) Drift of the Cotswold dipslope (Shotton *et al.*, 1980), which appear to consist entirely of weathered materials. Micromorphological studies of red-mottled paleo-argillic horizons in these deposits by Avery *et al.* (1972, 1982) and Bullock and Murphy (1979) have revealed evidence of more than one phase of reddening, separated by one or more phases of disruption or transport during which papules were formed. Bullock and Murphy also concluded that some pedological features in the soil they studied were emplaced as components of larger bodies which they termed fossil aggregates. The complex micro-fabrics of these paleo-argillic horizons thus seem to reflect at least two pre-Devensian phases of pedological reorganization which affected the source materials during warm stages before or after their final emplacement. In some cases (e.g. Isaac, 1981), an initial phase of Tertiary weathering may have contributed to the complexity of the micro-fabric. As on the Wolstonian and Anglian tills, however, features representing the latest phase of reddening seem to have originated *in situ*, probably during the Ipswichian stage.

SOILS WITH ARGILLIC B HORIZONS IN ENGLAND AND WALES

The recently published 1:250,000 soil-association maps (Soil Survey of England and Wales, 1983) show the generalized distribution of soils conforming to argillic and paleo-argillic classes, and those occupying significant areas are briefly described below, employing the textural terms clayey, loamy, etc. as defined by Avery (1980). Placement of the classes in the US system is indicated in parentheses.

Argillic pelosols (udalfs)

Brownish or reddish, slowly permeable, clayey soils with little gleying in the upper 40 cm, typically developed on initially calcareous argillaceous rocks such as Mercia mudstone and also on chalky till. The ordinary argillic B horizon has a coarse blocky or prismatic structure and is usually weakly expressed.

Argillic brown earths (udalfs)

More or less freely drained loamy, or loamy over clayey, soils with an ordinary argillic B horizon, normally brown or reddish. The parent materials include Devensian loess (brickearth), head and till; and older Quaternary or pre-Quaternary sediments with or without a cover of loamy head.

Argillic brown sands (arenic udalfs or udults)

Brown or reddish soils in coarse Quaternary or pre-Quaternary sediments. More than half the upper 80 cm is sandy or sandy-skeletal, and the argillic B horizon is usually represented by one or more brightly coloured bands in which strongly oriented clay coats and bridges the sand grains.

Paleo-argillic brown earths (udalfs or udults)

Brownish upper horizons over a strong brown to red, loamy or clayey paleo-argillic B horizon. As indicated above, they are typically developed partly or wholly in pre-Ipswichian Quaternary deposits, including quasi-residual materials such as Clay-with-flints, but also occur locally in weathering products of hard non-calcareous rocks.

Paleo-argillic podzols (orthods or humods)

A strongly coloured paleo-argillic B horizon beneath coarser textured E and B horizons characteristic of podzols, and are mainly restricted to existing or former heathland on high-level gravel deposits in the southern counties. Similar soils with gley features in and/or above the paleo-argillic B have been classed as Stagnogley-Podzols (Aquods).

Typical (argillic) stagnogley soils (aqualfs)

Strongly mottled with greyish and ochreous colours within 40 cm depth and have ordinary argillic B (Btg) horizons that are slowly permeable or immediately overlie impermeable substrata, implying that periodic wetness results primarily from impeded percolation of rain water during the winter half year. Normally loamy or loamy over clayey, these soils are widespead on Devensian tills in the Midlands and northern England and on older Quaternary and pre-Quaternary deposits further south. Most have composite profiles with upper horizons in originally coarser textured materials and are often difficult to distinguish from similar soils (e.g. Bardsey in Table 4.1) showing little evidence of clay translocation. Associated pelo-stagnogley soils, for example on Jurassic clays and Midland chalky tills, are clayey throughout and may also have an argillic B horizon, but have not been subdivided on this basis.

Paleo-argillic stagnogley soils (aqualfs or aquults)

Similar in most respects to typical stagnogley soils but differ in having thick, gleyed paleo-argillic B (Btg) horizons characterized, at least in the lower part, by dominantly strong brown or redder colours, or a reticulate pattern of red and grey mottles. They are confined to plateau sites on Wolstonian or older Quaternary sediments.

Argillic gley soils (aqualfs)

Also resemble the typical stagnogley soils in morphology but are developed in relatively pervious loamy, sandy or gravelly materials, including loess (brickearth) and Devensian river-terrace deposits which are low lying and/or overlie impervious substrata below root range. The horizons affected by seasonally rising ground water often contain abundant ferruginous segregations. Both typical stagnogley and argillic gley soils occur in catenary association with argillic brown earths, but those of the first group are much more extensive.

SIGNIFICANCE OF ARGILLIC HORIZONS IN RELATION TO QUATERNARY LANDSCAPE EVOLUTION

The authors of the US Taxonomy (Soil Survey Staff, 1975) believed that in humid temperate regions, 'the argillic horizon is mainly a mark of a stable surface and a seasonal moisture deficit'. Soil surveys in England and Wales have supported this conclusion to the extent that argillic B horizons have seldom been identified either in Flandrian sediments or in the perhumid uplands where average rainfall exceeds evapo-transpiration in every month of the year. But since around half the remaining area is occupied by soils with non-argillic B (cambic Bw or Bg, or podzolic Bs or Bh) horizons, or soils without B horizons, other factors are clearly involved, the most important of which is undoubtedly parent-material composition.

As in north-eastern USA, Ireland (Gardiner and Radford, 1980) and other parts of western Europe, the majority of soils with ordinary argillic B horizons in England and Wales have formed in or over initially calcareous Quaternary or pre-Quaternary sediments. They apparently represent the culminating stage of a developmental sequence involving solution and removal of carbonates, in which earlier stages are represented successively by rendzinas or pararendzinas (with A-C profile) and brown calcareous earths (with partially decalcified cambic horizon distinguished by colour and structure), or their gleyed analogues. Other soils with ordinary argillic B horizons, usually less-well developed, are chiefly in 'pre-weathered' non-calcareous sediments composed of 2:1 lattice clays and coarser fractions resistant to weathering. By contrast, soils in stony weathering products of crystalline rocks and hard non-calcareous Paleozoic sediments, for example in lowland Wales and south-west England, nearly all have cambic or podzolic B horizons showing evidence of accumulation of clay and iron oxides by weathering, though some, as indicated above, have illuviation argillans in underlying horizons containing less clay. Other substrates which appear to have retarded or prevented development of well-defined argillic horizons in the English lowlands include heavy clays and ferruginous weathering products of 'ironstones' as well as quartzose sands and sandstones and extremely calcareous materials such as chalk.

Hence, whilst the regular association of an argillic horizon with a particular geomorphic surface generally constitutes evidence of stability, at least over the last few thousand years, the absence of such a horizon does not necessarily imply that the surface is younger, even when some parts of the surrounding landscape have soils with argillic horizons. In the southern English chalklands, for example, the distribution of argillic soils is evidently determined mainly by the varying carbonate contents of the materials including fragmented chalk and flint, and loess which formed the surface layers at the close of the last major episode of periglacial instability, and to a lesser extent by the incidence of Flandrian erosion and deposition.

Particularly in long-cultivated areas, evidence of the latest stage of landscape evolution is afforded by accumulations of colluvium in valley bottoms and on footslopes, accompanied by signs of erosion on higher ground nearby. Where parent materials are relatively uniform, and post-Devensian soil formation has involved progressive leaching and clay translocation, as in west European loess landscapes, truncation or burial of the resulting argillic horizon can serve as a marker for estimating the degree and extent of erosion and aiding the recognition of Flandrian colluvial deposits (e.g. Kwaad and Mücher, 1977, 1979; Bolt *et al.*, 1980). Weir *et al.* (1971) showed that a leached profile in Devensian loess at Pegwell Bay, Kent, was buried by colluvium about 5000 years BP, by which time the loess had been decalcified to at least 2 m and an argillic horizon as well developed as in neighbouring unburied soils had formed. Surveys in the East Anglian Fenland (e.g. Seale, 1975) have similarly shown that soils with argillic horizons in Devensian fluvial deposits were buried beneath peat as a consequence of the mid-Flandrian marine transgression. Most of the clay translocation leading to development of ordinary argillic B horizons in originally calcareous Devensian deposits, and presumably in other materials also, therefore appears to have taken place during early Flandrian or even late Devensian times.

Soils with paleo-argillic B horizons have a much more restricted distribution, covering in all little more than 3 per cent of England and Wales (Mackney *et al.*, 1983). More than half this area is on Clay-with-flints and other locally derived superficial deposits over Chalk or older Cretaceous rocks, and the remainder chiefly in pre-Ipswichian glacial and river drift and in Carboniferous Limestone terrains including Mendip and the Derbyshire Dome. The rarity of paleo-argillic horizons in other areas unaffected directly by Devensian glaciation implies that erosion caused widespread removal or deep burial of pre-existing weathered mantles. Alternatively some of the existing surface soils may retain relict features which do not conform to the paleo-argillic horizon concept. Examples of the latter include soils with red mottled but apparently non-argillic subsurface horizons (Bg or BCg) over sandstone and shale in northern and south-west England (Bullock *et al.*, 1973; Harrod, 1978, 1981), podzols containing pedogenic haematite and kaolinite on Dalradian schists in Scotland (Stevens and Wilson, 1970) and gibbsite bearing soils on hornfelses in the Southern Uplands (Wilson and Brown, 1976). Deep bedrock alteration attributable to interglacial or Tertiary weathering has also been reported from numerous other localities in Highland Britain (e.g. FitzPatrick, 1963; Ball, 1964; Eden and Green, 1971; Hornung and Hatton, 1974), though many profiles appear to have been truncated prior to burial beneath less-weathered materials. Other soils which may have begun to form before the Devensian, but lack both argillic horizons and features hitherto identified as positive marks of pre-Devensian pedogenesis, include ferritic brown earths (Avery, 1980) on deeply weathered ironstones, shallow bright-brown or reddish soils (brown rendzinas) on more or less disturbed Jurassic limestones, and some initially non-calcareous or decalcified stagnogley soils on argillaceous rocks and clayey tills.

Even when allowance is made for the latter possibilities, however, it still seems that most soils on older Quaternary and pre-Quaternary formations have profiles showing approximately the same degrees of weathering and horizon development as those in Devensian sediments of similar composition, indicating that the associated ground surfaces are of comparable age. Thus, of the large total area more or less directly underlain by Anglian and Wolstonian chalky tills in East Anglia and the Midlands, more than one-third is covered by soils that remain calcareous to the surface, and only around 10 per cent by deeply leached soils with paleo-argillic B horizons. Similarly, many interfluves in the south-western peninsula, including relics of early or pre-Quaternary erosion surfaces as previously interpreted (e.g. Balchin, 1952), have shallow soils with cambic or podzolic B horizons over bedrock or head composed mainly of physically weathered rock and soils with reddish paleo-argillic B horizons are very limited in extent.

CONCLUSIONS

Recognition of argillic horizons is important in Quaternary studies, first because failure to do so can lead to misinterpretation of textural variations in near-surface deposits, and secondly because these horizons generally mark phases of ground-surface stability representing particular stages in the evolution of the landscapes in which they occur. Although pre-Devensian surfaces cannot as yet be dated with any certainty by intrinsic characteristics of the associated soil horizons, the recognition and mapping of soils with paleo-argillic B horizons has cast fresh light on the evolution of polycyclic landscapes like those of southern England. As noted by Catt (1979), more multidisciplinary case studies are needed to resolve the various problems of identification and interpretation that have arisen. Particle-size, mineralogical and micromorphological analyses of samples from carefully selected representative profiles will continue to play an essential role in such studies, but most progress is likely to come from intensive field investigations in which visually identifiable horizons are traced laterally in order to fully elucidate their stratigraphic and geomorphological relationships.

REFERENCES

Aubert, G. and Duchaufour, P. (1956). 'Projet de classification des sols', *Rapp. 6th int. Congr. Soil Sci. (Paris)*, E, 597–604.

Avery, B. W. (1980). 'Soil Classification for England and Wales (Higher categories)', *Soil Surv. Tech. Monogr.* No. 14, Harpenden.

Avery, B. W., Bullock, P., Catt, J. A., Newman, A. C. D., Rayner, J. H. and Weir, A. H. (1972). 'The soil of Barnfield', *Rothamsted Experimental Station Report for 1971*, Part 2, 5–37.

Avery, B. W., Bullock, P., Catt, J. A., Rayner, J. H. and Weir, A. H. (1982). 'Composition and origin of some brickearths on the Chiltern Hills, England', *Catena*, **9**, 153–174.

Avery, B. W., Stephen, I., Brown, G. and Yaalon, D. H. (1959). 'The origin and development of brown earths on Clay-with-flints and Coombe Deposits', *J. Soil Sci.*, **10**, 177–195.

Balchin, W. G. V. (1952). 'The erosion surfaces of Exmoor and adjacent areas', *Geogr. J.*, **118**, 453–476.

Baldwin, M., Kellogg, C. E. and Thorp, J. (1938). 'Soil classification', in *Soils and Men, US Dep. Agr. Yearbk. Agr.*, pp. 979–1001, US Govt. Printing Office, Washington DC.

Ball, D. F. (1964). 'Gibbsite in altered rock in North Wales', *Nature*, **204**, 673–674.

Beard, G. R. (1984). 'Soils in Warwickshire V: Sheet SP 27/37 (Kenilworth and Coventry South)', *Soil Surv. Rec.* No. 81, Harpenden.

Bolt, A. J. J., Mücher, H. J., Sevink, J. and Verstraten, J. M. (1980). 'A study on loess-derived colluvia in southern Limbourg (the Netherlands)', *Neth. J. agric. Sci.*, **28**, 110–126.

Bresson, L. M. (1976). 'Rubéfaction récente des sols sous climat tempéré humide', *Science du Sol* (No. 1), 3–22.

Brewer, R. (1964). *Fabric and Mineral Analysis of Soils*, Wiley, New York.

Brewer, R. and Sleeman, J. R. (1970). 'Some trends in pedology', *Earth Sci. Rev.* **6**, 297–335.

Brinkman, R. (1979). *Ferrolysis, a Soil-forming Process in Hydromorphic Conditions*, Agric. Res. Rep. 887, Pudoc, Wageningen.

Bullock, P. (1974). 'The use of micromorphology in the new system of soil classification for England and Wales', in *Soil Microscopy* (Ed. G. K. Rutherford), pp. 607–631, Limestone Press, Kingston, Ontario.

Bullock, P., Carroll, D. M. and Jarvis, R. A. (1973). 'Palaeosol features in northern England', *Nature, Phys. Sci.* **242**, 53–54.

Bullock, P. and Murphy, C. P. (1979). 'Evolution of a paleo-argillic brown earth (Paleudalf) from Oxfordshire, England', *Geoderma*, **22**, 225–252.

Catt, J. A. (1977). 'Loess and coversands', in *British Quaternary Studies: Recent Advances* (Ed. F. W. Shotton), pp. 221–230, Clarendon Press, Oxford.

Catt, J. A. (1979). 'Soils and Quaternary geology in Britain', *J. Soil Sci.* **30**, 607–642.

Chartres, C. J. (1980). 'A Quaternary soil sequence in the Kennet valley, Central Southern England', *Geoderma*, **23**, 125–146.

Chartres, C. J. and Whalley, W. B. (1975). 'Evidence of late Quaternary solution of Chalk at Basingstoke, Hampshire', *Proc. Geol. Ass.* **86**, 365–372.

Coope, G. R. (1977). 'Quaternary Coleoptera as aids in the interpretation of environmental history', in *British Quaternary Studies: Recent Advances* (Ed. F. W. Shotton), pp. 55–68, Clarendon Press, Oxford.

Duchaufour, P. (1982). *Pedology: Pedogenesis and Classification* (translated by T. R. Paton), George Allen and Unwin, London.

Eden, M. J. and Green, C. P. (1971). 'Some aspects of granite weathering and tor formation on Dartmoor, England', *Geogr. Annlr.* **53A**, 92–99.

FitzPatrick, E. A. (1963). 'Deeply weathered rock in Scotland, its occurrence, age and contribution to the soils', *J. Soil Sci.* **14**, 33–43.

Gardiner, M. J. and Radford, T. (1980). 'Soil associations of Ireland and their land-use potential', *Soil Survey Bull.* No. 36, An Foras Taluntais, Dublin.

Gibbard, P. L. (1982). 'Terrace stratigraphy and drainage history of the Plateau Gravels of north Surrey, south Berkshire and north Hampshire, England', *Proc. Geol. Ass.* **93**, 369–384.

Glinka, K. D. (1914). *Die Typen der Bodenbildung, ihre Klassifikation und geographische verbreitung*, Gebrüder Borntraeger, Berlin.

Harrod, T. R. (1978). 'Soils in Devon IV: Sheet SS30 (Holsworthy)', *Soil Surv. Rec.* No. 47, Harpenden.

Harrod, T. R. (1981). 'Soils in Devon V: Sheet SS61 (Chulmleigh)', *Soil Surv. Rec.* No. 70, Harpenden.

Holzhey, C. S. and Yeck, R. D. (1974). 'Micro-fabric of some argillic horizons in udic, xeric and torric environments in the United States', in *Soil Microscopy* (Ed. G. K. Rutherford), pp. 747–759, Limestone Press, Kingston, Ontario.

Hornung, M. and Hatton, A. A. (1974). 'Deep weathering in the Great Whin Sill, Northern England', *Proc. Yorks. Geol. Soc.* **40**, 105–114.

Horton, A., Worssam, B. C. and Whittow, J. B. (1981). 'The Wallingford Fan Gravel', *Phil. Trans. R. Soc., Ser B.* **293**, 215–255.

Isaac, K. P. (1981). 'Tertiary weathering profiles in the plateau deposits of East Devon', *Proc. Geol. Ass.* **92**, 159–168.

Jamagne, M. (1972). 'Quelques caractéristiques fondamentales des paléosols sur loess du nord de la France', *Pédologie, Gand*, **22**, 198–221.

Jones, R. J. A. (1983). 'Soils in Staffordshire III: Sheet Sk02/12 (Needwood Forest)', *Soil Surv. Rec.* No. 80, Harpenden.

Jongerius, A. (1970). 'Some morphological aspects of regrouping phenomena in Dutch soils', *Geoderma*, **4**, 311–332.

Kwaad, F. J. P. M. and Mücher, H. J. (1977). 'The evolution of soils and slope deposits in the Luxembourg Ardennes near Wiltz', *Geoderma*, **17**, 1–37.

Kwaad, F. J. P. M. and Mücher, H. J. (1979). 'The formation and evolution of colluvium on arable land in northern Luxembourg', *Geoderma*, **22**, 173–192.

Loveday, J. (1962). 'Plateau deposits of the southern Chiltern Hills', *Proc. Geol. Ass.* **73**, 83–102.

Mackney, D., Hodgson, J. M., Hollis, J. M. and Staines, S. J. (1983). 'Legend for the 1:250,000 soil map of England and Wales', *Soil Survey of England and Wales*, Harpenden.

McGregor, D. F. M. and Green, C. P. (1983). 'Post-depositional modification of Pleistocene terraces of the River Thames', *Boreas*, **12**, 22–33.

McKeague, J. A. (1983). 'Clay skins and argillic horizons', in *Soil Micromorphology: Vol. 2, Soil Genesis* (Ed. P. Bullock and C. P. Murphy), pp. 367–387, A. B. Academic Publishers, Berkhamsted.

Mitchell, G. F., Penny, L. F., Shotton, F. W. and West, R. G. (1973). 'A correlation of Quaternary deposits in the British Isles', *Geol. Soc. Lond. Spec. Rept.*, 4.

Paepe, R. (1968). 'Les sols fossiles Pleistocènes de la Belgique', *Pédologie, Gand*, **18**, 176–188.

Perrin, R. M. S., Davies, H. and Fysh, M. D. (1974). 'Distribution of late Pleistocene aeolian deposits in eastern and southern England', *Nature*, **248**, 320–324.

Pigott, C. D. (1962). 'Soil formation and development on the Carboniferous Limestone of Derbyshire: 1. Parent materials', *J. Ecol.* **50**, 145–156.

Robinson, G. W. (1932). *Soils, their Origin, Constitution and Classification*, Murby, London.

Ruhe, R. V., Hall, R. D. and Canepa, A. P. (1974). 'Sangamon paleosols of south-western Indiana, USA', *Geoderma*, **12**, 191–200.

Schwertmann, U., Murad, E. and Schulze, D. G. (1982). 'Is there Holocene reddening (hematite formation) in soils of axeric temperate areas?', *Geoderma*, **27**, 209–223.

Schwertmann, U. and Taylor, R. M. (1977). 'Iron oxides', in *Minerals in the Soil Environment* (Eds J. B. Dixon *et al.*), pp. 145–180, Soil Sci. Soc. Am., Madison, Wisconsin.

Seale, R. S. (1975). 'Soils of the Chatteris district of Cambridgeshire (Sheet TL38)', *Soil Surv. Spec. Survey*, No. 9, Harpenden.

Shotton, F. W., Goudie, A. S., Briggs, D. and Osmaston, H. A. (1980). 'Cromerian interglacial deposits at Sugworth, near Oxford, England, and their relation to the Plateau Drift of the Cotswolds and the terrace sequence of the Upper and Middle Thames', *Phil. Trans. R. Soc., Ser. B.* **289**, 55–86.

Soil Survey of England and Wales (1983). Soils of England and Wales: scale 1:250,000, Sheets 1–6, Ordnance Survey, Southampton.

Soil Survey Staff (1975). *Soil Taxonomy*, U.S. Department of Agriculture, Soil Conservation Service, Agriculture Handbook 436.

Stevens, J. H. and Wilson, M. J. (1970). 'Alpine podzols in the Ben Lawers Massif, Perthshire', *J. Soil Sci.* **21**, 85–95.

Sturdy, R. G. and Allen, R. H. (1981). 'Soils in Essex IV: Sheet TM12 (Weeley)', *Soil Surv. Rec.* No. 67, Harpenden.

Sturdy, R. G., Allen, R. H., Bullock, P., Catt, J. A. and Greenfield, S. (1979). 'Paleosols developed on chalky boulder clay in Essex', *J. Soil Sci.* **30**, 117–137.

Thorez, J., Bullock, P., Catt, J. A. and Weir, A. H. (1971). 'The petrography and origin of deposits filling solution pipes in the Chalk near South Mimms, Hertfordshire', *Geol. Mag.* **108**, 413–423.

Torrent, J., Schwertmann, U., Fechter, H. and Alterez, F. (1983). 'Quantitative relationships between soil color and haematite content', *Soil Sci.* **136**, 354–358.

Van Vliet-Lanoë, B. (1985). 'Frost effects in soils', in *Soils and Quaternary Landscape Evolution* (Ed. J. Boardman), Wiley, Chichester (in press).

Weir, A. H., Catt, J. A. and Madgett, P. A. (1971). 'Postglacial soil formation in the loess of Pegwell Bay, Kent (England)', *Geoderma*, **5**, 131–149.

Whitfield, W. A. D. and Beard, G. R. (1980). 'Soils in Warwickshire IV: Sheet SP29/39 (Nuneaton)', *Soil Surv. Rec.* No. 66, Harpenden.

Wilson, M. J. and Brown, C. J. (1976). 'The pedogenesis of some gibbsitic soils from the Southern Uplands of Scotland', *J. Soil Sci.* 513–522.

Soils and Quaternary Landscape Evolution
Edited by J. Boardman
© 1985 John Wiley & Sons Ltd.

5

^{14}C Dating of Palaeosols, Pollen Analysis and Landscape Change: Studies from the Low- and Mid-Alpine Belts of Southern Norway

CHRISTOPHER J. CASELDINE AND JOHN A. MATTHEWS

ABSTRACT

^{14}C dating and pollen analysis of palaeosols are used to reconstruct Holocene arctic-alpine landscapes and environmental change. Evidence is presented from soils buried during the 'Little Ice Age' beneath end moraines in front of two southern Norwegian glaciers. Estimates are made of, (a) the time elapsed since burial of palaeosols, (b) the date of initiation of soil formation (absolute soil age), and (c) the rate of soil development. The nature and origin of soil-pollen assemblages are discussed. The relative proportions and concentrations of pollen taxa indicate local plant community changes, which in turn suggest oscillations in the altitude of sub-alpine, low-alpine and mid-alpine vegetation belts over about the last 5000 years. Age/depth gradients within buried horizons clarify both the advantages and the limitations of palaeosols in establishing moraine chronologies and glacier variations. Neoglacial temperatures are estimated based on the geomorphological, pedological and palaeobotanical evidence associated with the palaeosol sites. In conclusion it is emphasized that soils are a widespread and neglected source of palaeoenvironmental information in the alpine zone.

INTRODUCTION

The understanding of landscape change in arctic-alpine environments in middle latitudes relies heavily on extrapolation from various kinds of record at lower altitudes. Organic production is generally low above the sub-alpine woodland and decreases with increasing altitude, whilst peat accumulations are extensive only near the tree line. Combined with the absence of tree growth, these features of the alpine zone all militate against the detailed reconstruction of palaeoenvironmental change. Thus, although research on vegetation history and the study of tree-line variations has a long history in Scandinavia (Hustich, 1948, 1966; Hyvärinen, 1976; Karlén, 1976; Eronen, 1979; Moe, 1979; Kullman, 1979, 1980; Eronen and Hyvärinen, 1981; Hafsten, 1981a,b), information derived from the alpine zone itself remains relatively sparse.

Glacial deposits provide the main exception to this rule. Studies on end moraine sequences deposited in front of Scandinavian glaciers indicate complex Holocene glacier variations caused by climatic changes in the alpine zone (Karlén, 1973, 1979, 1982; Karlén and Denton, 1976; Griffey, 1976; Griffey and Matthews, 1978; Griffey and Worsley, 1978; Matthews, 1980, 1981; Matthews and Shakesby, 1984). Nevertheless, the establishment of an accurate chronology of Holocene glacier variations in Scandinavia is limited by the low organic status of sediments buried beneath the moraines and, in particular, by a necessary dependence upon the ^{14}C dating of palaeosols. In an environment where soils provide the most widely available source of organic matter, the importance of palaeosols for absolute dating cannot be overestimated. The role of palaeosols in the study of landscape change depends, however, on their usefulness not only for ^{14}C dating but also as a source of palaeoenvironmental evidence.

In this paper, two ^{14}C-dated palaeosols from beneath end moraines in southern Norway are used to examine landscape change over about the last 5000 years. Detailed ^{14}C dating provides a unique framework within which such changes can be discussed. Apart from confirming the importance of the climatic deterioration experienced during the 'Little Ice Age', the palaeosols provide evidence for the way vegetation communities and soils within sub-alpine and alpine environments responded to earlier Holocene climatic variations. Consideration is also given to the technical implications of the results for the ^{14}C dating and pollen analysis of soils, and assessments are made of the implications for Neoglacial glacier and climatic variations.

STUDY AREA AND SITE DETAILS

Detailed consideration is given here to two sites within the Jotunheimen/ Jostedalsbreen region, an area of glaciated mountains and plateaux rising to 2469 m (Figure 5.1). The first site lies close to the glacier Haugabreen, to the west of the Jostedalsbreen ice cap, at an altitude of about 660 m. The second site is in eastern Jotunheimen, near the glacier Vestre Memurubreen, at an altitude of about 1480 m. Both sites lie above the upper limit of birch (*Betula pubescens*) woodland, which rises across the region from west to east.

The birch woodland forms the sub-alpine belt in Scandinavia above which the arctic-alpine landscape has traditionally been sub-divided, on the basis of vegetational differences, into low-, mid- and high-alpine belts (Nordhagen, 1943; Dahl, 1956, 1975a). The Haugabreen site lies immediately above the birch woodland, just within the low-alpine belt, whereas the Memurubreen site is

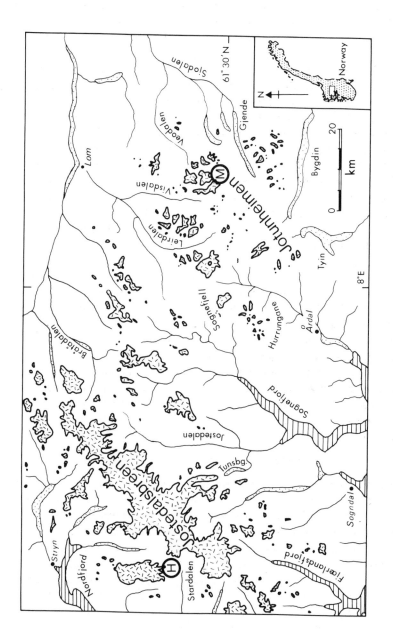

Figure 5.1 The location of Vestre Memurubreen (M) and Haugabreen (H), within the Jotunheimen/Jostedalsbreen region of southern Norway. Glaciers (flecked), lakes (stippled) and fjords (horizontal shading) are shown

at least 100 m above the lower limit of the mid-alpine belt. The former site is characterized by an oligotrophic dwarf-shrub heath, dominated by *Vaccinium myrtillus* and *V. uliginosum*; the latter site by a lower growing 'grass' heath community, dominated by *Juncus trifidus*, *Salix herbacea* and *Carex bigelowii*. Both sites are freely draining. Based on adjusted data from neighbouring meteorological stations (Bruun, 1967), mean annual temperatures are probably between +1 and +2 °C at the Haugabreen site and between −2 and −3 °C at the Memurubreen site.

Excavations beneath the outermost Neoglacial end moraine ridges in front of the glaciers revealed intact and *in situ* palaeosols that were laterally continuous with the present-day soils beyond the moraines (Figure 5.2). The palaeosols are thus overlain by Neoglacial tills and are developed on pre-Neoglacial land surfaces (also composed of till). All the tills are acidic in reaction, being derived from gneissic bedrock. Independent historical and lichenometric dating evidence indicates that the moraines were deposited by the glaciers at their Neoglacial limits reached, for both sites, in the 'Little Ice Age', probably about AD 1750 (Matthews, 1980, 1981, 1982). The pre-Neoglacial tills are assumed to have been deposited by the Weichselian inland ice sheet, which dissipated about 9000 ^{14}C yr BP (Andersen, 1980). Thin, incipient soils on the Neoglacial moraine ridges have therefore developed over about 230 years whereas the present-day soils beyond the moraines, and the palaeosols beneath them, developed over a much longer period.

At Haugabreen, the palaeosol is a typical Humo-ferric Podzol. Beneath an organic (FH) horizon of variable thickness up to 14.5 cm, this palaeosol possesses a 5 cm-thick eluvial (Ea) horizon overlying an illuvial zone comprising upper Bh and lower Bs horizons of combined thickness 30 cm. The Memurubreen palaeosol is a Brown soil, equally well developed but with less distinct horizons. The upper 10 cm of this palaeosol is a dark-brown sandy silt beneath which about 7 cm of olive-brown soil merges into the underlying grey till. Further details of the palaeosols, including full profile descriptions and photographs of the sites, may be found in Matthews (1980, 1981, 1982), Ellis and Matthews (1984) and Caseldine (1983b, 1984). The classification (Ellis, 1979) and environmental relationships (Ellis, 1980a) of similar soils have been discussed with reference to the Okstindan Mountains of northern Norway, while the development of a similar podzolic profile has been described by Alexander (1982) from the vicinity of the glacier Engabreen, also in northern Norway.

RADIOCARBON DATING

Methods and results

Intensive dating has been carried out at the Haugabreen site where the moraine ridge is about 2 m high (Matthews and Dresser, 1983). Field sampling for ^{14}C

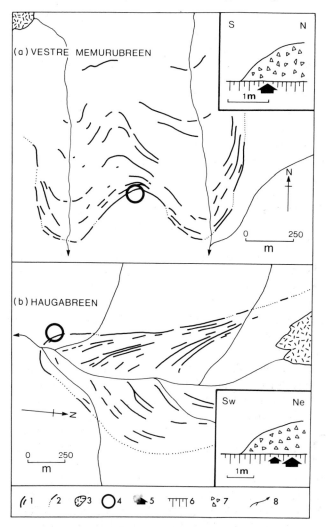

Figure 5.2   Study sites and moraine cross sections (inset) at (a) Vestre Memurubreen
and (b) Haugabreen: 1, end moraine ridges; 2, 'Little Ice Age' and Neoglacial glacier
limit ca AD 1750; 3, present glacier margin; 4, site of excavation; 5, stratigraphic position
of ^{14}C-dated samples; 6, palaeosol and its extension into the present-day soil; 7,
Neoglacial till; 8, meltwater stream

assay was carried out from the exposed surface of the buried podzol in
an open pit after excavation to a depth of about 2 m and removal of 1.3 m
of till overburden. Thin slices of 5 mm or 10 mm thickness were taken
with a stainless steel palette knife from an area of 0.3 m × 0.3 m where the
organic-rich FH horizon was 14.5 cm thick. Great care was taken to avoid

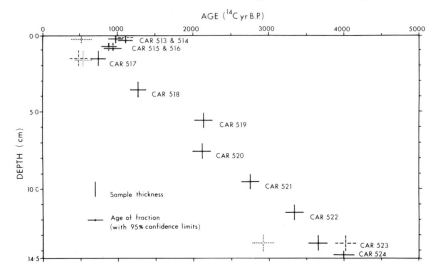

Figure 5.3 ^{14}C dates (uncalibrated) in relation to depth within the buried FH horizon of the Haugabreen palaeosol. Dated fractions are: 'humic acids' (——), 'fulvic acids' (. . . .) and 'fine residual' (– – –) (after Matthews and Dresser, 1983). Reproduced by permission of the Swedish Research Councils

contamination of samples; they were wrapped in aluminium foil and, after return to the UK, were deep frozen prior to pretreatment.

Pretreatment followed these stages (Matthews, 1980).

1   Physical separation of rootlets in a wet recycling sieving system.
2   Acid chemical extraction, which involved boiling the <2 mm fraction with 10 per cent HCl for 10 min and separation of the acid-soluble *'fulvic acids'* by centrifugation and filtration.
3   Two successive alkali chemical extractions, each involving boiling the acid-insoluble residue with 0.5 per cent NaOH for 10 min and separating the alkali-soluble fraction by centrifugation. The alkali-soluble/acid-insoluble *'humic acids'* were then recovered by acid precipitation using concentrated HCl.
4   Separation of the *'fine residual'* fraction (<250 μm) retained by sieving after acidification of the alkali-insoluble residue.

Results are given in Figure 5.3, which shows uncorrected ^{14}C ages for the three fractions in relation to depth within the FH horizon. Important features of the results include, (a) the order of magnitude differences in age within the horizon, ranging from 485±60 ^{14}C yr (CAR-517C) to 4020±70 ^{14}C yr (CAR-523C), (b) the steep overall increase in age with depth, (c) significantly different ages between fractions at the same depth, and (d) the age/depth reversal in the uppermost 1.0 cm of the palaeosol. After eliminating the dates from the

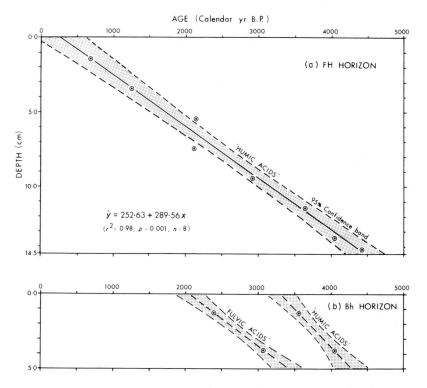

Figure 5.4 ^{14}C dates (calibrated) from the Haugabreen palaeosol: (a) 'humic acid' fractions within the FH horizon from 1.0–14.5 cm depth; (b) 'humic acid' and 'fulvic acid' fractions within the Bh horizon at two depths. Linear age/depth gradients are shown with 95 per cent confidence intervals (after Ellis and Matthews, 1984). Reproduced by permission of the Regents of the University of Colorado

uppermost centimetre (on the grounds of probable contamination), the linear nature of the age/depth gradient for 'humic acids' is shown in Figure 5.4a in which sample age has been calibrated according to Clark (1975). Probable contamination of the uppermost 1.0 cm is suggested not only by the age/depth reversal but also by significantly lower values of total oxidizable carbon at this level (see Figure 5.6).

Calibrated dates from the Bh horizon (Ellis and Matthews, 1984), based on 2.5 cm-thick samples where the horizon was 5 cm thick and using an identical pretreatment, are shown in Figure 5.4b. The most important features of these dates from the Bh horizon are, (a) the greater difference in age between fractions, (b) the absence of ages younger than about 2000 ^{14}C yr (c) the presence of an age/depth gradient, and (d) the lack of evidence for any material older than that found in the FH horizon.

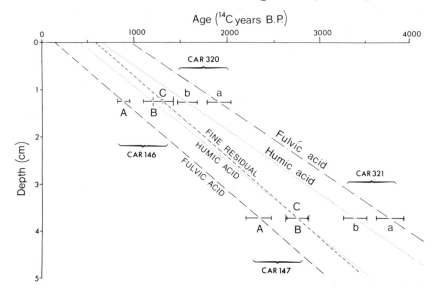

Figure 5.5 ^{14}C dates (uncalibrated) from the uppermost 5 cm of the buried Brown soil at Vestre Memurubreen (lower-case letters) compared with further dates from the Haugabreen palaeosol where the FH horizon was 5 cm thick (capital letters). Dated fractions are: 'fulvic acids' (Aa,– . – . –); 'humic acids' (Bb, ...); 'fine residual' (C– – – –). Possible linear age/depth gradients are indicated (after Matthews, 1980, 1981). Reproduced by permission of Springer-Verlag and the publishers of *Geografiska Annaler*

The dating at Vestre Memurubreen has been less detailed. Samples of thickness 2.5 cm, taken from the uppermost 5 cm of the buried Brown soil have been dated using the same pretreatment procedure (Matthews, 1981). Insufficient carbon remained in the 'fine residual' fraction to permit dating in a conventional counter. Nevertheless, 'fulvic acid' and 'humic acid' fractions from this site show comparable steep age/depth gradients (Figure 5.5) although it should be noted that the 'fulvic acids' are older than the 'humic acids' at the Memurubreen site. Additional dates from the Haugabreen podzol where the FH horizon was 5 cm thick (Matthews, 1980) are also shown in Figure 5.5. Differences between these dates and the more intensive dating framework provided by Figure 5.3 will be discussed later.

Technical implications

The interpretation of ^{14}C dates from palaeosols is beset with many problems which have been discussed in detail in Matthews (1985). Even if contamination can be discounted, the ^{14}C age of a buried soil is only its *apparent mean residence time* (AMRT) (Campbell *et al.*, 1967; Scharpenseel, 1971; Scharpenseel and Schiffmann, 1977). This quantity may be far removed from either the time elapsed since the soil was buried or its absolute age (the date at which soil development commenced at the site) (Geyh *et al.*, 1971; Gerasimov, 1974; Stout *et al.*, 1981).

The usefulness of ^{14}C dates from palaeosols for reconstructing landscape change depends on how closely the time elapsed since burial and absolute soil age can be estimated. Because the soil contained some relatively old organic matter at the time of burial, a ^{14}C date from a palaeosol provides only a *maximum* estimate of the time elapsed since then. The AMRT of a buried soil also provides only a *minimum* estimate of absolute soil age because it is not composed entirely (or in some cases even partially) of material surviving from the commencement of soil formation at the site. Closest approximations to the time elapsed since burial will therefore be provided by the *youngest* uncontaminated fractions in a palaeosol; similarly, closest estimates of absolute soil age are dependent on isolating and dating the *oldest* uncontaminated fractions.

Application of these principles to the detailed results from the FH horizon at Haugabreen (Figure 5.3) is instructive, particularly when viewed in the light of the independent evidence for burial of the palaeosol about 230 years ago. Bulk sampling of whole horizons, sampling from uncontrolled depths within horizons, or dating eroded soils are all likely to produce large errors. Although age/depth gradients have been reported from many different types of soil (Scharpenseel, 1975), previous data on within-horizon age variation has been slight and likely errors appear to have been grossly underestimated.

The age/depth gradient in Figure 5.3 and Figure 5.4a seems to reflect the steady accumulation of plant remains during development of this organic-rich horizon (Matthews and Dresser, 1983; Ellis and Matthews, 1984). The differences in age between fractions provide little if any evidence of contamination effects; they can be explained in terms of normal decomposition, humification and translocatory processes in the soil prior to burial. Precise definition of the age/depth gradient and extrapolation to the presumed former surface of the soil (zero depth in Figure 5.4a) suggests, therefore, that close estimates of *time elapsed since burial* can be made provided the ^{14}C dating is sufficiently sophisticated. Other than by defining age/depth gradients by intensive dating, thin sampling *near* the surface of well-preserved horizons would appear to be the only way of obtaining close maximum estimates of time elapsed since burial from palaeosols of this type. The age/depth gradient found at the Memurubreen site suggest that a similar approach may be required where other, less organic, soils are involved.

The oldest dated fractions, 4020 ± 70 ^{14}C yr BP (CAR-523C) at Haugabreen and 3810 ± 75 ^{14}C yr BP (CAR-321A) at Memurubreen, are much younger than the assumed *age of the terrain* on which they have developed. Nevertheless, these dates do provide minimum estimates, and dates from deeper in the soil profiles

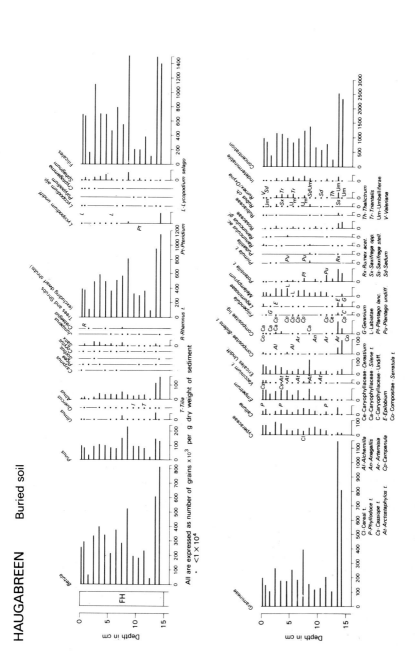

Figure 5.6 Pollen diagrams from the Haugabreen palaeosol: relative pollen diagram and concentration diagram expressed as the number of grains per gram (after Caseldine, 1983b). Reproduced by permission of Universitetsforlaget

may provide closer approximations. Much depends on the turnover rate of soil carbon prior to burial, particularly at depth in the profiles. Although this is relatively low in arctic-alpine environments, whether any resistant fractions survive from the time of regional deglaciation of the terrain is problematic.

For several reasons, the *date of initiation of the FH horizon* itself may be closely approximated by the dates presently available from the Haugabreen site. First, the organic rich nature of the horizon, with recognizable birch-bark fragments near its base, suggests that some organic matter at the base of the horizon may be true relics of horizon initiation. Second, the age/depth relationship for 'humic acids', with no evidence for a departure from linearity near the base of the horizon, suggests a constant accumulation rate following sudden initiation of horizon formation. Third, the degree of age-differentiation between fractions at depth in the horizon is not as great as would be expected if decomposition and translocation were dominant over accumulation of organic matter. Fourth, there is no evidence from the Bh horizon dates for the presence of older material. Although it is possible that the Bs horizon contains somewhat older material (Guillet, 1972) this too probably formed after the initiation of the FH horizon (Ellis and Matthews, 1984).

Taking into account, (a) the age of CAR-523C, which suggests that the 'fine residual' fraction may be about 500 ^{14}C years older than the 'humic acids' at the same depth, and (b) the width of the confidence interval in Figure 5.4a, at least the FH horizon at Haugabreen is considered to have been initiated before 4500 Calendar yr BP, possibly shortly before 5000 Calendar yr BP. Thus is can be claimed that the procedures adopted in this study have significantly improved the precision with which both time elapsed since burial and absolute soil age can be estimated, although conclusions regarding the latter must be expressed with less certainty.

POLLEN ANALYSIS

Methods and results

Sixteen contiguous samples of maximum thickness 1.0 cm were taken from the FH horizon at Haugabreen. These included sub-samples of the ^{14}C-dated samples and additional samples at intermediate depths. Similarly, fourteen samples were taken from the Memurubreen palaeosol but in this case they were taken close to the adjacent ^{14}C-dated profile. All samples were prepared using standard procedures (Moore and Webb, 1978). Initial maceration in NaOH was followed by boiling in HF to remove silicates. Prior to treatment, tablets of 'exotic' *Eucalyptus* pollen were added to known volumes and dry weights of material (Stockmarr, 1971). For all samples, a count of at least 500 Total Land Pollen (TLP) plus exotic pollen was made ensuring that at least 100 'exotic' pollens were counted. Details of pollen preservation (Cushing, 1967) were also recorded and analyses were made of soil pH and carbon content.

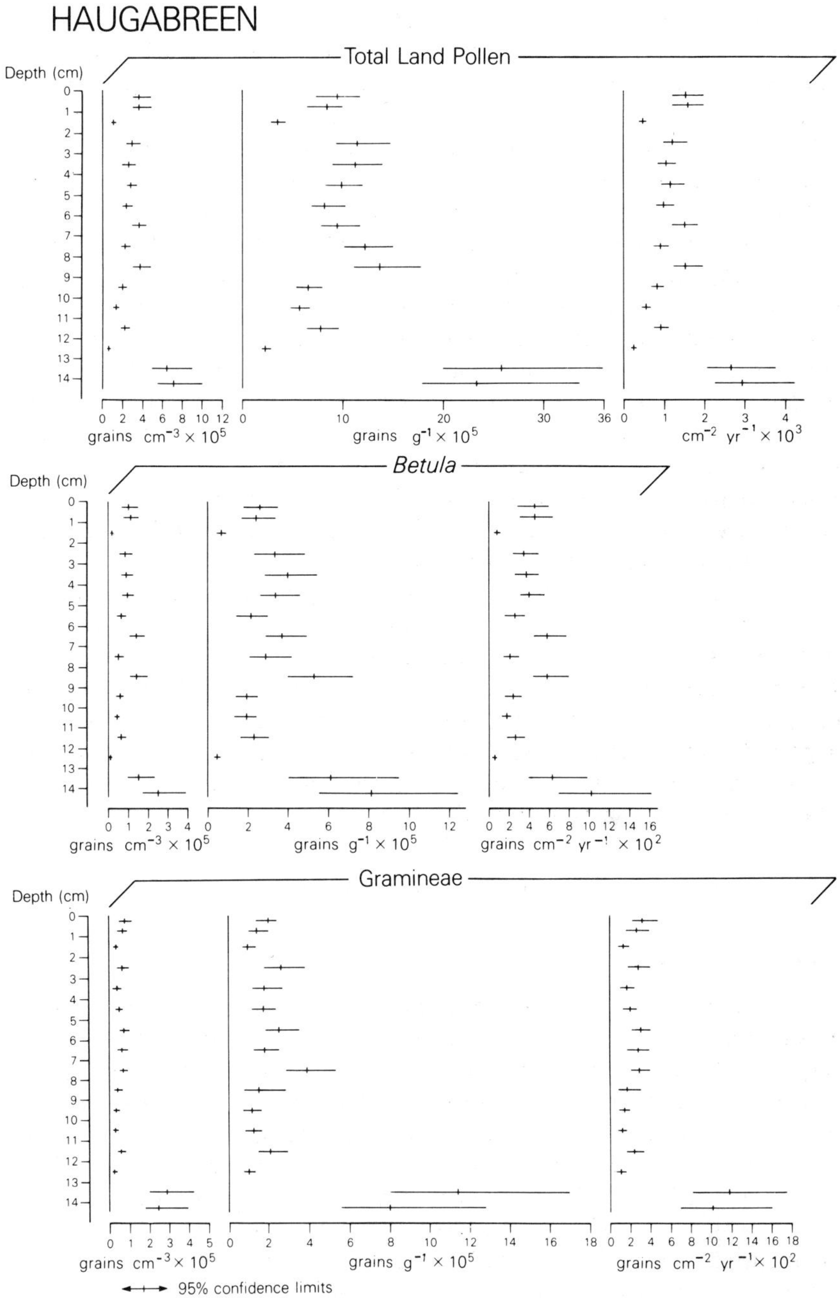

HAUGABREEN

Figure 5.7   Pollen incorporation rates for selected taxa from the Haugabreen palaeosol (after Caseldine, 1983b). Reproduced by permission of Universitetsforlaget

Soils and Quaternary landscape evolution

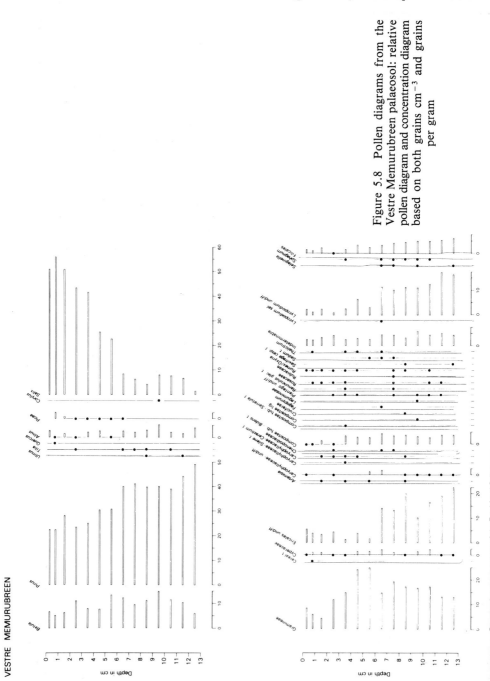

Figure 5.8 Pollen diagrams from the Vestre Memurubreen palaeosol: relative pollen diagram and concentration diagram based on both grains cm^{-3} and grains per gram

VESTRE MEMURUBREEN

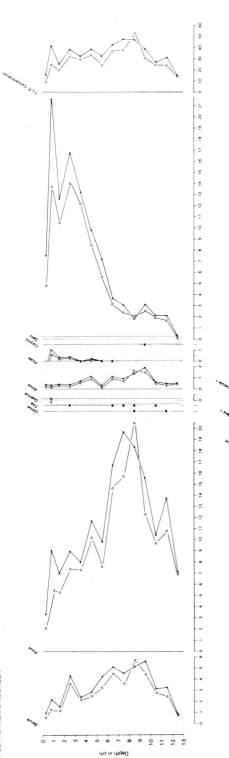

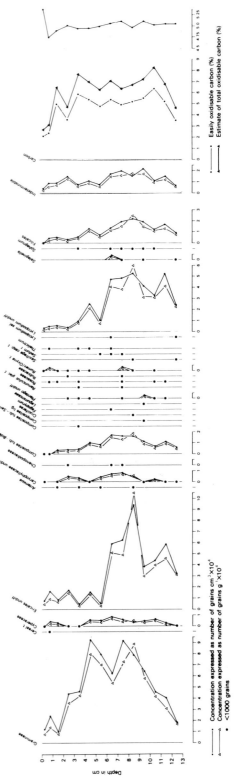

Pollen diagrams for the two palaeosols, based on both relative amounts and concentration values, are presented in Figures 5.6–5.8. A major recurrent problem was the determination of accurate values for fossil pollen concentration where this varied considerably even within the same horizon (cf. Cruickshank and Cruickshank, 1981; Caseldine, 1983a). This may result in counts falling outside the range of percentages for the amount of 'exotic' pollen advised by Bonny (1972) and explains the wide variations in the widths of the confidence intervals in the lowermost levels of Figure 5.7. Derivation of the values for pollen incorporation rates (pir) at Haugabreen have been discussed in Caseldine (1983b) and are commented upon in the next section. Pollen assemblage zones have not been defined for either site because changes in soil pollen diagrams appear to be continuous and zone boundaries would be artificial.

Interpretation of these results requires separation of pollen derived locally from that of more distant origin. The importance of *Pinus*, and to a lesser extent *Betula* at the Memurubreen site is clearly a result of long-distance pollen transport. Within the Non-Arboreal Pollen (NAP) record, however, there is a clear change in the dominant taxa up the profile (Figure 5.8). Below 6 cm, pollen of *Ericales* undiff. (including *Vaccinium* and *Empetrum*) and *Gramineae*, and spores of *Lycopodium* undiff. (including both *L. alpinum* and *L. clavatum*) are dominant, whereas above this level *Salix* increases dramatically. All these taxa are likely to be predominantly of local origin and hence may reflect vegetation change at or around the site.

At Haugabreen, the distinction between local and non-local pollen is not so clear cut, due to the proximity of the woodland and to the abundance of *Betula*. Nevertheless, two findings are of particular importance: first, the variability in the curves for *Betula*, with significantly high concentrations in the basal levels and significant minima at depths of 1.0–2.0 cm and of 12.0–13.0 cm (Figure 5.7); and second, changes in NAP taxa with *Gramineae* and *Potentilla* type pollen being replaced by *Cyperaceae*, ericaceous taxa and *Melampyrum* above 13.0 cm (Figure 5.6).

All of these features of the pollen diagrams, further details of which are given in Caseldine (1983b, 1984), may be interpreted in terms of, (a) local vegetation changes, and (b) oscillations in the altitude of the sub-alpine, low-alpine and mid-alpine vegetation belts (see below).

Technical implications

There has been considerable debate concerning the origin of pollen assemblages in soils and the validity of soil-pollen assemblages as representatives of former vegetation communities. It is now generally recognized that although most soil horizons incorporate pollen of varying age it is sometimes possible to isolate elements from different communities and hence derive valuable evidence concerning vegetational, pedological and environmental history (e.g. Dimbleby,

1962; Vasari, 1965; Munaut, 1967; Havinga, 1974; Andersen, 1979; Cruickshank and Cruickshank, 1981; Caseldine, 1983b). Despite the obvious value of such studies little is known about the processes behind pollen incorporation into soils, although a number of workers have examined aspects of this problem (Dimbleby, 1957, 1961; Munaut, 1967; Walch *et al.*, 1970; Havinga, 1974; Andersen, 1979). The value of the palaeosols discussed here lies in the extremely detailed dating framework within which pollen variations can be examined.

The pir derived for the Haugabreen profile and referred to earlier were calculated by determining the pollen concentration in the soil (both in grains g^{-1} dry weight, and in grains cm^{-3}) and then calibrating the result using the age/depth curve to give a figure for number of pollen grains incorporated per ^{14}C year. The following discussion provides the background against which these results can be evaluated.

For any soil on a level site the following state should apply for pollen incorporation *at the surface.*

$$pir = pi - (l) - (d) + (a_b) \qquad (1)$$

where pir = pollen incorporation rate
pi   = pollen influx at the surface
l = pollen loss through biological, chemical or mechanical degradation
d = down profile movement of pollen by percolation, translocation or mixing processes
a_b = pollen added from below, mainly by faunal mixing.

Thus the amount of pollen being incorporated into the uppermost layer of the soil will depend not only on the amount of pollen arriving at the surface but also on the level of biological and chemical activity within the surface layers. At Haugabreen, where all the evidence points to a high degree of stratification in the FH horizon, pollen loss, down profile movement of pollen and mixing processes are assumed to be very small and possibly negligible (Caseldine, 1983b). In organic horizons the difference between pir and pi may therefore be slight, particularly as overall pollen loss would be unlikely to exceed that at the surface of bogs or in traps, traditional sources for determining pollen influx values. In mineral soils (e.g. the Brown soil at Memurubreen) such a close relationship between pir and pi cannot be expected in view of enhanced rates of biological and chemical activity. Nevertheless, in the alpine zone, this activity is likely to be restricted seasonally and to be generally less than at comparable lower altitude sites. Thus some degree of stratification may also exist in the Memurubreen palaeosol, a suggestion supported by the steep age/depth gradient there.

For conditions *within a soil horizon*, equation (1) may be modified to take the following form:

$$pir = rp - (l) - (d) + (a_{ab}) \qquad (2)$$

where a_{ab} = pollen added either from above or from below by percolation, translocation or mixing processes
rp = resident pollen.

Here, the dominant element will vary in different soil types. In the Haugabreen example, where the FH horizon developed in a manner comparable to a peat with incremental upward development, the values for (l), (d) and (a_{ab}) can be virtually ignored. The calculated value for pir will therefore approximate to rp, pollen derived from the period when that part of the organic horizon was at or near the surface. As already argued, this should closely reflect contemporary pollen influx. With the ^{14}C dates available from the FH horizon, it is therefore possible to establish, within reasonable margins of error, a detailed picture of pollen influx on to the soil surface.

In mineral soils, while it is possible, in theory, to apply equation (2), the complex nature of pollen addition and loss within horizons, the unknown mobility of pollen within the soil profile, and the greater continuing influence of surface-pollen influx, renders such a model of limited value. However, if the ideas of Dimbleby (1957, 1961) are accepted, some pollen will become 'resident' within horizons, the age of this rp increasing with depth. Although pollen loses will be much higher than in organic horizons, these may be broadly estimated where, as at Memurubreen, it is possible to derive estimates of absolute soil age.

At the Memurubreen site, the oldest date so far obtained from the soil of 3810 ± 75 ^{14}C yr BP (Figure 5.5) is the minimum period over which the palaeosol developed. The maximum period is *ca.* 9000 ^{14}C yr. By comparing values for pollen concentration in the soil with that expected under comparable vegetation communities (using modern influx data and taking into account the presence of more than one vegetation community at the site) at least 15–20 per cent and possibly up to 70–75 per cent of pollen appears to have been lost.

In neither palaeosol is there any strong relationship between the amount of pollen damage and soil carbon content or pH. Correlations between percentage damaged pollen and total oxidizable carbon are weak (Haugabreen $r = 0.42$, ns; Memurubreen $r = 0.40$, ns). In the Brown soil there is a strong inverse relationship ($r = -0.87$, $p < 0.01$) between carbon and the amount of degraded pollen (*sensu* Cushing, 1967) as a percentage of damaged pollen, but such a relationship is not apparent in the FH horizon at Haugabreen. In both soils damage is largely mechanical in origin, as manifested by breakage or crumpling (Figure 5.9). The details of pollen preservation do not allow the identification of pollen from different sources, such as rp, for although preservation data were collected at taxon level few patterns emerged and those that could be seen were largely affected by pollen type.

HAUGABREEN

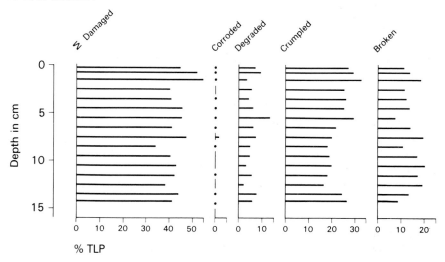

VESTRE MEMURUBREEN

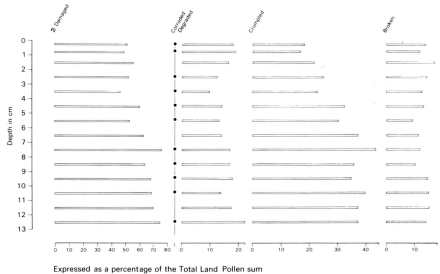

Expressed as a percentage of the Total Land Pollen sum
• <1%

Figure 5.9 Comparison of pollen preservation from the Haugabreen and the Vestre Memurubreen palaeosols (after Caseldine, 1983b). Reproduced by permission of Universitetsforlaget

IMPLICATIONS FOR LANDSCAPE CHANGE

Moraine chronology

The results indicate both the potential and the dangers of using palaeosols for dating end moraines and hence glacier variations. Detailed ^{14}C dating at Haugabreen and Vestre Memurubreen has produced results fully consistent with the independent evidence for burial of the palaeosols in the 'Little Ice Age'. If age/depth gradients had not been identified in the buried soils, however, overestimates of moraine age by up to 4000 ^{14}C years would have been possible. The information now available on ^{14}C age variations within these palaeosols provides, therefore, persuasive reasons for doubting many previous estimates of *moraine age* based on palaeosol dating.

In southern Norway, the work of Griffey and Matthews (1978) on the ^{14}C dating of palaeosols at other Jotunheimen glaciers, has been reinterpreted in the light of the age/depth gradients described above (Matthews, 1980). Griffey and Matthews interpreted ^{14}C ages of up to 2830 ± 130 ^{14}C yr BP (Birm-720B) as direct evidence for Neoglacial moraine ridges dating from before the 'Little Ice Age'. They recognized pre-'Little Ice Age' Neoglacial glacier maxima dating from about 1300 Calendar yr BP at Leirbreen and about 2700 Calendar yr BP at Styggedalsbreen. Their interpretation was all the more convincing because similar glacier expansions had been recognized elsewhere in Europe. However, the possibility of steep age/depth gradients in samples of thickness 2.5 cm had not been taken sufficiently into account. The revised interpretation of a 'Little Ice Age' date for the moraines at Leirbreen and Styggedalsbreen has since been further supported by lichenometric evidence from eighty glaciers and boulder weathering data from thirty-four glaciers, which has shown that only a small minority (< 10 per cent) of Jotunheimen glaciers are likely to have reached their Neoglacial maxima before the 'Little Ice Age' (Matthews and Shakesby, 1984). The study included Leirbreen and Styggedalsbreen.

A larger number of Neoglacial glacier expansion episodes have been recognized in northern Scandinavia, where the glacier variation chronology rests heavily on ^{14}C dating of palaeosols from beneath moraines (Karlén, 1982). However, as, (a) it has been demonstrated that age differences with depth in a single horizon can encompass almost the whole of the Neoglacial time scale, and (b) age/depth gradients are unlikely to be unique to southern Norwegian palaeosols, much of the Scandinavian glacier variation chronology must be open to question. The precision with which individual moraines have been dated would appear to cast doubt on the timing of most of the major expansion episodes that have been recognized (cf. Karlén, 1981; Matthews, 1982).

[14]C dates from well-developed palaeosols, such as those of 3170 ± 120 ^{14}C yr BP (Birm-498) from Steikvassbreen in the Okstindan mountains of northern Norway (Griffey, 1976) and of 2970 ± 95 ^{14}C yr BP (St-7192) from Mikkajekna in Swedish Lappland (Karlén, 1981) must be considered with great caution. At

these and other sites in the north, steep age/depth gradients may possibly have led to the overestimation of moraine age. In contrast, incipient soils that have developed over a short period cannot contain steep age/depth gradients and may therefore be expected to yield ^{14}C ages that approximate burial events more closely. Examples are provided by the dates of 245 ± 65 ^{14}C yr BP (SI-1321B) from Årjep Ruotesjekna in Lappland (Karlén and Denton, 1975) and of 240 ± 100 ^{14}C yr BP (Birm-497) from Steikvassbreen (Griffey and Worsley, 1978). Such thin soils may, however, be subject to greater risk of contamination or other problems associated with recently deglaciated terrain (Sutherland, 1980). There are also considerable problems with calibrating such recent dates as these (Stuiver, 1978).

Vegetation history

The relationship of the study sites to the altitudinal limits of birch woodland and to the boundaries of the vegetation belts in the alpine zone can be used in conjunction with the results of pollen analysis and ^{14}C dating to examine changes in the location of these belts during the Neoglacial period. Although the sensitivity of the altitudinal limits of pine and birch to climatic variations is well known in Scandinavia, this contrasts markedly with current views on the stability of the vegetation communities in the alpine zone (Dahl, 1975b).

At Memurubreen, there has been a single well-defined change in the local vegetation community during the period of soil development. The present vegetation cover is a 'grass' heath typical of the mid-alpine belt in Jotunheimen. The *Salix* dominated pollen assemblage of the upper levels appears to be representative of this community. In the lower levels of the soil the pollen assemblage is more characteristic of the dwarf-shrub heaths of the low-alpine belt. It is unlikely that such a change would occur as a result of pedological change as the dwarf-shrub species are not dominants in the mid-alpine belt and are rarely found there. Thus it seems likely that this substitution of communities occurred as a result of some form of climatic change allowing a downward displacement of vegetation belts within the alpine zone. The ^{14}C dates currently available from the Memurubreen site are not sufficient for provision of an absolute time scale for these changes.

Despite difficulties in the separation of pollen of local origin from long-distance pollen at the Haugabreen site, it has been argued by Caseldine (1983b) that changes in both the relative proportion and the pollen incorporation rate for *Betula* represent changes in the altitude of the upper limit of the sub-alpine belt. Two periods of woodland recession have been identified. The most recent of these was at the onset of the 'Little Ice Age', possibly around the thirteenth century AD. The earlier recession occurred between about 3600 ^{14}C yr BP and 3300 ^{14}C yr BP. Prior to about 3600 ^{14}C yr BP it is probable that birch woodland occupied the site, but it did not return to the site subsequently. Except at the base of the FH horizon, the dominantly local NAP in the Haugabreen record is representative of species commonly found in the low-alpine, dwarf-shrub heath communities, although it should be recognized that these species occur today within the sub-alpine woodland.

In the alpine zone of southern Norway the results of this study show Neoglacial vegetation changes in some detail, suggesting a reappraisal of the supposed stability of the alpine communities. Although particularly resilient communities, or sites lying at the heart of vegetation belts (and therefore buffered from changes near their altitudinal limits) are likely to yield relatively uniform pollen records, even in these cases oscillations representative of community dynamics may possibly be identified.

Soil development

Detailed age/depth gradients within palaeosols are useful as a basis for inference about soil development. The possible extension of the arguments from the FH horizon at Haugabreen to the soil profile as a whole has been considered by Ellis and Matthews (1984), on the basis of the dates from the Bh horizon (Figure 5.4b) and micromorphological analysis of the soil profile.

Illuviation of organic matter probably commenced in the Bh horizon shortly after the FH horizon began to accumulate. This inference is based on statistically indistinguishable ages for similar fractions at the base of the two horizons (Figure 5.4). Large differences between the ages of fractions at similar depths within the Bh horizon appear to reflect the translocation of material with a wide range of ages from the FH horizon. The absence of dates younger than 2340 ± 65 ^{14}C yr BP from the Bh horizon is similarly interpreted as reflecting large apparent mean residence times of organic fractions inherited from the FH horizon. The age/depth gradient in the Bh horizon seems to reflect an upward movement of the zone of maximum deposition of illuvial material (cf. Aaltonen, 1939). Ellis and Matthews (1984) concluded that not only the FH horizon but also the whole podzol profile may have been initiated suddenly in response to a deteriorating climate (see below).

Other attempts to infer rates of podzolic soil development in Scandinavian arctic-alpine environments have been few in number (Alexander and Worsley, 1973; Ellis, 1980b) and have been made without the benefit of known age/depth gradients. Although podzolic soil development can be much more rapid at lower altitudes, well-developed podzols of the Haugabreen type would appear to require many thousands of years to develop in an arctic-alpine environment.

In the absence of any ^{14}C dating or micromorphological evidence as to the nature of the soil at the Haugabreen site prior to the onset of FH horizon formation, it can only be assumed that pedological or other changes initiated at or before about 5000 Calendar yr BP were sufficiently drastic to remove evidence of any pre-existing soil. The absence of traces of this pre-existing soil may be due, in part, to genetic similarity to the Humo-ferric Podzol that developed subsequently. This interpretation is in agreement with Ellis (1980b) who suggested, in the context of the similar environment of Okstindan, that recognizable podzolic soils may have developed by about 8000 ^{14}C yr BP. Whatever the cause, the absence of ^{14}C ages in excess of 5000 years from the FH horizon means that it is not possible to obtain from ^{14}C dating evidence alone, a close estimate of the *age of the land surface* on which the Haugabreen palaeopodzol began to develop.

Closer estimates may be possible at the Vestre Memurubreen site if new ^{14}C dates can be obtained from deeper in the buried Brown soil profile. The climate at 1480 m may be sufficiently severe for the preservation of resistant organic fractions at depth in the soil profile. Furthermore, uninterrupted soil development may have occurred slowly at this site since regional deglaciation about 9000 ^{14}C yr BP. A date of 8350 ± 120 ^{14}C yr BP (T-406) reported by Østrem (1965) from a depth of 20 cm in a present-day soil at an altitude of 1900 m near Gråsubreen, about 18 km to the north, provides some support for this assertion.

Climatic inferences

Climatic inferences can be made on the basis of the geomorphological, pedological and palaeobotanical evidence that is available from the palaeosol sites.

As the ^{14}C dates indicate that both Haugabreen and Vestre Memurubreen expanded to their Neoglacial limits during the 'Little Ice Age', it follows that climatic *deteriorations* over the whole Neoglacial period were unlikely to have exceeded those of the 'Little Ice Age'. Summer temperature estimates for southern Norway since the 'Little Ice Age' glacier maximum of between -2.0 and $+1.0$ °C relative to the present, with an overall warming of about 1.0 °C (Matthews 1976, 1977), may therefore apply to earlier Neoglacial climatic deteriorations.

The presence of low-alpine pollen at the Memurubreen site prior to the 'Little Ice Age' shows that temperatures were probably higher than today. Minimum and maximum estimates of these former higher temperatures are obtainable from possible downward displacements of the relevant vegetation belts. The mid-alpine belt extends at least 100 m below the altitude of the Memurubreen site. The present sub-alpine birch woodland is found, in favourable locations, a further 200 m below this. As there is no evidence in the pollen record for the tree line having approached the palaeosol site, growing season temperatures during the most *favourable* period for development of the palaeosol are likely to have been between $+0.7$ and $+2.1$ °C relative to the present (assuming a temperature gradient of 0.7 °C per 100 m).

At Haugabreen, two dated intervals of relative warmth can be inferred from upward extensions of the woodland limit. The most favourable interval occurred between FH horizon initiation at or before about 5000 ^{14}C yr BP and about 3600 ^{14}C yr BP, when macrofossil as well as microfossil evidence suggests that birch woodland grew at the palaeosol site. Temperatures at that time were therefore higher than they are today but were probably lower than those characteristic of the warmest period of the Holocene (cf. Hafsten, 1981a). The second interval of relative warmth is inferred to have occurred between about 3300 ^{14}C yr BP and the beginning of the 'Little Ice Age'.

The apparently sudden onset of FH horizon itself provides the earliest dated evidence for climatic change at Haugabreen. Although relatively steep age/depth gradients were found in the FH horizon where it was only 5 cm thick (Figures 5.4a and 5.5), predicted ages at the base of this horizon are compatible with a climatic deterioration large enough to initiate development almost simultaneously on different microtopographic positions (Matthews and Dresser, 1983). The woodland recession at about 3600 ^{14}C yr BP could have been a response to the same general deterioration, possibly related to the end of the Postglacial warmth period (Hafsten, 1969; Hafsten and Solem, 1976). This is supported by evidence that the Tunsbergdalsbreen outlet of the Jostedalsbreen ice cap was smaller than it is today during the interval 8083 ± 100 to 3855 ± 55 ^{14}C yr BP (Mottershead *et al.*, 1974; Mottershead and Collin, 1976). The timing of the woodland recession at around the thirteenth century AD, which marks the onset of the 'Little Ice Age' in southern Norway, is in fairly close agreement with evidence for the development of Omnsbreen (Elven, 1978) when temperatures slightly higher than today were followed by a rapid and strong deterioration of climate in the fourteenth century AD.

Assuming climate has continuously varied throughout the Neoglacial period, it is concluded that only major shifts that either crossed particular thresholds or induced a cumulative response in the vegetation have been registered in the palaeosol record. Observations on the effect of the post-'Little Ice Age' climatic amelioration at both Memurubreen and Haugabreen support this view. At Memurubreen, the site has not been recolonised by low-alpine communities, suggesting that either a climatic change of the order of $+ 1\,°C$ has not been sufficient, or that the vegetation communities respond slowly to improving climate. Similarly, the Haugabreen site has not been characterized by a local revertence to birch woodland, although there is an added complication here in the possibility of local grazing pressure preventing woodland regeneration.

CONCLUSIONS

Despite the many problems involved in the study of landscape change in the alpine zone, the results presented here suggest that significant advances are possible both in ^{14}C dating and in the reconstruction of palaeoenvironments from palaeosols. The implications of the work extend beyond the immediate Jotunheimen/Jostedalsbreen region and are relevant to palaeosols buried before the Neoglacial period.

Although moraine chronologies from other areas, especially from the Alps (Patzelt and Bortenschlager, 1973; Patzelt, 1974; Schneebeli and Röthlisberger, 1976; Röthlisberger *et al.*, 1980) may be better controlled, the limitations of the ^{14}C dating of palaeosols, as demonstrated here, are almost certainly a factor in the apparent lack of synchroneity in Holocene glacier variations world-wide (Grove, 1979; Porter, 1981; Williams and Wigley, 1983). Previous studies may have overestimated the number of glacier expansion episodes through the absence of detailed information on ^{14}C age/depth gradients. Such information is essential if the timing of geomorphological events is not to be confounded with AMRT effects in well-developed palaeosols. The full implications for the Holocene moraine chronology of Scandinavia have yet to be resolved, as independent lichenometric evidence indicates the widespread occurrence of pre-'Little Ice Age' Neoglacial moraines in northern Sweden and northern Norway (Karlén, 1973, 1975, 1979; Matthews and Shakesby, 1984).

Palaeosols dating from earlier in the Quaternary may well possess similar steep age/depth gradients to those reported here, provided the palaeosols developed over a long enough period of time. Where palaeosols developed over a shorter time interval, such as in the Allerød or Bølling Chronozones, steep age/depth gradients are less likely as AMRT effects cannot be very large.

In palaeosols with steep age/depth gradients, the considerable potential for tracing and dating the initiation, development and termination of soil formation is not without problems, some of which have been identified or clarified as a result of our studies. These problems include, first, the necessity for precision sampling of thin soil slices, with sophisticated pretreatment procedures and a large number of ^{14}C age determinations. Second, although accurate dating of the termination of soil formation appears to be feasible in some palaeosols, the uppermost layer may have been contaminated during burial even where the surface of the palaeosol has been preserved intact. Third, most soils do not develop by simple upwards accumulation of organic matter. There appears, however, to be an *element* of upward development even in many mineral soils (as illustrated by the Brown soil at Memurubreen and the Bh horizon at Haugabreen), which holds the key to the reconstruction of palaeoenvironmental change as well as the study of soil development. Fourth, although the date of initiation of soil formation may be closely approximated in favourable situations by dating the oldest organic fraction remaining near the lowest fringes of the soil profile, this ^{14}C age may not be a close approximation of the age of the land surface on which the soil developed. This is related to the problem recognized at Haugabreen of the possibility of a pre-existing soil at this site prior to the onset of formation of the dated FH horizon.

In view of the lack of other direct evidence from the alpine zone, the pollen analytical results from Haugabreen and Memurubreen must lead to questioning the general stability of Scandinavian alpine vegetation communities (Dahl, 1975b). This view of general stability may well apply to the most resilient communities and also to the heart of vegetation belts, but it is likely that some communities are sensitive to change. It should be recognized that changes may be relatively slow, the substitution of the communities at Memurubreen probably representing one single community change over at least the last 4000 years.

Three further problems remain to be clarified. The first involves the precise environmental factors responsible for the changes that have been identified. Although we have emphasized possible temperature change, other factors or factor complexes are almost certainly involved, either directly or indirectly. One of the most important of these in the alpine zone is snow cover, which generally increases in depth, duration and areal extent with increasing altitude. The second problem is the rate of response of the various communities to environmental change and the possible differential response to climatic deterioration and amelioration. It seems that climatic deteriorations are relatively quickly registered, particularly where the birch-tree line is involved, and they may have long-lasting effects resulting in disequilibrium between climate and vegetation at high altitudes. Some communities may be more resistant to change except when critical thresholds are crossed, events that may have been experienced on only a very few occasions in the Holocene. Again, in the Late Weichselian and early Holocene this could account for the registration or non-registration of events such as the Older Dryas and Preboreal climatic deteriorations.

The third problem is the scale of the effect, if any, of the proximity of glaciers to the palaeosol sites. There is a possibility, particularly in the case of variations in the limit of birch woodland at Haugabreen, that the expansion or recession of the glaciers is at least a contributing factor in the vegetation changes. If this is so, the pollen record in such palaeosols could be turned to advantage in a similar way to pollen records preserved in the peat bogs associated with moraines in the Alps. In the Alps it has been shown that Holocene climatic deteriorations are well represented in pollen profiles only if the bogs investigated are close to the tree line or in the vicinity of glaciers (Bortenschlager, 1970 a and b; Patzelt and Bortenschlager, 1973; Patzelt, 1974).

Finally, we would emphasize that, with well-preserved organic matter in its soils, there is much to be learned in the alpine zone without having to rely on scattered and more traditional peat and lake sites for evidence of landscape change. Indeed, with the potential for ^{14}C dating and palaeoenvironmental analysis based on surface as well as buried soils, the alpine zone could, in some respects become a model for the elucidation of relationships between climate and landscape.

ACKNOWLEDGEMENTS

This paper forms part of the continuing research programme of the Jotunheimen Research Expeditions. We would like to thank all those who assisted in the field, and the many organizations and individuals who helped financially or otherwise with various aspects of the work.

REFERENCES

Aaltonen, V. T. (1939). 'Zur stratigraphie des podsolprofils II', *Communicationes Instituti Fenniae*, **17**, 4, 1–133.

Alexander, M. J. (1982). 'Soil development at Engabredal, Holandsfjord, north Norway', in *Okstindan Research Project Preliminary Report 1980 and 1981* (Ed. J. Rose), pp. 1–23, Birkbeck College, University of London, and University of Reading.

Alexander, M. J. and Worsley, P. (1973). 'Stratigraphy of a Neoglacial end moraine in Norway', *Boreas*, **2**, 117–142.

Andersen, B. G. (1980). 'The deglaciation of Norway after 10,000 B.P.', *Boreas*, **9**, 211–216.

Andersen, S.Th. (1979). 'Brown earth and podzol: soils genesis illuminated by microfossil analysis', *Boreas*, **8**, 59–73.

Bonny, A. P. (1972). 'A method for determining absolute pollen frequencies in lake sediments', *New Phytol.* **71**, 393–405.

Bortenschlager, S. (1970a). 'Waldgrenz- und Klimaschwankungen im pollenanalytischen Bild des Gurgler Rotmooses', *Mittl. Ostalp. Dinar. Ges. f. Vegetationskunde*, **11**, 19–26.

Bortenschlager, S. (1970b). 'Neue pollenanalytische untersuchungen von Gletschereis und Gletschernahen mooren in den Ostalpen', *Zeitschr. f. Gletscherk. und Glazialgeol.* **6**, 107–118.

Brunn, I. (1967). *Standard Normals 1931–1960 of the Air Temperature in Norway*, Det Norske Meteorologiske Institutt, Oslo.

Campbell, C. A., Paul, E. A., Rennie, D. A. and McCallum, K. J. (1967). 'Factors affecting the accuracy of the carbon-dating method in soil humus studies', *Soil Sci.* **104**, 81–85.

Caseldine, C. J. (1983a). 'Problems of relating site and environment in pollen analytical research in South West England', in *Site, Environment and Economy* (Ed. B. Proudfoot), Brit. Arch. Reports. Int. Series, **173**, 61–71.

Caseldine, C. J. (1983b). 'Pollen analysis and rates of pollen incorporation into a radiocarbon-dated palaeopodzolic soil at Haugabreen, southern Norway', *Boreas*, **12**, 233–246.

Caseldine, C. J. (1984). 'Pollen analysis of a buried arctic-alpine Brown soil from Vestre Memurubreen, Jotunheimen mountains, Norway: evidence for recent high altitude vegetation change', *Arctic and Alpine Res.* **16**, (in press).

Clark, R. M. (1975). 'A calibration curve for radiocarbon dates', *Antiquity*, **49**, 251–266.

Cruickshank, J. G. and Cruickshank, M. M. (1981). 'The development of humus-iron podzol profiles, linked by radiocarbon-dating and pollen analysis to vegetation history', *Oikos*, **36**, 238–253.

Cushing, E. J. (1967). 'Evidence for differential pollen preservation in late Quaternary sediments in Minnesota', *Rev. Palaeobot. Palynol.* **4**, 87–101.

Dahl, E. (1956). 'Rondane: mountain vegetation in southern Norway and its relation to environment', *Skrifter utgift av det Norske videnskaps-academi i Oslo. Matematisk-naturvidenskapelig klasse*, **3**, 1–374.

Dahl, E. (1975a). 'Flora and plant sociology in Fennescandian Tundra areas', in *Fennoscandian Tundra Ecosystems*, Pt. 1 (Ed. F. E. Wiegolaski), pp. 62–67, Springer-Verlag, Berlin.

Dahl, E. (1975b). 'Stability of tundra ecosystems in Fennoscandia', in *Fennoscandian Tundra Ecosystems*, Pt. 2 (Ed. F. E. Wiegolaski), pp. 231–236. Springer-Verlag, Berlin.

Dimbleby, G. W. (1957). 'Pollen analysis of terrestrial soils', *New Phytol.* **56**, 12–28.

Dimbleby, G. W. (1961). 'Soil pollen analysis', *J. Soil Sci.* **12**, 1–11.

Dimbleby, G. W. (1962). 'The development of British heathlands and their soils', *Oxford Forestry Memoir*, **23**.

Ellis, S. (1979). 'The identification of some Norwegian mountain soils types', *Norsk Geogr. Tidsskr.* **33**, 205–212.

Ellis, S. (1980a). 'Soil-environmental relationships in the Okstindan Mountains, north Norway', *Norsk Geogr. Tidsskr.* **34**, 167–176.

Ellis, S. (1980b). 'Physical and chemical characteristics of a podzolic soil formed in Neoglacial till, Okstindan, Northern Norway', *Arctic and Alpine Res.* **12**, 65–72.

Ellis, S., and Matthews, J. A. (1984). 'Pedogenic implications of a ^{14}C-dated palaeopodzolic soil at Haugabreen, southern Norway', *Arctic and Alpine Res.* **16**, 77–91.

Elven, R. (1978). 'Subglacial plant remains from the Omnsbreen glacier area, south Norway', *Boreas*, **7**, 83–89.

Eronen, M. (1979). 'The retreat of pine forest in Finnish Lappland since the Holocene climatic optimum: a general discussion with radiocarbon evidence from subfossil pines', *Fennia*, **157**, 93–114.

Eronen, M. and Hyvärinen, H. (1981). 'Subfossil pine dates and pollen diagrams from northern Fennoscandia', *Geol. För. i Stockholm Förh.* 437–445.

Gerasimov, I. P. (1974). 'The age of recent soils', *Geoderma*, **12**, 17–25.

Geyh, M. A., Benzler, J-H. and Roeschmann, G. (1971). 'Problems of dating Pleistocene and Holocene soils by radiometric methods', in *Palaeopedology Origin, Nature and Dating of Palaeosols* (Ed. D. H. Yaalon) pp. 63–75. Int. Soc. Soil Sci. and Israel Universities Press.

Griffey, N. J. (1976). 'Stratigraphical evidence for an early Neoglacial glacier maximum of Steikvassbreen, Okstindan, north Norway', *Nor. Geol. Tidsskr.* **56**, 187–194.

Griffey, N. J. and Matthews, J. A. (1978). 'Major Neoglacial glacier expansion episodes in southern Norway: evidence from moraine ridge stratigraphy with ^{14}C dates on buried palaeosols and moss layers', *Geogr. Annlr.* 60A, 73–90.

Griffey, N. J. and Worsley, P. (1978). 'The pattern of Neoglacial glacier variations in the Okstindan region of northern Norway during the last three millennia', *Boreas*, **7**, 1–17.

Grove, J. M. (1979). 'The glacial history of the Holocene', *Progress in Physical Geography*, **3**, 1–50.

Guillet, B. (1972). *Relation Entre l'Histoire de la Végétation et la Podzolisation dans les Vosges*, PhD Thesis, Université de Nancy.

Hafsten, U. (1969). 'A proposal for a synchronous subdivision of the late Pleistocene period having global and universal applicability', *Nytt Mag. Bot.* **16**, 1–13.

Hafsten, U. (1981a). 'An 8000 years old pine trunk from Dovre, South Norway', *Norsk Geogr. Tidsskr.* **35**, 161–165.

Hafsten, U. (1981b). 'Palaeo-ecological evidence of a climatic shift at the end of the Roman Iron Age', *Striae*, **14**, 58–61.

Hafsten, U. and Solem, T. (1976). 'Age, origin and palaeo-ecological significance of blanket bogs in Nord-Trøndelag, Norway', *Boreas*, **5**, 119–142.

Havigna, A. J. (1974). 'Problems in the interpretation of pollen diagrams of mineral soils', *Geologie Mijnb.* **53**, 449–453.

Hustich, I. (1948). 'The Scotch pine in northernmost Finland and its dependence on the climate in the last decades', *Acta Bot. Fenn.* **42**, 1–75.

Hustich, I. (1966). 'On the forest-tundra limit and the northern tree lines', *Ann. Univ. Turku,* AII, 36, 7–47.

Hyvärinen, H. (1976). 'Flandrian pollen deposition rates and tree-line history in northern Fennoscandia', *Boreas,* **5**, 163–176.

Karlén, W. (1973). 'Holocene glacier and climatic variations, Kebnekaise mountains, Swedish Lappland', *Geogr. Annlr.* **55A,** 29–63.

Karlén, W. (1975). 'Lichenometrisk datering i norra Skandinavien—metodens tillförlitlighet och regionala tillämpning', *Forskningsrapport, Naturgeografiska Institutionen, University of Stockholm,* **22**, 1–70.

Karlén, W. (1976). 'Lacustrine sediments and tree-limit variations as indicators of Holocene climatic fluctuations in Lappland', *Geogr. Annlr.* **58(A)**, 1–34.

Karlén, W. (1979). 'Glacier variations in the Svartisen area, northern Norway', *Geogr. Annlr.* **61(A)**, 11–28.

Karlén, W. (1981). 'A comment on John A. Matthews's article regarding ^{14}C dates of glacier variations', *Geogr. Annlr.* **63(A)**, 19–21.

Karlén, W. (1982). 'Holocene glacier variations in Scandinavia', *Striae,* **18**, 26–34.

Karlén, W. and Denton, G. H. (1976). 'Holocene glacier variations in Sarek National Park, northern Sweden', *Boreas,* **5**, 25–56.

Kullman, L. (1979). 'Change and stability in the altitude of the birch tree-limit in the Southern Swedish Scandes 1915–1975', *Acta Phytogeogr. Suecica,* **65**.

Kullman, L. (1980). 'Radiocarbon dating of subfossil Scots pine (*Pinus sylvestris* L.) in the southern Swedish Scandes', *Boreas,* **9**, 101–106.

Matthews, J. A. (1976). ' "Little Ice Age" palaeotemperatures from high altitude tree growth in S. Norway', *Nature,* **264**, 243–245.

Matthews, J. A. (1977). 'Glacier and climatic fluctuations inferred from tree-growth variations over the last 250 years, central southern Norway', *Boreas,* **6**, 1–24.

Matthews, J. A. (1980). 'Some problems and implications of ^{14}C dates from a podzol buried beneath an end moraine at Haugabreen, southern Norway', *Geogr. Annlr.* **62(A)**, 185–208.

Matthews, J. A. (1981). 'Natural ^{14}C age/depth gradient in a buried soil', *Naturwissenschaften,* **68**, 472–474.

Matthews, J. A. (1982). 'Soil dating and glacier variations: a reply to Wibjörn Karlén', *Geogr. Annlr.* **64(A)**, 15–20.

Matthews, J. A. (1985). 'Radiocarbon dating of surface and buried soils: principles, problems and prospects', in *Geomorphology and Soils* (Eds K. Richards, S. Ellis and R. Arnett) George Allen & Unwin, London (in press).

Matthews, J. A. and Dresser, P. Q. (1983). 'Intensive ^{14}C dating of a buried palaeosol horizon', *Geol. För. i Stockholm Förh.* **105**, 59–63.

Matthews, J. A. and Shakesby, R. (1984). 'The status of the "Little Ice Age" in southern Norway: relative-age dating of Neoglacial moraines with Schmidt hammer and lichenometry', *Boreas,* **13**, 333–346.

Moe, D. (1979). 'Tregrense-fluktuasjoner på Hardangervidda etter siste istid', in *Førtiden i Søkelyset. [14]C datering gjennom 25 år* (Eds R. Nydal, S. Westin, U. Hafsten and S. Gulliksen) pp. 199–208, Laboratoriet for Radiologisk Datering, Trondheim.

Moore, P. D. and Webb, J. A. (1978). *An Illustrated Guide to Pollen Analysis,* Hodder & Stoughton, London.

Mottershead, D. N., Collin, R. L. and White, I. D. (1974). 'Two radiocarbon dates from Tunsbergdalen', *Nor. geol. Tidsskr.* **54**, 131–134.
Mottershead, D. N. and Collin, R. L. (1976). 'A study of Flandrian glacier fluctuations in Tunsbergdalen, southern Norway', *Nor. geol. Tidsskr.* **56**, 413–436.
Munaut, A. V. (1967). 'Recherches paléo-écologiques en Basse et Moyenne Belgique, *Acta Geogr. Louv.* 6.
Nordhagen, R. (1943). 'Sikilsdalen og Norges fjellbeiter', *Bergens Mus. Skr.* 22.
Østrem, G. (1965). 'Problems of dating ice-cored moraines', *Geogr. Annlr.* **47(A)**, 1–38.
Patzelt, G. (1974). 'Holocene variations of glaciers in the Alps', *Colloques Int. du Cent. Natl. de la Rech. Sci.* **219**, 51–59.
Patzelt, G. and Bortenschlager, S. (1973). 'Die postglazialen Gletscher- und Klimaschwankungen in der Venedigergruppe (Hohe Tauern, Ostalpen)', *Zeitschr. f. Geomorphol. Neue Folge, Suppl. Bd.* **16**, 25–72.
Porter, S. C. (1981). 'Glaciological evidence of Holocene climatic change', in *Climate and History: Studies on Past Climates and their Impact on Man* (Eds T. M. L. Wigley, M. J. Ingram and G. Farmer) pp. 82–110, Cambridge University Press, Cambridge.
Röthlisberger, F., Haas, P., Holzhauser, H., Keller, W., Bircher, W. and Renner, F. (1980). 'Holocene climatic fluctuations—radiocarbon dating of fossil soils (fAH) and woods from moraines and glaciers in the Alps', *Geographica helv.* **35**, 21–52.
Scharpenseel, H. W. (1971). 'Radiocarbon dating of soils—problems, troubles, hopes', in *Paleopedology, Origin, Nature and Dating of Paleosols* (Ed. D. H. Yaalon), pp. 77–88, Int. Soc. Soil Sci. and Israel Universities Press, Jerusalem.
Scharpenseel, H. W. (1975). 'Natural radiocarbon measurements on humic substances in the light of carbon cycle estimates', in *Humic Substances: Their Structure and Function in the Biosphere* (Eds D. Povoledo and H. L. Goltermann) pp. 281–292, Centre for Agricultural Publishing and Documentation, Wageningen.
Scharpenseel, H. W. and Schiffmann, H. (1977). 'Radiocarbon dating of soils, a review', *Zeitschr. Pflanzenernährung Düngung Bodenkunde,* **140**, 159–174.
Schneebeli, W. and Röthlisberger, F. (1976). '8000 Jahre Walliser Gletschergeschichte', *Die Alpen, Zeitschr. des Schweizer Alpen Club,* **52**, 1–151.
Stockmarr, J. (1971). 'Tablets with spores used in absolute pollen analysis', *Pollen Spores,* **13**, 615–621.
Stout, J. D., Goh, K. M. and Rafter, T. A. (1981). 'Chemistry and turnover of naturally occurring resistant organic compounds in soil', in *Soil Biochemistry* (Eds E. A. Paul and J. N. Ladd) pp. 1–73. Marcel Dekker, New York and Basel.
Stuiver, M. (1978). 'Radiocarbon timescale tested against magnetic and other dating methods', *Nature,* **273**, 271–274.
Sutherland, D. G. (1980). 'Problems of radiocarbon dating deposits from newly deglaciated terrain: examples from the Scottish Lateglacial', in *Studies in the Lateglacial of North-West Europe* (Eds J. J. Lowe, J. M. Gray and J. E. Robinson) pp. 139–149. Pergamon, Oxford.
Vasari, Y. (1965). 'Studies on the vegetational history of the Kuusamo district (North East Finland) during the Late-Quaternary period. IV. The age and origin of some present day vegetation types', *Ann. Bot. Fenn.* **2**, 248–273.
Walch, K. M., Rowley, J. R. and Norton, N. J. (1970). 'Displacement of pollen grains by earthworms', *Pollen Spores,* **12**, 39–44.
Williams, L. D. and Wigley, T. M. L. (1983). 'A comparison of evidence for Late Holocene summer temperature variations in the northern hemisphere', *Quaternary Res.* **20**, 286–307.

Soils and Quaternary Landscape Evolution
Edited by J. Boardman
© 1985 John Wiley & Sons Ltd.

6

Frost Effects in Soils

B. VAN VLIET-LANOË

ABSTRACT

Evidence of frost action in paleosols is very common in temperate regions; two types are considered.

First, it may be inherited from a former disturbance of the parent material and can influence pedogenesis. A good example of this is the periglacial type of fragipan (former permafrost table) developed in loessic material. Also inherited polygonal patterns, geliflucted ground mass with oriented porosity, suprapermafrost silt accumulations are all elements important enough to influence profile development.

Secondly, frost may have disturbed the pedological organization of the profile on a macroscopic or a microscopic scale. Disturbance is mainly associated with deep seasonal frost. It ranges from macroscopic cryoturbation (differential frost heave) and thrusting, to microscopic disturbances.

Ice lensing, soil heaving, sorting of stones and particles and translocation mechanisms are reviewed and discussed from a pedological viewpoint. A series of features, generally considered by soil scientists to be the result of biological or geochemical processes, are related to frost action.

Application of these observations leads to the possible reconstruction of climate based on paleosols with similar accuracy to that based on palynology. This leads also to understanding the formation of surface soils in temperate regions of western Europe in a chronostratigraphic manner.

INTRODUCTION

In temperate and arctic or alpine soils, frost as well as rainfall and vegetation are important pedogenic agents. The effects of frost have been recognized often in fossil and buried soils, but in many surface soils such effects, although clearly visible, have been largely neglected. A wide range of phenomena due to frost occur in soils, from faint features like platy structure or uplifted stones, to more prominent ones such as ice-wedge casts, and the results of cryoturbation and gelifluction. Frost effects, unlike those created by desiccation alone, can be active on a scale varying from a millimetre to several metres.

All these phenomena are related to one dominant process, the segregation of ice in the soil, usually in the form of lenses, and the heaving and associated strains resulting from both freezing and thawing. Other processes such as thermal cracking, desiccation network development and sliding give secondary associated features.

In the soil, frost dynamics are controlled by texture (in the pedological sense), specifically the ability of the soil to retain capillary and adsorbed water, and by local drainage conditions. Ice segregation is controlled by the equilibrium between the thermal deficit at the freezing plane and the thermal input transferred by water (heat of crystallization and soil heat). The greater and steadier the water supply, the more important is the amount of segregated ice. Optimum conditions for ice lensing are represented by loamy textured sediments in the capillary fringe of a water table.

ICE LENSING AND RELATED PROCESSES IN SOILS

Ice lensing

The commonest form of segregated ice is that of ice lenses growing parallel to the thermal gradient in the sediment, usually roughly parallel to the soil

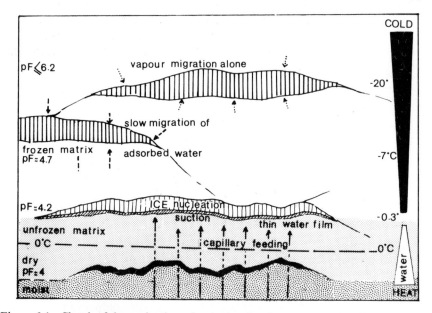

Figure 6.1   Sketch of the nucleation of an ice lens in a loamy textured sediment. Suction of capillary water induced by the thermal gradient produces desiccation of the underlying sediment with opening of shrinking cracks. Ice lenses will nucleate in these cracks. ▥, ice; ▨, water; ▣, moist, unfrozen matrix; ▨, dry, unfrozen matrix; ■, air; □, frozen

surface. Kokkonen (1927) was the first to observe ice lensing, later to be explained by Taber (1929, 1943).

In a slowly freezing soil characterized by a dominant silty texture and also good water reserves, ice lenses nucleate usually in medium-size pores (around 50 μm in diameter) at a temperature slightly below 0 °C (Figure 6.1) behind the freezing line (Williams and Wood, 1982). The ice lenses commonly form a series, one below the other, reflecting change of thermal/hydraulic balance at the freezing line or, more exactly, at the ice nucleation line. As long as the pore-water pressure in the unfrozen soil remains higher (pF $\leqslant$ 4) than the pore pressure at the ice/water interface, the ice lens will continue to grow. The initial lens cannot grow beyond a certain point during the first stage because of the desiccation of the underlying unfrozen soil as water is extracted by suction induced by ice crystallization. Palmer (1967) showed that for further ice lenses to nucleate the thermal diffusivity of the soil must exceed the water diffusion coefficient. At this moment, the break below the ice lens in the capillary water supply (pF = 4) prevents thermal input at the freezing plane and cold penetrates deeper into the desiccated soil until the thermal equilibrium is restored. Thus, a new ice lens is able to nucleate, for example in an open desiccation fissure parallel to but below the freezing line.

If, on the contrary, the water diffusion coefficient is equal to the thermal diffusivity, as in steady water supply conditions (Williams and Wood, 1982), ice crystals grow continuously and form large ice bodies as may occur above a water table (Mackay, 1971). Pipkrakes or ice needles also occur in such conditions and although considered as extrusion ice by many authors, their mechanism of nucleation and growth is only a particular case of segregated ice (Figure 6.7).

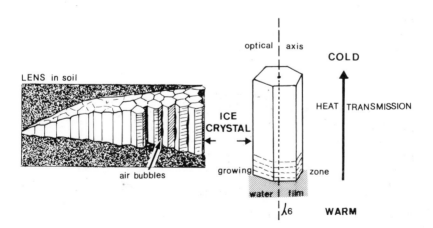

Figure 6.2   Crystallographic pattern of an ice lens, related to its thermal conductivity

Ice crystals forming ice lenses nucleate first orthogonal to the fissure walls as hexagonal columns, growing in the direction of the water supply and heat influx. The thermal conductivity of the crystal is high and follows the optical axis (λ6, Figure 6.2) (Shumskii, 1964). As ice is nearly four times more conductive than water, it means that in a frozen state, a wet peat, a shallow lake or a soil is highly conductive to cold or to heat loss.

Once the capillary feed is blocked, ice lenses can continue to grow by slow migration of still unfrozen adsorbed water (Anderson and Morgenstern, 1973; Burt and Williams, 1976) along the face of mineral grains until the pressure in the finest pores is equal to that at the ice/water interface. Suction at the ice/water interface is sufficient to induce this adsorbed water displacement as shown by a detectable soil heave occurring at temperatures as low as $-4\,°C$ in loamy textured sediments and $-20\,°C$ in clays. If the temperature drops sufficiently, theoretically all hygroscopic water can also be used not only by slow migration but also by vaporization because the vapour tension of ice is much lower than that of water at the same temperature.

The growing ice lenses result in soil heaving, termed secondary heaving in the literature. This heaving is closely controlled by soil texture and by the quantity of water, in other words by the drainage conditions, as illustrated in Table 6.1.

The importance of soil texture to susceptibility of soils to ice lensing has been studied by Beskow (1935), Csathy and Thowsend (1962) and by Williams (1966) who stress the importance of pore size in assessing frost susceptibility. It is very important to determine the effective frost susceptibility of a sediment as will be discussed later.

Table 6.1   Relation between frost susceptibility, texture and drainage conditions (Research Institute of Glaciology, Lanchou, China, 1975). The highest numbers (III and IV) correspond to the most sensitive conditions to frost heave

Texture	Water (%)	Depth of water table	Frost susceptibility classes
Gravel	—	—	I
Fine sand	$W < 14$	$> 1.5\,m$	I
		$\leqslant 1.5$	II
Silt	$14 \leqslant W < 18$	> 1.5	II
		$\leqslant 1.5$	III
	$W \geqslant 18$	> 1.5	III
		$\leqslant 1.5$	IV
Clay	$W > Wp + 9$	not considered	IV
	$Wp + 9 \geqslant W > Wp + 5$	> 2.0	III
	$Wp + 5 \geqslant W > Wp$	$\leqslant 2.0$	III
		> 2.0	II
	$W \leqslant Wp$	$\leqslant 2.0$	II
		> 2.0	I

W: total water content, Wp: water content at plastic limit.

Penner (1959) stressed that ice lensing may not always be associated with saturated soils and therefore saturated hydraulic conductivity. Soils developed above a permanent water table will be in steady state of water supply and only thick ice bodies will grow. In contrast, soils affected by a perched permafrost water table will be more favourable to ice lensing because of the limited heat supply. From field observations and thermal survey, it is clear that waterlogged soils on permafrost are less favourable to regular ice lensing and heave than unsaturated soils located in the lower part of the capillary fringe of a water table (pF around 2.5) (Van Vliet-Lanoë, 1983).

During thaw, successive water migrations occur but in a reverse order with an hysteresis component (Burt and Williams, 1976). The speed of thawing is directly related to the amount of segregated ice and thermal conditions: the higher the ice content, the slower and later the melting of the soil. This phenomenon, associated with the high thermal conductivity of frozen ground, explains why in the discontinuous permafrost zone, permafrost is restricted to local humid depressions and does not exist in valleys with a permanent water table, except in highly frost-susceptible sediments.

Refreezing ice

A process particularly important in fine-grained sediments is the migration of meltwater by way of water films to colder horizons, following the surface of mineral or organic particles (adsorbed water) and the interface between ice crystals (Williams and Wood, 1982). This phenomenon was first observed by Globus in 1962 and demonstrated by Burt and Williams in 1976. It results from the inversion of the thermal gradient during the early stages of thawing and is responsible for heave of the soil (termed tertiary heave) apparently as important as the well-known freezing heave. This has been observed in the field during the melting season in hummocky soils (Mackay, 1981), in snow patches (Woo and Heron, 1981) and in temporary hydrolaccoliths or ice blisters (Van Vliet-Lanoë, 1983). It results from an increase of the segregated ice content in the subsurface horizons. This process is also able to enhance cryoturbation processes or stone heaving and also explains the large amount of segregated ice in the top horizon of permafost as observed by Hoekstra (1969) and the congelation ice in thermal cracks that form ice wedges (Shumskii, 1964). Nevertheless, this process is limited, in so far as significant amounts of ice lenses can reduce noticeably the permeability of the frozen soil to meltwater at temperatures just below 0 °C (Burt and Williams, 1976).

Aggregate stability and vesicles

For several reasons, aggregates created by ice lensing are generally very stable once formed. So, it is very surprising that soils affected by frost very frequently include vesicles. These observations are conflicting and have to be discussed.

Stability and compaction

Platy aggregates created solely by desiccation (pF: 4.2) have not been compacted and therefore are very unstable upon remoistening (Miller, 1971). In contrast, those created at depth by ice lensing under natural conditions are extremely firm and resistant to collapse even if not cemented by amorphous material as can frequently occur below arctic podzols. The mechanical action of the growing ice can exert a surface pressure of over 14 kg cm^{-2} (Taber, 1929). This reduces the internal porosity of aggregates, particularly the medium-size pores. When viewed through a microscope, the reorientation of fine particles parallel to the aggregate surface is evident (Van Vliet-Lanoë, 1976; Van Vliet-Lanoë and Langohr, 1981), forming a *neostrian* (Brewer, 1976), also termed stress cutan.

Ultra-desiccation is also an important factor. Pissart (1970) stressed that it could result in a break in capillary menisci (pF: 4.2). In fact, desiccation induced by frost works further by exploiting an important part of the adsorbed water. It is possible that near the upper layers of the permafrost, the winter temperature does not drop sufficiently to exploit all adsorbed water and reach the theoretical pF of 6.2 as stressed by Schenk (1968). Nevertheless, cryodesiccation is a very important factor in structural stabilization. In most cases a pF of at least 4.7 is reached. Following recent works of Tessier (1978, 1984), an originally fresh sedimented clay is able to keep a state of microaggregation related to the highest desiccation it has undergone since its deposition. This phenomenon explains why aggregates formed by frost are very difficult to disperse. The Btx horizon developed in clay loams of the Belgian Ardennes (Langohr and Van Vliet-Lanoë, 1979) is a good example of it.

The action of solutes also stabilizes aggregates. During freezing a salt concentration can occur due to the suction existing at the freezing plane which facilitates flocculation of clay particles and even salt precipitation as gypsum or calcite. Many of these precipitates disappear during thawing. In the field and in experiments iron precipitation by frost, as proposed by Krivan (1958) and Bertouille (1972), has never been observed by us. Precipitates of iron hydroxides exist but are mainly of bacterial origin; most of the iron present in such an active soil is in the Fe^{2+} state, or complexed with organic matter, or adsorbed at the surfaces of clay minerals. One feature that may occur, associated with accumulation of solutes at the top of permafrost (Van Vliet-Lanoë, 1976; McKeague, 1981) is the synthesis of amorphous silicates resulting from vertical leaching of weathering products. This does not result from chemical weathering produced by frost ultra-desiccation (as proposed by Tyutyunov (1964) where, at low temperatures, the kinetics of chemical reactions are extremely slow (Arrhenius equation)), but rather mostly from biochemical weathering close to the soil surface (Ellis, 1983; Van Vliet-Lanoë, 1983).

The very stable aggregates are frequently reworked in fluvial or colluvial deposits forming soft pellets.

Structural collapse and vesicles

Vesicles are common in a wide range of soils from the subtropics to the arctic. They are characteristic of soils with a discontinuous plant cover, e.g. agricultural soils, and generally result from the collapse of soil aggregates under the influence of splash and upon wetting (Volk and Geyger, 1970; Miller, 1971; De Ploey and Mücher, 1981).

Various hypotheses have been used to explain their peculiar abundance in arctic soils. Fitzpatrick (1956) proposed that vesicles were the result of air expelled from soil water during freezing, a theory that has been accepted by most authors since then (Chandler, 1972; Bunting, 1977; Harris and Ellis, 1980). However, according to Brewer (1976) vesicles can result from growing ice lenses during freezing of the soil surface. In fact, air entrapped in the ice crystals exists but in such small amounts that it can not explain the large volume of air contained in the entire vesicle.

Vesicles are very common in all frost-susceptible soils, both in natural exposures and in experiments and are found in the upper 5–10 cm in both patterned and non-patterned ground and in gelifluction lobes. They are especially

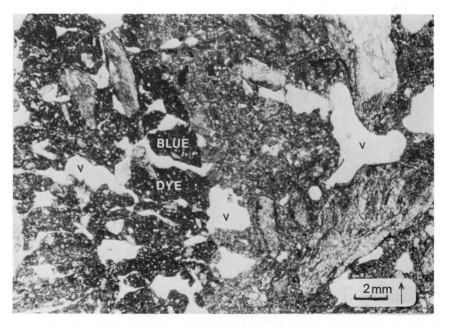

Figure 6.3   Vesicle formation in a field experiment (Chambeyron, French Alps). The blue dye on the left side of the picture acts like a structure stabilizer and prevents collapse of the aggregates on thawing. Fissures created by ice lensing remain visible to the left, though in natural soil (right) they are replaced by well-rounded vesicles. V, void (from Van Vliet-Lanoë *et al.*, 1984)

well developed in small depressions or beneath snow patches where soils are likely to be supersaturated with water when thawing occurs. In some cases, for example overlying a perched water table on permafrost, vesicles can occur in deeper horizons. Their size is generally larger in natural exposures than in experiments, where especially in the finest textured sediments, they can sometimes reach up to 2–3 cm in length (tubular forms). In thixotropic material, they are well rounded though in a more clayey or stable matrix they remain mammillated (Figures 6.3, 6.4).

Vesicles result from air expulsion confined by structural collapse that occurs during thawing (Van Vliet-Lanoë *et al.*, 1984). The reason for this is that during ice segregation soil aggregates are dried (ice $pF \geqslant 4.7$) and air is held in the sediment between ice lenses and also between ice crystals (Figure 6.2). Usually, when thawing starts from the surface, melting water confines the soil air at depth which has to be expelled. Air bubbles try to escape but need to overcome the hydrostatic pressure and the high viscosity of cold water (1.6 times higher than at 20 °C). This means that the energy needed by the bubbles to reach the soil surface is much more destructive to soil aggregates in arctic than in temperate or warm conditions.

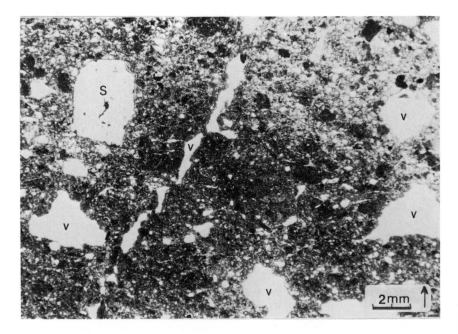

Figure 6.4   Desiccation crack in glacio-lacustrine mud (on a former solifluxion lobe, Spitsbergen); it was used as an escape route by soil air during thawing. The crack is cut into segments by traces of ice lenses which are laterally displaced by frostcreep.
V, void; S, mineral skeleton (from Van Vliet-Lanoë *et al.*, 1984)

Thawing produces an important bubbling at the surface of the temporarily soupy soil which is more intense if the ice content is high and thawing rapid. The high dielectric properties of meltwater can partially explain the superficial instability of surface aggregates. Because aggregates close to the surface have been weakly compacted by rapid migration of the freezing line, their shape and characteristics are similar to those resulting solely from desiccation (Van Vliet-Lanoë *et al.*, 1984). By contrast, aggregates at depth are compacted by a long process of segregation. Using the Poiseuille law, it can be assumed that stability of aggregates in cold water is very high due to the high viscosity of water; it is also clear that the less porous and more compacted aggregates at depth are a thousand times more stable than those occurring near the soil surface (square radius of the main porosity). In fact, wide open fissures left by intense ice segregation are extremely stable and once formed are recorded in soils for millenia. These open fissures have been confused with collapsed vesicles by Bunting (1977) and Harris and Ellis (1980). In deep horizons only a superficial collapse of aggregates along vertical cracks used as a main escape route by the soil air has been observed (Figure 6.4). This is particularly clear, for example, in bleached cracks of fragipan soils or at the base of injections in cryoturbated soils.

Interaggregate porosity

As already noted ice lensing creates porosity which remains after thawing of the ice. If the horizon is close to the soil surface, thaw consolidation is restricted and the material becomes highly porous after one or a few cycles of freeze–thaw. At depth, the situation is little different, the settling of the aggregates being closer as a result of overburden pressure; in this case the porosity is fissure-like and discontinuous. From the experiments performed in Caen, it is clear that after a first freeze–thaw cycle, the hydraulic conductivity of the sediment increases considerably, especially in originally uncompacted clay and clay loam. Usually, the situation is almost stable after three to four cycles as a result of the mechanical compaction from ice-lens growth and ultra-desiccation. From this, it is clear that the effects of frost susceptibility of a material need to be determined after several cycles to be closer to natural conditions, and not one alone.

Stone lifting, particle sorting and illuviation

Stone lifting and particle sorting

Sorting of particles and stones is one of the most striking features of frost-affected soils. The process can give rise to macroscopic features such as those observed in patterned ground or to microscopic ones. Stone lifting and expulsion from the soil is characteristic of all frost-susceptible soils and results from a sequence of processes acting in freezing soils.

The first to occur is a *vertical uplift* resulting from the heaving of the soil matrix following ice lensing, as proposed by Beskow (1935) and demonstrated by Kaplar (1965). As the nucleation front reaches the stone, (Figure 6.5) a pore appears immediately above it, resulting in local depletion in water supply. In the second phase, the stone is raised by heaving of the matrix with a consequent pore formation at the base of the stone. Finally, segregated ice nucleates in the basal pore. This basal growth of segregated ice is able to push upwards against the stone. It acts principally on fragments in the centimetre size range in cases where thermal conductivity of the stone is greater and its porosity lower than that of the moist soil matrix; then ice can nucleate earlier in the basal pore and locally deplete the water supply for its own use. If the stone is as porous and conductive as the soil matrix, this process is considerably limited or absent. Nucleation and development of ice lenses is clearly visible in the field of frozen soils and after thawing the effects remain visible in profiles (Taber, 1943) and in thin sections (Figure 6.6) (Van Vliet-Lanoë, 1976, 1982; Fox, 1979). Pipkrake heaving of stones is one particular example (Figure 6.7).

Uplifting is very effective on elongate flat stones with rather high specific surfaces, i.e. a high *adhesion surface* to the frozen matrix, compared to blocks and pebbles. Stones, originally slightly tilted, may become vertical. It results

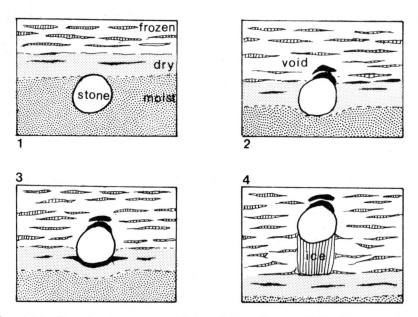

Figure 6.5   Stone-heaving processes (adapted from Kaplar, 1965). The symbols are identical to those of Figure 6.1

from the action of a torque applied on the extremities of the stone as reproduced experimentally by Pissart (1969). From experimental work performed in Caen, field observations in the Alps, and the geotechnical work of Penner (1974) on foundations, stones or blocks are raised in a few cycles. When thawing occurs, the refreezing at depth of melting water which percolates along vertical cracks favours the process: the upward injection of stones in crack networks as observed in sorted polygons is accelerated by it.

Figure 6.6 Uplifted stones in fragipan horizon of Belgian Ardennes. Notice the pore (V) left by ice lensing at the base of shale fragments. In the field, this horizon was associated with buried sorted polygons

If the matrix is thixotropic, the gap left by the ice collapses during thawing, and although the stone falls, it never settles into its original position. The finer the matrix and therefore the more stable upon remoistening, the closer the stone remains to its original position prior to thawing. Only flat-lying stones return to their original position though frequently tilted. If the stone reaches the soil surface and ice segregation is not particularly significant, the stone will settle on thawing to a position close to its original. Sediments which have been plastically deformed by stone extrusion (Figure 6.7) retain evidence of the phenomena in the form of a depression around the stone; if the fragment is coarse, lateral sliding of the matrix may also be involved. This is a significant process and affects elements from as small as 0.5 mm in length to blocks of

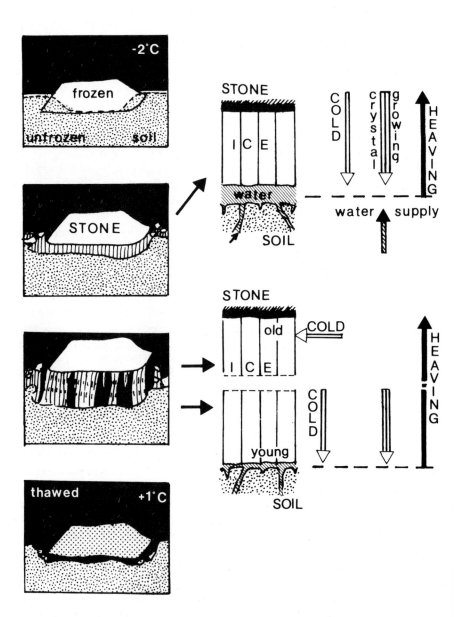

Figure 6.7   Stone lifting by pipkrake growth, a particular case of stone uplifting. At the crystallization line, the optical axis of the ice needles remains parallel with the transmission of heat. Symbols as in Figure 6.1

bedrock. If a frost-susceptible matrix exists between the pebbles of a fluvial or marine gravel, all the pebbles may become vertically oriented.

In areas of patterned ground if stones have a higher thermal diffusivity than soil matrix then thawing first occurs adjacent to stones, allowing them to slide to the edge of the stony border of the sorted ground (Washburn, 1956; Schunke, 1975). Other authors, e.g. Pissart (1966) and Whittow (1968) advocate pipkrakes as the origin of the sorting. Most authors agree that fine particles migrate ahead of the freezing line as demonstrated by Corte (1962, 1966). Based on our microscopic observations, although this theory is valid for loose, coarse grained sediments, in other cases sorting is mostly related to the specific *adhesion surface* of the stone and its ability to be uplifted in the soil or engulfed by the growing ice. Macroscopic sorted accumulations of stones are mostly the result of a relative displacement of coarse-grained material with respect to fine matrix, with large fragments being quickly displaced relative to fines, as shown experimentally by J. P. Coutard (pers. comm.).

Another process evident in north-west Spitsbergen (Van Vliet-Lanoë, 1983) is the continual creation of fine material by frost shattering (Lautridou, 1982) especially when the stones are in contact with a mud boil as on the inner edge of a sorted circle. Frost shattering is able to enhance sorting. As observed many times in the field, granular disintegration by frost shattering of coherent rocks embedded in fine grained matrix is common in fluviatile loam or glacio-marine clay-loams. The effectiveness of this underground frost shattering is less than at the soil surface; as the soil matrix is usually more porous than the stone, this is generally desiccated by the soil cryo-suction (Lautridou and Ozouf, 1982). During snow melting, the removal of the fines by rill wash action may also emphasize sorting.

Microscopically, experimental observations of Corte (1962, 1966) have been confirmed by those of Rowell and Dillon (1972). Fine particles with a high specific surface, especially clay minerals, are able to migrate ahead of growing ice crystals, whereas coarser particles are rapidly trapped by the ice crystals and remain stationary in the ice lens. At this scale, it is the thickness of the adsorbed water layer which prevents the entrapping process: if the water is firmly bonded, for example to a clay mineral, the suction needed to extract this layer is greater than the suction at the water/crystal interface. Therefore, the particle is propelled in front of the growing crystal. However, for quartz grains or coarser particles, this layer is thin, weakly bonded and is rapidly extracted by the growing crystal with the result that the particle is enclosed in it. It has also been shown that particles in ice can migrate under a temperature gradient (Hoekstra and Miller, 1965; Römkens and Miller, 1973). This movement of fines is related to the thickness of the adsorbed layer and is directed towards the warmer side of the crystal.

Stone heaving and micro-sorting, working together on a small scale, are both able to produce the microfabric typical of repeated freeze-thaw (Figure 6.8).

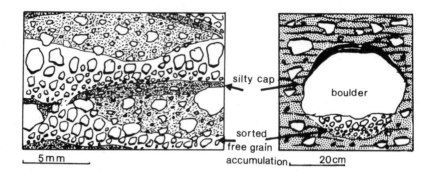

Figure 6.8 Particle sorting at microscopic and macroscopic scales in a till from the French Vosges. This sorting with reverse grading is typical at alternating freeze–thaw processes

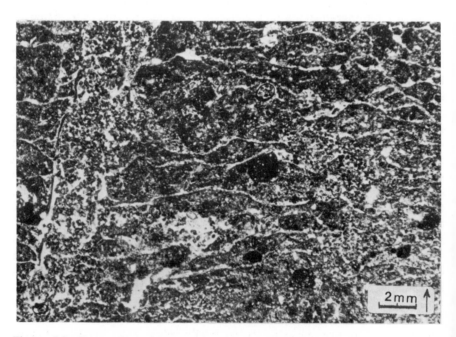

Figure 6.9 Freeze–thaw fabric in clay loam (Fairbanks, Alaska). Fecal pellets (Enchytreids) in the platy aggregates represent biological activity. Fecal pellets associated with present biological activity are visible in tubules on the left side of the picture

Such processes can explain the reversed sorting of ice-expelled skeleton and fine grains capping aggregates which is observed microscopically and it may also explain reverse sorting of leached gravels and sands occurring in the same profile at the base of a boulder below which an important mass of segregated ice was seasonally growing (see also Figure 6.5).

Some authors, such as Mermut and St Arnaud (1981) and Fedoroff *et al.* (1981), advocate liquefaction on thawing to explain the capping by fines. However, such cappings are very stable due to their ultra-desiccation, resisting cryoturbation and mass wasting stresses. Freeze–thaw microfabrics can incorporate fecal pellets (Figure 6.9) formed during summer time which are not destroyed but only plastically deformed. In French alpine soils with an unstable soil matrix the cappings are resistant to collapse on thawing and are commonly overlaid with vesicles.

Capping and illuviation

Some authors have advocated electrophoresis induced by frost to explain the migration of fine particles towards the warmer part of the soil (Bertouille, 1972). However, experiments by Brewer and Haldane (1957) have clearly shown that clay suspensions migrate in the direction of water suction which in freezing materials, pure sands excepted, normally occurs in the opposite direction, towards the cold part. As the viscosity of cold water is very high, it is assumed that clay migration induced by electrophoresis is insignificant compared to that induced by cryo-suction. Thus it is difficult to explain the accumulation of fine particles to form massive or banded Bt illuvial horizons in temperate and arctic soils by frost activity alone.

An additional factor which enhances sorting is that, on thawing, meltwater, including snow meltwater, percolates deep into the soil profile. The high dielectric properties of such waters allow some dispersion of weakly stabilized aggregates, usually those lying close to the soil surface or along frost drying cracks; this leads to macro- and microerosion in the soil profile. This facilitates accumulation at depth of fine particles ranging from clays, if the pH is neutral to weakly acid, to coarse silts and more exceptionally to fine sands. We compare this process of translocation to an *internal colluviation*.

These translocated particles accumulate in coarse pores (Figure 6.10) such as occur in granitic sands (Van Vliet-Lanoë and Valadas, 1983), in tills (Collins and O'Dublain, 1980) or in cracks in fragipan horizons of the loessic belt (Van Vliet-Lanoë and Langohr, 1981). Accumulation occurs generally at the top of permafrost (Corte, 1963) or at a discontinuity as at the boundary between two different textures (Van Vliet-Lanoë, 1983). In the case of permafrost, a compact layer forms, especially below a semi-permanent snow patch (Van Vliet-Lanoë and Valadas, 1983) and has been observed in many fossil slope formations, for example in Scotland (Romans *et al.*, 1966; Romans and

Figure 6.10   Silt and silt-size clay accumulation with cross bedding in coarse pores of
a bedded granitic sand (western Vosges, France)

Table 6.2   Cumulative granulometric curves of supra-permafrost silt accumulation (a,
b) resulting from internal translocation, compared with superficial colluviation (c) (from
Van Vliet-Lanoë, 1983)

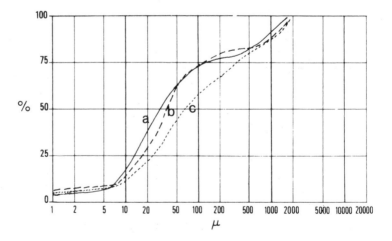

Robertson, 1974), in the Vosges and in the north-eastern part of the French Central Massif. This silt accumulation has frequently been compacted by permafrost aggradation and for this reason has also been described as a fragipan (e.g. Fitzpatrick, 1956, 1974; Romans *et al.*, 1966; Payton, 1983; Van Vliet-Lanoë and Valadas, 1983).

An additional phenomenon, attributed to frost action by Konichev *et al.* (1973) and Morozova (1972), is the occurrence of microscopic circles of skeleton grains, corresponding in thin section to the surface of the peds (plain light). This phenomenon is related to frost micro-sorting, somewhat disturbed by cryoturbation. Similar features are frequently observed but only in transmitted polarized light. In these cases they seem to be connected with former bioturbation or bioturbation alternating with frost activity. Indeed, burrowing animals are frequently active in low arctic and mountain environments.

STRUCTURES AND STRUCTURAL PATTERNS
CREATED BY ICE LENSING

Deep seasonal freezing

Platy and prismatic structures

In homogeneous loamy textured sediments under continuous cooling and a sufficient supply of water, ice lenses are closely spaced near the soil surface and become more widely spaced with depth, reflecting a decrease in the rate

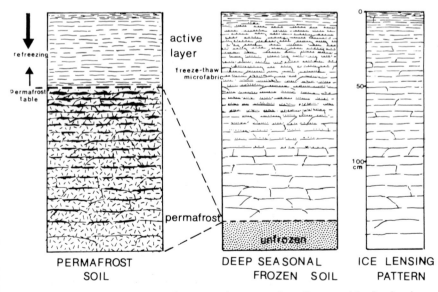

Figure 6.11 Different types of structural patterns in soils caused by ice lensing

of cooling (Shumskii, 1964). After the ice melts, the soil close to the surface has a bladed structure usually thickening to a platy structure at depth (Figure 6.11). This structural sequence is common in soils and in weathered or porous rocks such as limestone that have undergone deep seasonal freezing (Bertouille, 1972).

If lithic discontinuities or stratification exist, they can influence the location of ice lensing by local changes in hydraulic conductivity and porosity. For example, in a thinly stratified sand, ice lensing usually follows the bedding. In uncompacted sediments in the form of proglacial varves, traces of ice lenses are slightly tilted (2–3 °C). The location of the ice lenses is influenced by the orientation of the thermal gradient (isotherms) rather than by the stratification.

Pores and discontinuities created by the first cycle of ice lensing guide the location of subsequent ice segregation cycles. The sizes of platy aggregates range from a few tenths of a millimetre near the surface to several centimetres thick at depth. Ice lenses usually develop in existing desiccation cracks which are open to below the freezing line. The roughness of crack walls is smoothed by pressure exerted by ice crystal growth. Crack or fissure walls are no longer complementary as is the case in desiccation cracks (shrinkage). After thaw settlement or consolidation, fissures are discontinuous and vary in width (Figure 6.6) both at macro- and microscopic scale. They are particularly visible in poorly drained soils and in horizons affected by refreezing ice (Van Vliet-Lanoë *et al.*, 1984).

This bladed to platy structure corresponds under the microscope to the isoband fabric of Dumanski (1964) and Dumanski and St Arnaud (1966). It was also observed by Fitzpatrick (1956, 1974), Haesaerts and Van Vliet (1973), and Bunting and Fedoroff (1974).

In the field the occurrence of thinly bladed structures thickening with depth is indicative of soil surface stabilization under a cold climate between two phases of sedimentation. This occurs frequently in loessic and fluviatile sediments. In many soils of Belgium and, northern and western France, this feature may be indicative of a polysequum or of buried horizons. If the original water reserve of the sediment is limited the structure gradually vanishes with depth.

In clay loams or clays, platy structures are often replaced by short prisms, a consequence of a rapid exhaustion of the capillary water supply in such a texture (Van Vliet-Lanoë *et al.*, 1984). This prismatic pattern is frequent in poorly drained depressions and is often observed in alluvial plains (Kuznecova, 1973) or in the minerogenic core of palsas (Svenson, 1964; Van Vliet-Lanoë, 1976; Gangloff and Pissart 1983). This feature occurs in quickly frozen soils as observed on a site exposed to wind in north-western Spitsbergen, and in experiments when a break in the water supply occurs (Van Vliet-Lanoë *et al.*, 1984).

Freeze–thaw structure

When a soil is affected by alternating freezing and thawing, the fabric resulting from ice lensing evolves by erosion and plastic deformation, the platy aggregates becoming lens shaped. A *capping of segregated fine particles* progressively develops on the top face of the aggregates due to successive phases of particle sorting by ice and of particle translocation on thawing. In the field the capping is clearly visible because of a higher proportion of clays and silt-size quartz than in the soil matrix. The capping is usually composed of silt and clay-size particles which microscopically show a weak birefringence (Fedorova and Yarilova, 1972) if secondary staining by organic compounds or bacterial iron precipitation is absent. The capping may also be composed of humic particles, fecal pellets and

Figure 6.12   Freeze–thaw structural pattern associated with a frostcrack in a humic sand (early Glacial deposit, Fontainebleau, France). The whitish lines correspond to the accumulation of sorted free grains (skeletan)

clay micro-aggregates. Some coarser elements such as mica flakes or quartz grains can sometimes be included as fragments of soil coatings (papules) or other components (sclerotes, pollen grains). These inclusions are usually flat lying.

At the same time, clean loose skeleton grains (skeletan, Brewer 1976) appear between the peds, progressively forming whitish loose coatings, of ice-expelled grains. As with the capping, fragments of soil components can also be included (Figure 6.20). As this loose coating shows a reverse-graded bedding (Figure 6.8) (Van Vliet-Lanoë, 1976; Curmi, 1979) this type of fabric is called a *sorted platy structure* and is considered by the author as typical of alternating freezing and thawing. It corresponds to the banded fabrics of Dumanski (1964) and the silt droplets of Romans *et al.* (1966).

Microscopically, this fabric is well known to soil scientists working in boreal and low arctic environments (Dumanski and St Arnaud, 1966; Bjorkhem and Jongerius, 1974; Fedorova and Yarilova, 1972) but has also been observed in temperate alpine mountains by the author and is also frequently preserved in fossil periglacial deposits (Fitzpatrick, 1974; Romans and Robertson, 1974; Van Vliet-Lanoë, 1976; Van Vliet-Lanoë and Flageollet, 1981) (Figure 6.12).

In imperfectly drained soils in which ice lensing is effective, freeze–thaw fabric develops rapidly. In the field, the fabric requires many cycles to develop; it is incipient after 100 cycles and fairly well developed after 1000 (Van Vliet-Lanoë, 1983). Sorting is much more effective in experiments as a consequence of better controlled thermal and hydraulic parameters (Van Vliet-Lanoë *et al.*, 1984). Dumanski (1964) produced such a fabric in forty-five cycles in a water-saturated experiment. In Caen, a weaker development in twenty-eight cycles was produced but in a larger size (5 m × 5 m) experiment and under imperfectly drained conditions. Grading of the sediment is also very important. The fabric develops more quickly in poorly graded sediments such as glacial till or colluvial sandy loam than in well-graded loess or fluvial sand (Van Vliet-Lanoë and Langohr, 1981; Van Vliet-Lanoë *et al.*, 1984). However, Dumanski (1964) found that the fabric does not relate to percentage of sand, silt, clay or carbon. Mineralogical and chemical composition of the soil matrix is also important (Van Vliet-Lanoë, 1976), amorphous clays or very stable organic matter completely preventing this phenomenon.

Permafrost soils and fragipan

The structural pattern in permafrost differs slightly from the previous type and occurs in fossil and arctic soils. The pattern is more clearly seen in fossil soils as the climatic event was simpler and shorter in time (about 1000 years, Haesaerts and Van Vliet, 1973) than that in arctic soils where Late Glacial and Holocene climatic changes are thought to have occurred during soil development and give rise to complex structural patterns.

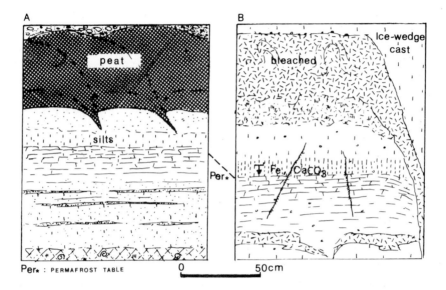

Figure 6.13   Buried soils with traces of a permafrost table. (A) poorly drained fluviatile calcareous sandy loam (Somme valley, France); (B) imperfectly drained loess, associated with an ice-wedge cast network of 30 m diameter (Harmignies, Belgium, 25 000 BP) Note the shearing planes. Per*, permafrost table; ▤, traces of ice lensing (medium to coarse platy structure); ▨, loam, unbleached; ▨, loam, bleached; ▨, organic matter

The permafrost table is usually fairly distinct due to clear structural change, marked by a change in the size of the platy structures (Figures 6.13 and 6.14) resulting from the autumn refreezing at the top of permafrost table. Below it, except in waterlogged soils, the structural pattern is similar to that in deep seasonally frozen soils.

In the upper layer of permafrost, ice lensing is generally more intense due to the refreezing process. After thaw consolidation, aggregates are extremely firm, with a high bulk density ($\geqslant 1.7$) and between them occur discontinuous open fissures equant pores with smooth, curved surfaces (Figure 6.6) (Van Vliet-Lanoë, 1976). If an ice-wedge cast is associated with the soil, the horizon corresponding to the upper permafrost does not cross the infilling (Haesaerts and Van Vliet, 1973) (Figure 6.14).

These characteristics are identical to those of *fragipan horizons* developed in loams in western Europe and the USA and sometimes attributed to periglacial phenomena. This resemblance together with the extent of decalcification of the loess in Belgium suggests a common genesis (Van Vliet-Lanoë and Langohr, 1981) (Figure 6.15) for both structural patterns. Recently Langohr (1984) suggested the term *consolidated horizon* instead of fragipan to avoid confusion

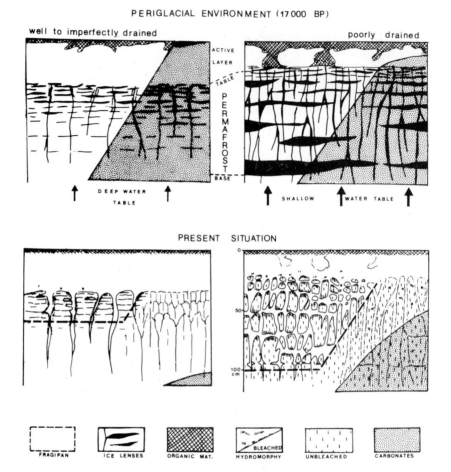

Figure 6.14   Relationship between Late Glacial permafrost soils and present fragipan soils (adapted from Van Vliet-Lanoë and Langohr, 1981). In poorly drained sites, ice lensing prisms develop instead of platy structure. If the sediment is leached, compaction and traces of ice lensing are preserved in further soil development in form of fragipan horizon

with similar forms reviewed by Smalley and Davin (1982). This term can also be extended to supra-permafrost silt accumulation.

These horizons are often associated with a polygonal network. In Pleniglacial soils, the network is due to shearing planes commonly developing in the upper permafrost horizon similar to those in sands described by Bertouille (1972). In Late Glacial soils, the present 'fragipan' soils, the polygonal network results from soil shrinkage following freeze desiccation. The more the sediment was initially saturated with water, the smaller the network (Czeratzki, 1956;

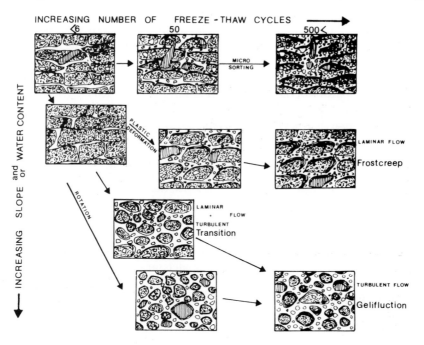

Figure 6.15   Evolution of the fabric produced by ice lensing relative to increasing slope and increasing number of freeze–thaw cycles. Sandy-loam texture

Van Vliet-Lanoë, 1980; Langohr, 1984). This agrees well with the observations of Svenson (1964) and Mackay (1974) in active permafrost. Once open, network fissures are usually exploited by ice lensing particularly in imperfectly and poorly drained sites. Other features that are generally observed in these soils include bleaching of the surface layer especially in non-calcareous material, associated with seasonal waterlogging and the chelate character of mosses and algae humus, and an accumulation of organic matter (Pettapiece, 1974, Van Vliet-Lanoë, 1983), silt (Corte, 1963, Fedoroff, 1966) iron and carbonates (Van Vliet-Lanoë, 1976) and organo-mineral complexes (McKeague, 1981).

SLOPE DEPOSITS: FROM FROSTCREEP TO GELIFLUCTION

If frost and thaw are acting on gentle slopes (2 °–17 °), initially the general fabric of the sediment is identical to that on flat surfaces, but a progressive sliding of the sediment occurs in the downslope direction.

According to Williams (1959) total displacement by solifluction is characterized by the summation of small displacements along sliding planes corresponding to the upper face of the melting ice lenses. Higashi and Corte

(1971) considered frostcreep to be limited to the soil surface and the layer directly underneath: soil particles as well as stones are lifted by ice needles normal to the soil surface (pipkrakes) and deposited after thawing under the influence of gravity a little further downslope. This feature is only a particular case of the process proposed by Williams.

Most authors consider gelifluction as a mass displacement restricted to soil sliding on permafrost.

Most slope movements involving sedimentary masses show traces of ice lensing in the form of platy structures roughly parallel to the soil surface (Benedict, 1970; Harris, 1981a; Van Vliet-Lanoë *et al.*, 1980; Van Vliet-Lanoë, 1982; Van Vliet-Lanoë and Valadas, 1983). In the field (active and fossil deposits) and in experiments where ice lensing has developed the sliding occurs sheet by sheet in a downslope direction with the shearing or slipping plane corresponding to one of the fissures created by the ice. This slipping is associated with minute lateral displacement which appears during thawing without marked plastic deformation of aggregates. This can be observed in soils with bleached tongues or in granitic sands with veins or joints. Small lateral discontinuities are clearly visible in the field and are usually accentuated near the top of the formation. Displacement takes place as slow laminar flow and increases more or less regularly in an upward direction with a surface displacement ranging from 2 to 10 mm a cycle.

This process is responsible for the terminal curvature observed in granitic sands (Dylik, 1967). From field observations in the same granitic massif, terminal curvature is more abrupt on southern exposed slopes than on northern slopes (Van Vliet-Lanoë *et al.*, 1980). This feature, associated with platy aggregates which thicken with depth, results from deep seasonal freezing. This process is independent of sorting of the platy aggregates, but, as on flat surfaces, sorting develops as the number of freeze–thaw cycles increases. In platy structured granitic sands, Curmi (1979) observed a well-sorted skeletan, though this was not interpreted as being due to frost action.

The terminal curvature or the minute lateral displacements associated with platy structure correspond to an *initial frostcreep fabric* (Van Vliet-Lanoë, 1982) (Figures 6.4, 6.15).

If the displacement takes place over a longer distance and period, aggregates become asymmetric due to a weak plastic deformation, and the capping of fines thickens downslope (Van Vliet-Lanoë, 1982). This structure is very common in many solifluction lobes, heads (Harris, 1981a and b), creep, elongated patterned ground such as sorted stripes, or in step mudboils, corresponding macroscopically to platy structure (Van Vliet-Lanoë, 1983). It corresponds to *frostcreep sensu-stricto*. Usually at this stage the mineral particles (Harris, 1981a) and stones (Smith, 1956) are elongated downslope in the direction of the greatest flow dynamics (weakest resistance).

Change in the organization of the fabric occurs if the speed of displacement increases due to steeper slopes or to more water available. The interstitial water pressure increases due to hydrostatic loading and reduces the friction between aggregates and stones which begin to roll resulting in the formation of fines capping the sides of stones or aggregates (Van Vliet-Lanoë, 1982). This feature which initially occurs in some laminae of the flow is a *transitional form between frostcreep and gelifluction* (Figure 6.15). The features are often visible in the field, the capping generally being lighter in colour than the matrix. Frostcreep and gelifluction fabrics may be observed in contiguous laminae of the mass flow in form of platy and granular structures. Differences in fabric reflect small variations in aggregation related to texture and lateral inflow of water.

If the speed increases again, a new stage is reached when upper horizons of the ground mass flow about 20–50 cm per cycle downslope. Laminar flow sheets tend to disappear and all the particles are rolled over each other. This type of flow is compared with turbulent flow: it is *gelifluction sensu-stricto* (Van Vliet-Lanoë, 1982). In the field, it is characterized by granular structure and a random orientation of small- and medium-size stones.

Harris (1981b) in his excellent synthesis of periglacial mass wasting adopted a similar nomenclature based on a geotechnical approach. The speed difference forecast by Dylik (1967) represents the difference between a water saturated mass movement and an oversaturated one. The gelifluction fabric is usually developed above permafrost but local topographic conditions favourable to water concentration such as the presence of shallow bedrock or highly frost susceptible material can also promote this process in non-permafrosted slopes (Van Vliet-Lanoë and Valadas, 1983). This can occur for example in the southern Alps, during a very humid spring.

Another phenomenon generally occurring near the base of slopes, commonly in sediments with a low-bearing capacity, is *mudflow*. The flow is suddenly initiated when hydrostatic pressure exceeds the liquid limit and displacement occurs on a scale of metres. It is not specific to the periglacial environment but is favoured by a high content of segregated ice, a rapid thawing or an abundant water supply (snow bank or heavy rainfall). If the winter temperature does not drop too low, aggregation is poor and easy to destroy. Mudflows lack characteristic structure and fabric; rather they are massive, hard when dry, and frequently contain numerous vesicles. Under the microscope, remnants of broken cappings mixed with the ground mass or preserved on mineral grains can be observed sometimes. This pattern is rather rare in fossil deposits because of the superposition of other features.

Gelifluction or mudflow fabrics are rapidly formed if the original content of fine particles is high or is subsequently increased by lateral migration. In granitic sands a few metres displacement is needed to develop a frostcreep fabric and about fifty for a transitional form, whereas in illuvial horizons of buried soils only 50 cm are needed to develop frostcreep, and thixotropic mudflow can occur a few metres further downslope.

Harris (1981b) considers the term solifluction to be particularly useful in describing topographic forms such as solifluction lobes, terraces and sheets where initially the relative importance of gelifluction and frostcreep is unknown. In active lobes in north-western Spitsbergen and the French Alps, frostcreep, the laminar flow responsible for relatively flat lobes, is clearly the dominant process. Sometimes a superficial and/or frontal reduction of the speed due to plant cover produces a frontal swell of the lobe or at least some superficial waves. Thixotropic lobes are less frequent and in the Alps occur only in exceptional lithic and climatic conditions. True gelifluction lobes are also rather infrequent, with only flat very elongate forms existing like those downstream of glacial ice margins or below important semipermanent snow patches, showing a gelifluction fabric in their upper 20 to 30 cm. Below this depth the fabric developed is of a frostcreep type (Van Vliet-Lanoë, 1983).

CRYOTURBATION AND PATTERNED GROUND

Introduction

Many hypotheses have been advanced to explain cryoturbation and patterned ground but only two are commonly cited.

The first explains involutions as due to load-casting effects on thawing. Kuenen (1958) produced pocket or drop-like shapes by flowage of water-saturated sediments. Two layers are unstable if the upper has a greater specific weight than the lower due to some differences in packing and water content. In such a case, increase in pore pressure associated with stress will produce liquefaction (Anderson and Bjerrun, 1967). This hypothesis has been supported by the experiments of Cegła and Dżulinski (1970). In a recent paper Vandenberghe and Van den Broek (1982) also use it to explain convolutions existing in Late Glacial deposits near the northern Belgian border. They measured the density of water-saturated sediments of various textures and showed that there was a very narrow range of values (1.98–2.19 g cm^{-3}).

The other hypothesis, proposed by Sharp (1942), uses differential frost heaving to explain cryoturbation. He invoked differential frost heaving of contiguous sediments of various textures. As a result of sediment heaving caused by ice segregation, sediments are compressed and injected either upwards or laterally. Additional perturbations are also thought to occur by liquefaction during thawing. This hypothesis has been supported by Dylikowa (1961), Corte (1962), Washburn (1973) and Pissart (1976).

Observations

From our field and micromorphological observations at the Gåsbu site in north-western Spitsbergen, the second hypothesis is more acceptable. The Gåsbu sequence is a twentieth-century aeolian deposit lying on glacio-marine sands. The site is imperfectly drained with the top of the water table at 50 cm and the top of permafrost at 80 cm. Involutions are found only in the upper part of the active layer. In this profile, the moist aeolian layer is prevented by rapid superficial consolidation from upward heaving from the outset of freezing and can only heave downwards. It exerts a stress which forces the underlying sediment to be progressively injected upwards through a network of desiccation and thermal cracks. These injections affect various textured sediments ranging from coarse sand to fine humic silt. From this observation it is clear that the hypothesis of load casting does not fit this case. Micromorphological observations on the injected material and in the prismatic structural units of aeolian sediments revealed traces of ice lensing parallel to the soil surface (Figure 6.16). This also proves that liquefaction did not occur during thawing although the possibility remains of liquefaction occurring in loose sands. Based on the position of the perched water table and the thermal survey for the entire profile during the year 1982–83, it is evident that conditions were optimal for significant heave to occur from late August to early November. Because the base of the aeolian sand was dated at about 1917, this profile has possibly undergone between 65 and 100 cycles of freeze–thaw.

Figure 6.16 Upward injection of humic silt (left) in a coarse sand (right). Gåsbu site (north-west Spitsbergen). Notice the traces of ice lensing orthogonal to the injection and the uplifted stone (S); V, void

Such injections have been reproduced experimentally by Coutard and Mücher (1984) along shearing planes and desiccation cracks by twenty-eight cycles of gentle freezing and thawing in an imperfectly drained laminated loam. Experimental work performed by Pissart (1982) demonstrated that after twenty-two freeze–thaw cycles bulb-like deformation of loamy cylindrical units (7 cm in diameter) entrapped in ice-cemented gravel produced a shape very similar to the aeolian prismatic units found at the Gåsbu site. In these experiments, ice lensing remained mostly parallel to the soil surface.

Micromorphological analysis of the cylindrical units from the Pissart experiments (Van Vliet-Lanoë, 1982) revealed that platy aggregates occurring close to the faces of the cylinders became rounded by plastic deformation. Similar deformation has been observed in injected loams of the Coutard and Mücher experiment (1984) and in fossil involutions in France and Belgium (Van Vliet-Lanoë and Coutard, 1984). Injections, as well as platy and granular structures, are commonly observed in the field. Because aggregates formed by ice in loamy textured soils have a high degree of stability it is easier to follow fabric evolution produced by lateral stresses. Granular structures have been reported commonly in the literature and often attributed to cryoturbation (Fedoroff, 1966; Morozova, 1965, 1972; Crampton, 1977; Konichev *et al.*, 1973; Fox and Protz, 1982; Van Vliet-Lanoë, 1982) but the process leading to their formation was never clearly explained. In fact, from measuring stresses during freezing and thawing, Mackay and McKay (1976) showed that stresses were not important during freezing but much more sensitive during thawing due to the occurrence of refreezing ice at depth. This explains why platy aggregates evolve to granular ones. This process is rather rapid and in most cases the skeleton ice-expulsion is still undeveloped; in long-term cryoturbations it could be the origin of the microcircles of skeleton grains observed by Morozova (1972).

All fossil involutions observed show traces of oxidation-reduction, indicating that they were active at the capillary fringe of a water table when summer temperatures were not low enough ($\geqslant 4\,^{\circ}$C) to inhibit bacterial iron precipitation. Goździk (1964) found that these forms are located on low terraces or in small depressions. Mapping of cryoturbation and patterned ground in relation to drainage conditions (Van Vliet-Lanoë, 1983) shows that although some true mudboils are located in poorly drained sites, most are in imperfectly drained sites together with sorted polygons, sorted circles and step-like forms. Similar sites exist in non-permafrost areas such as the Alps.

Genesis

These observations allow optimum conditions for the development of frost-induced involutions to be defined.

If materials of various frost susceptibilities are superimposed or interbedded and if they are located in the capillary fringe of a water table, with or without permafrost, they are under alternating freezing and thawing conditions, potential sites for involutions (Figure 6.17).

The following sequence of events can be reconstructed (Van Vliet-Lanoë and Coutard, 1984): cryo-desiccation and thermal cracking of surface layer(s), shearing in deeper horizons; prevention of upwards heaving by superficial consolidation from the outset of freezing; progressive deformation and upward injection of various textured substrata along vertical cracks as a result of heaving stresses. Once this initial pattern is achieved (Figure 6.17), the genesis of

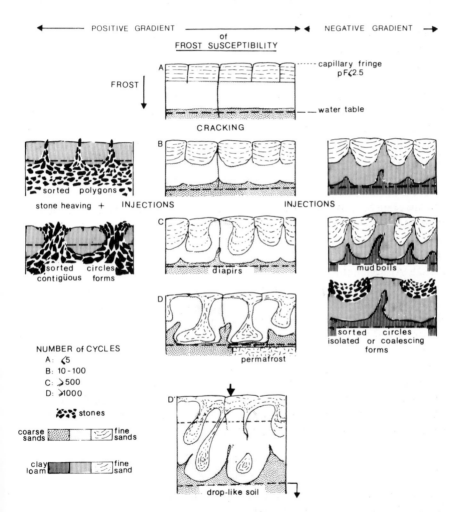

Figure 6.17 The evolution of patterned ground and involutions on imperfectly to poorly drained sites (adapted from Van Vliet-Lanoë and Coutard, 1984). A positive gradient means an increase of frost susceptibility towards the surface

patterned ground can be explained using the *gradient of frost susceptibility* (Van Vliet-Lanoë and Coutard, 1984). If the *gradient* is *positive*, it means that the frost susceptibility increases towards the surface. The injected material forms a diapir and progressively squeezes the surface prismatic units which evolve to a drop-like form and slowly migrate downwards. Initial stratification, if present, is folded but only weakly perturbated. When the drop-like form reaches the top of the water table, water-supply conditions for heaving become constant and flat-based pockets occur because large ice bodies impede further downwards migration of heave. Although claimed by Gullentops and Paulissen (1978), permafrost is not necessary to explain this process. Permafrost is, however, necessary to explain the presence of a water table in well-drained sites (Figure 6.17D). If the *gradient* is *negative*, the injected material heaves much more than the surface prismatic units and succeeds in bursting the soil surface, particularly due to the heave produced by water freezing at depth from the outset of thawing: a mud boil appears. If the surface material is stony, isolated or coalescent sorted circles occur and are progressively enlarged by the heave of the central boil. Sorted polygons and circles are, on the contrary, related to a positive gradient (Figure 6.17C—left). With better drainage, cryoturbated sands evolve into small hummocks with the upward heaving being less impeded at the surface. Mud boils and circles also evolve into high-centred forms as observed by Zoltai and Tarnocai (1981) in Northern Canada and by ourselves in north-western Spitsbergen (1983). This evolution is also confirmed by the experiments of Pissart (1982).

Most of the forms observed in the field are related to differential frost heave and result from the summation of minute displacements as observed by Pissart (1982). Although this can explain most of the observed involutions and patterned ground, nevertheless, in valleys with very saturated sandy soils, dewatering and loadcasting may produce similar forms. In clayey soils, differential swelling and shrinking of clays leads to converging forms (Knight, 1980), but in this case traces of ice segregation are lacking.

Diagenesis of buried forms

The theory of differential frost heaving provides an understanding of the climatic and hydrological reasons for the diagenesis of weakly active and totally inactive forms.

1 Active forms are susceptible to fossilization in a particular stage of evolution by severe climatic change.
2 Fossil forms remaining close to the topographical surface may be periodically reactivated during cold stadials or even winters if hydraulic conditions favourable to their formation are restored.
3 Sand wedges or ice-wedge casts in alluvial deposits are susceptible to

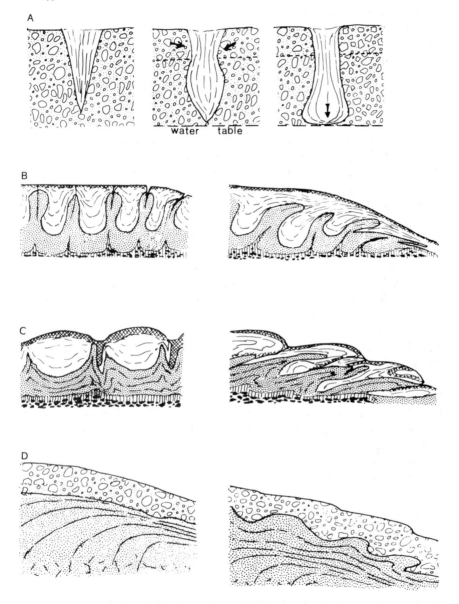

Figure 6.18   Deformation of periglacial structures by differential heaving and frost creep. (A) Sand wedge in fluviatile gravel. The deformation is associated with raising of watertable (north Belgium). (B) and (C) Deformation of involutions and hummocks by frost creep following valley incision and a lowering of the water table (north-west Spitsbergen). (D) Deformation of a complex periglacial slope deposit by frost creep remobilization (south-east of Central Massif, France)

evolution—forming pockets or 'cauldron' forms if the frost susceptibility gradient is positive (Figure 6.18A).

4 Initial involutions resulting from dewatering, load casting, karst or thermokarst are also likely to evolve in the same way, if affected by further deep seasonal freezing.

5 Burial of active forms by fluvial or aeolian sedimentation may induce the superimposition of new involutions on the still active ones. This can explain the complexity of polygenetic sequences in valley infillings.

6 Valley incision associated with a lowering of the water table can induce a change in the profile dynamics, frost creep supplanting differential frost heaving (Figure 6.18B and C).

Clay illuviation and its relationship with frost

As previously discussed, soils affected by repeated freeze–thaw are sensitive to internal colluviation during thawing and this is reinforced by snow melting. With lowering of the transport capacity of percolating water, fine sands and silts are first deposited, followed at depth by clay-size particles. This means that if the clays are potentially easy to disperse because of their base saturation or their incorporation in organo-mineral complexes (Guillet *et al.*, 1981), fine clay accumulation is able to take place in such an environment. If the acidity is too high, only flocules of aluminized clays move down in pseudo-silt form. This explains the difference in reaction of soils or sediments to particle translocation as a consequence of their state of leaching and base content. In such an environment in a saturated state (e.g. in young sediments), well-oriented, fine-clay coatings may develop, often locally interstratified with coarser material. In an unsaturated situation, coarse clay interstratified with silts accumulates in the form of poorly oriented coatings; in this case the occurrence of fine clay signifies a rejuvenation of the profile.

  If clay translocation occurs within the active layer, clay particles forming a 'summer' coating, as observed in some Spitsbergen soils (developed on limestone), are quickly integrated into the soil matrix by frost stresses and sorting to form a stress cutan or a capping, creating at various depths (5 cm to deeper horizons) a Bt horizon without coatings. Conversely, if translocation of clay occurs below the depth of winter frost penetration, coatings on pores and aggregates are preserved. This has been clearly observed in the form of thick, massive and well-sorted argillans below frostcreep deposits.

  Clay accumulation in soils affected by deep seasonal frost has been observed by McKeague *et al.* (1983), Fedorova and Yarilova (1972), Sokolov (1980). Sokolov *et al.* (1976). Clay accumulation occurs initially in the form of ferriargillans (Cailler, 1977) and, under conditions of thawing, translocation of fines seems to be important. Classical clay migration can also occur later during the summer.

When climatic conditions change and the active layer includes a classical Bt horizon with well-oriented clay coatings, ice lensing and ultra-desiccation are responsible for breaking the cutans into angular papules which are quickly rounded after a few freeze–thaw cycles as reproduced experimentally in the Caen laboratory. This feature is commonly observed in many fossil European loess soils (Jamagne, 1972) (Figure 6.19). Cryo-desiccation associated with the mechanical cutting by ice needles and lenses disrupts clay cutans into small pieces and loosens them from the matrix or the sand grains. Once formed, these papules are very stable (Pede and Langohr, 1983) and survive mass wasting, thixotropic collapse or colluviation processes.

After many freeze–thaw cycles sorting affects skeleton grains as well as papulized coatings. This effect is visible between peds and in bleached tongues of Alfisols and Ultisols but can also be responsible for the progressive deepening of the top of the Bt horizon, in the same way as the action of heavy rainfall in tropical regions (P. Faivre, pers. comm.). Clay and silt accumulating in large pores and cracks (tongues) contrast in frost susceptibility with adjacent soil material, especially when this process is acting in a former Bt horizon. This contrast explains macro- and microcryoturbation in soils such as the granular structure observed in bleached tongues, their deformation and enlarging by differential heave (Figure 6.20).

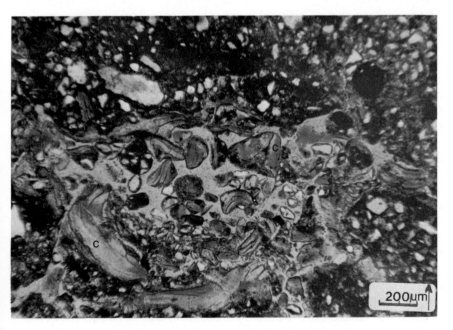

Figure 6.19   Cryoturbated and papulized pore ferriargillan from a buried paleosol in loess of Normandy (St Romain profile, France)

A final cryoturbation phenomenon affects organic matter in soils. Ice needles and lenses occurring in plant litter or in buried organic horizons cut plant tissues into minute fragments. This process is clearly observed in Spitsbergen soils (Van Vliet-Lanoë, 1983) where faunal activity and mulching are practically non-existent. It produces coarse silt-size particles which can be translocated, and accumulate deeper in the soil as described by Pettapiece (1974) and Ellis (1983).

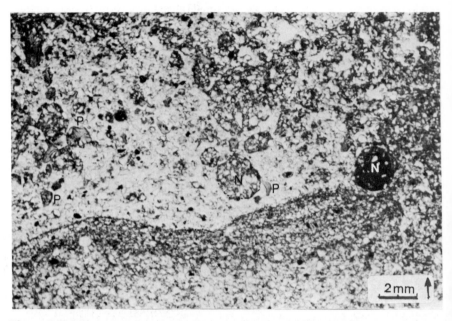

Figure 6.20   Freeze–thaw microfabric in a cryoturbated tongue of a buried paleosol in loess of Normandy (St Romain profile, France). Notice the thick capping and the sorted free-grain accumulation including papules (P) and soil nodules (N)

CONCLUSIONS: FROST STRUCTURES — A VALUABLE CLIMATIC INDICATOR

Macro- and microscopic observations can be used as climatic indicators in soils particularly as pollen is usually poorly preserved or absent. Structural patterns created by ice lensing are excellent indicators of the location of a former permafrost table. They are more common and as reliable as ice-wedge casts or flat-based involutions in a deeply drained environment. The depth of the permafrost table reconstructed in this way in Belgian tundra gley soils (Figure 6.13, 25 000 BP) (Van Vliet-Lanoë, 1976) and in fragipan soils of the same area (Figure 6.14, 17 000 BP) (Van Vliet Lanoë and Langohr, 1981) fits well with drainage conditions commonly observed in the Arctic. These depths seem abnormally shallow to scientists working in high latitudes. During the cold periods of the Late Glacial, the insolation rate was much less than today (Broecker and Van Donk, 1980). The Pleniglacial was characterized by cool summers and the winters were not too cold (Berger, 1983) as in north-western Spitsbergen today, but during Lateglacial the summers were warm and the winters very cold as in the continental Arctic today (Berger, 1983). This is reflected in the permafrost and fragipan soils already described. Ice wedges and shearing planes dominate in Pleniglacial soils, though, in the same region, in fragipan soils drastic winter cooling enhanced cryo-desiccation and thermal cracking.

Sorted or non-sorted platy structures, granular cryoturbation structures, frost cracks, involutions in low topographic conditions are all features indicative of seasonal frost. The thickness of the platy aggregates can provide an estimate of soil truncation by erosion or position an older soil surface in a complex profile.

On the basis of these climatic indicators and their relationship to sediment-ological and pedological events, climatic curves almost as accurate as those recorded by palynological analysis can be proposed (Haesaerts and Van Vliet-Lanoë, 1981). In addition, phenomena occurring in soils which have been attributed to the 'acid degradation' of soils (podzol flour in the literature) can be shown to result from alternating freezing and thawing and associated

Figure 6.21 Paleoargilic horizon with traces of ice lensing enhanced by humic infiltration from a complex podzol (Lessay, Normandy, France). This profile is associated with large sand and pebble wedges. A fossil alios (A) has been broken by cryoturbation and new humic matter (H) infiltrates along vertical cracks

phenomena (sorting and internal colluviation) (Langohr and Pajares, 1983; Van Vliet-Lanoë and Langohr, 1983); it may be enhanced by synchronous and younger hydromorphic clay destruction (Cailler, 1977). This is a great help in establishing chronostratigraphy and profile genesis of many surface soils (Figure 6.21) which began to develop in early Lateglacial time, after the main loess deposition, or sometimes even in the Interpleniglacial (Van Vliet-Lanoë and Langohr, 1981). This will facilitate a better understanding of Lateglacial and Holocene landscape evolution (Velichko and Morozova, 1976; Langohr, 1984). Based on such observations, surface soil chronostratigraphy is not only valuable in loessic areas but also in coarser textured materials such as heads and granitic sands with a Late Glacial aeolian admixture such as occurred in the French Vosges, in some sectors of the French Central Massif, in Brittany, Cotentin and in Cornwall (UK), where aeolian deposits have been described by many authors.

ACKNOWLEDGEMENTS

We would like to express our thanks to Douglas Fisher for suggesting ways of improving the original text both in language and content.

REFERENCES

Andersen, A. and Bjerrun, L. (1967). 'Slides in subaqueous slopes in loose sand and silt', in *Marine Geotechnique* (Ed. A. F. Richards), pp. 221–222, Urbana University, Illinois Press.

Anderson, D. M. and Morgenstern, N. R. (1973). 'Physics, chemistry and mechanics of frozen ground: A review', North American contribution, Second International Permafrost Conference, Yakoutsk, USSR. July 1983, pp. 257–288, Washington, DC, National Academy of Sciences.

Benedict, J. B. (1970). 'Down slope soil movement in a Colorado alpine region: rates, processes, and climatic significance', *Arctic and Alpine Research*, 2, 165–226.

Berger, A. L. (1983). 'Approche astronomique des variations paléoclimatiques: les variations mensuelles et en latitude de l'insolation de −130,000 à −100,000 et de −30,000 à aujourd'hui', *Bulletin de L'Institut Géologique du Bassin d'Aquitaine*, 34, 7–25.

Bertouille, H. (1972). 'Effet du gel sur les sols fins', *Revue de Géomorphologie Dynamique*, 2, 71–84.

Beskow, G. (1935). 'Tjälbildningen och tjällyftningen med särskild hänsyn till vägar och jarnägar', *Sveriges Geologiska Undersökning*, Arsbok 26, ser. C, no. 375.

Bjorkhem, U. and Jongerius, U. (1974). 'Micromorphological observations in some podzolized soils from Central Sweden', in *Soil Microscopy* (Ed. G. K. Rutherford), pp. 320–332, Limestone Press, Kingston, Ontario.

Brewer, R. (1976). *Fabric and mineral analysis of soils*, Krieger, New York.

Brewer, R. and Haldane, A. D. (1957). 'Preliminary experiments in the development of clay orientations in soils', *Soil Science*, 84, 301–309.

Broecker, W. S. and Van Donk, J. (1970). 'Insolation changes, ice volumes and the ^{18}O record in deep sea cores', *Rev. Geophysic and Space Physics*, 8, 169–198.

Bunting, B. and Fedoroff, N. (1974). 'Micromorphological aspects of soil development in the canadian High Arctic', In *Soil Microscopy* (Ed. G. K. Rutherford), pp. 350–365, Limestone Press, Kingston, Ontario.

Bunting, B. (1977). 'The occurrence of vesicular structures in arctic and subarctic soils', *Zeitschrift für Geomorphologie*, **21**, 87–95.

Burt, T. P. and Williams, P. J. (1976). 'Hydraulic conductivity in frozen soils', *Earth Surface Processes and Landforms*, **1**, 349–360.

Cailler, M. (1977). *Etude chronoséquentielle des sols sur les terrasses alluviales de la Moselle. Genèse et évolution des sols lessivés glossiques.* Thèse de Spécialité Université de Nancy.

Cegła, J. and Džuĺinski, S. (1970). 'Uklady niestatecznie wartwowane i ich wystepowanie w srodowisku peryglacjalnym', *Acta Universitatis Wratislaviensis*, no. 124, Studia Geograficzne **13**, 17–42.

Chandler, R. J. (1972). 'Periglacial mudslides in Vestspitsbergen and their bearing on the origin of fossil 'solifluction' shears in low angled clay slopes', *Quarterly Journal of Engineering Geology*, **5**, 223–241.

Collins, J. and O'Dublain, T. (1980). 'A micromorphological study of silt concentrations in some Irish podzols', *Geoderma*, **24**, 215–224.

Corte, A. (1962). 'Vertical migration of particles in front of a moving freezing plane', *Journal of Geophysical Research*, **67**, (3), 1085–1090.

Corte, A. (1963). 'Relationship between four ground patterns, structure of the active layer and type and distribution of ice in permafrost', *Biuletyn Peryglacjalny*, **12**, 7–90.

Corte, A. (1966). 'Particle sorting by repeated freezing and thawing', *Biuletyn Peryglacjalny*, **15**, 175–240.

Coutard, J. P. and Mücher, H. M. (1984). 'Deformation of laminated silt loam due to repeated freezing and thawing', *Earth Surface Processes and Landforms*, (in press).

Crampton, C. B. (1977). 'A study of the dynamics of hummocky microrelief in the Canadian North', *Canadian Journal of Earth Sciences*, **14**, 639–649.

Csathy and Thowsend, D. L. (1962). 'Pore size and field frost performance of soils', *Highway Research Board Bulletin*, **331**, 67–80.

Curmi, P. (1979) 'Genèse d'une structure litée à granulométrie hétérogène dans une arène granitique', *C.R. Acad. Sci. de Paris, Série D*, **288**, 731–733.

Czeratzki, V. (1956). 'Bödems Struktur Bildug bei Wechselnder Befeuchtung und Trocknung (model Versuche)', Instit. für Wissenchaftlichen Film, Göttingen.

De Ploey, J. and Mücher, H. M. (1981). 'A consistency index and rainwash mechanisms on Belgian loamy soils', *Earth Surface Processes and Landforms*, **6**, 319–330.

Dumanski, J. A. (1964). *A Micropedological Study of Eluviated Horizons*, Unpublished Masters Thesis, University of Saskatchewan.

Dumanski, J. A. and St Arnaud, R. J. (1966). 'A micropedological study of eluviated horizons, *Canadian Journal of Soil Science*, **46**, 287–292.

Dylik, J. (1967). 'Solifluction, congelifluction and related slope processes', *Geografiska Annaler*, **49**, 167–177.

Dylikowa, A. (1961). 'Structures de pression congélistatiques et structures de gonflement par le gel près de Katarzynów près de Lódz', *Bulletin Société des Lettres de Lódz*, **12**, (9), 1–23.

Ellis, S. (1983). 'Micromorphological aspects of arctic-alpine pedogenesis in the Okstindan Montains, Norway', *Catena*, **10**, (1–2), 133–147.

Fedoroff, N. (1966). 'Les cryosols', *Science du Sol*, **2**, 77–110.

Fedoroff, N., De Kimpe, C., Page, F. and Bourbeau, G. (1981). 'Essai d'interprétation des transferts sous forme figurée dans les podzols du Québec Méridional à partir de l'étude micromorphologique des profils', *Geoderma*, **26**, 25–45.

Fedorova N. N. and Yarilova, A. (1972). 'Morphology and genesis of prolonged seasonally frozen soils in Western Siberia', *Geoderma*, 7, 1–13.

Fitzpatrick, E. A. (1956). 'An indurated soil horizon formed by permafrost', *Journal of Soil Science*, 7, 248–257.

Fitzpatrick, E. A. (1974). 'Cryons and Isons', *Proc. North Engl. Soils Discuss. Group*, 11, 31–43.

Fox, C. (1979). *The Soil Micromorphology and Genesis of the Turbic Cryosols from the Mackensie River Valley and Yukon Coastal Plain*, Unpublished PhD Thesis, University of Guelph, Ontario.

Fox, C. (1983). 'Micromorphology of an Orthic Turbic Cryosol—a permafrost soil', in *Soil Micromorphology* (Eds. P. Bullock and C. P. Murphy), pp. 699–705. AB Academic Publishers, Berkhamsted.

Fox, C. and Protz, R. (1982). 'Definition of fabric distributions to characterize the rearrangement of soil particles in the Turbic Cryosols, *Canadian Journal of Soil Science*, 61, 29–34.

Gangloff, P. and Pissart, A. (1983). 'Evolution géomorphologique et palses minérales près de Kuujuaq (Fort Chimo, Québec)', *Bulletin de la Société Géographqie de Liège*, (in press).

Goździk, J. (1964). 'Repartition topographique des structures périglaciaires', *Biuletyn Peryglacjalny*, 14, 215–249.

Guillet, B., Rouiller, J. and Vedy, J. C. (1981). 'Dispersion et migration de minéraux argileux dans les podzols. Contribution des composés organiques associés, leur rôle sur la forme et l'état de l'aluminium', in *Migrations organo-minérales dans les sols tempérés*, Colloque International CNRS, Nancy 1979, Edition du CNRS, pp. 46–56.

Gullentops, F. and Paulissen, E. (1978). 'The drop soil of Eisden type', *Biuletyn Peryglacjalny*, 27, 105–115.

Haesaerts, P. and Van Vliet, B. (1973). 'Evolution d'un permafrost dans les limons du Dernier Glaciaire à Harmignies', *Bulletin de l'Association Francaise pour l'Étude du Quaternaire*, 3, 151–164.

Haesaerts, P. and Van Vliet-Lanoë, B. (1981). 'Phénomènes periglaciaires et sols fossiles observés à Maisières-Canal, à Harmignies et à Rocourt', *Biuletyn Peryglacjalny*, 28, 291–325.

Harris, C. (1981a). 'Microstructure in solifluction sediments from South Wales and North Norway', *Biuletyn Peryglacjalny*, 28, 221–226.

Harris, C. (1981b). 'Periglacial mass-wasting: a review of research', *BGRG Research Monograph*, 4, Geobooks, Norwich.

Harris, C. and Ellis, S. (1980). 'Micromorphology of soils in soliflucted materials, Okstindan, Northern Norway', *Geoderma*, 23, 11–29.

Higashi, A. and Corte, A. (1971). 'Solifluction: a model experiment', *Science*, 171, 480–482.

Hoekstra, P. (1969). 'Water movement and freezing pressures', *Soil Science Society of America, Proceedings*, 33, 512–518.

Hoekstra, P. and Miller, R. D. (1965). 'Movement of water in a film between glass and ice', *U.S. Army Cold Regions Res. Engineering Lab. Res. Rept.* 153.

Jamagne, M. (1972). 'Some micromorphological-aspects of soils developed in loess deposits of Northern France', in *Soil Micromorphology* (Ed. S. Kowalinski), pp. 554–582, Panstwowe Wydawnictwo Naukowe, Warszawa.

Kaplar, C. W. (1965). 'Stone migration by freezing of soil', *Science*, 149, 1520–1521.

Knight, M. J. (1980). 'Structural analysis and mechanical origins of gilgai at Boorook, Victoria, Australia', *Geoderma*, 23, 245–283.

Konichev, V. N., Faustova, M. A. and Rogov, V. V. (1973). 'Cryogenic processes as reflected in ground microstructure', *Biuletyn Peryglacjalny*, 22, 213–219.

Kokkonen, P. (1927). 'Beobachtungen über die Struktur des Bodenfrostes', *Acta Forestalia Fennica*, 30, 1–55.

Krivan, P. (1958). 'Tundrenerscheinungen mit Eislinsen und Eisblättigkeit in Ungarn', *Acta geologica*, 5, 323–334.

Kuenen, P. H. (1958). 'Experiments in Geology', *Geological Society Glasgow, Trans.*, 23, 1–28.

Kuznecova, T. P. (1973). 'Special features of cryolithogenesis in the alluvial plains, Central Yakutia', *Biuletyn Peryglacjalny*, 22, 221–231.

Langohr, R. (1984). 'The extension of permafrost in Western Europe in the period between 18,000 years and 10,000 years B.P. (Tardiglacial): information from soil studies', *Proceedings of the 4th International Permafrost Conference, Fairbanks, Alsaska, 1983.*

Langohr, R. and Van Vliet-Lanoë, B. (1979). 'Clay migration in well to moderately well drained acid brown soils of the Belgian Ardennes. Morphology and clay content determination', *Pédologie*, 29, 367–385.

Langohr, R. and Pajares, G. (1983). 'The chronosequence of pedogenic processes in Fraglossudalfs of the Belgian loess belt', in *Soil Micromorphology* (Eds P. Bullock and C. P. Murphy), pp. 503–510, AB Academic Publishers, Berkhamsted.

Lautridou, J. P. (1982). 'La fraction fine des débris de gélifraction expérimentale', *Biuletyn Peryglacjalny*, 29, 78–85.

Lautridou, J. P. and Ozouf, J. C. (1982). 'Experimental frost shattering. 25 years of research at the Centre de Géomorphologie du CNRS', *Progress in Physical Geography*, 6, (2), 215–232.

Mackay, J. R. (1971). 'The origin of massive icy beds in permafrost, western arctic coast Canada', *Canadian Journal of Earth Science*, 8, 397–422.

Mackay, J. R. (1974). 'Reticulate ice veins in permafrost; Northern Canada', *Canadian Geotechnical Journal*, 11, 230–237.

Mackay, J. R. (1981). 'Active layer slope movement in a continuous permafrost environment, Garry Island, N.W.T. Canada', *Canadian Journal of Earth Science*, 18, 1666–1680.

Mackay, J. R. and McKay, D. K. (1976). 'Cryostatic pressure in non-sorted circles mud hummocks, Inuvik, N.W.T. Canada', *Canadian Journal of Earth Science*, 13, 889–897.

McKeague, J. A. (1981). 'Organo-mineral migrations: some examples and anomalies in Canadian soils', in *Migrations organo-minerals dans les sols temperes*, Colloque International CNRS, Nancy 1979, Edition du CNRS, pp. 341–348.

McKeague, J. A., MacDougall, J. I. and Miles, N. M. (1973). 'Micromorphological, physical and mineralogical properties of a catena of soils from Prince Edwards Island in relation to their classification and genesis', *Canadian Journal of Soil Science*, 53, 281–295.

Mermut, A. R. and St Arnaud, R. J. (1981). 'Microband fabric in seasonally frozen soils', *Soil Sci. Soc. Am. J.*, 45, 578–586.

Miller, D. E. (1971). 'Formation of vesicular structure in soils', *Soil Science Society of America, Proceedings*, 35, 635–643.

Morozova, T. D. (1965). 'Micromorphological characteristics of pale yellow permafrost soils in Central Yakoutia in relation to cryogenesis', *Soviet Soil Science*, 7, (12), 1333–1342.

Morozowa, T. D. (1972). 'Micromorphological peculiarieties of the fossil soils and some problems of paleogeography of the Mikulino (Eemian) Interglacial on the Russian Plain', in *Soil Micromorphology* (Ed. S. Kowalinski), pp. 595–606, Panstwowe Wydawnictwo Naukowe, Warszawa.

Palmer, A. C. (1967). 'Ice lensing, thermal diffusion and water migration in freezing soil', *Journal of Glaciology*, **6**, (47), 681–694.

Payton, R. (1983). 'The micromorphology of some fragipans and related horizons in British soils with particular reference to their consistence', in *Soil Micromorphology* (Eds P. Bullock and C. P. Murphy), pp. 317–336, AB Academic Publishers, Berkhamsted.

Pede, K. and Langohr, R. (1983). 'Microscopic study of pseudo-particles in dispersed soil samples', in *Soil Micromorphology* (Eds P. Bullock and C. P. Murphy), pp. 265–272, AB Academic Publishers, Berkhamsted.

Penner, E. (1959). 'The mechanics of frost heaving in soils', *Highway Research Board Bulletin*, **225**, 1–22.

Penner, E. (1974). 'Uplift forces on foundations in frost heaving soils', *Canadian Geotechnical Journal*, **11**, (3), 323–338.

Pettapiece, W. W. (1974). 'A hummocky permafrost soil from the Subarctic of the North Western Canada and some influence of fire', *Can. Jour. Soil. Sci.*, **54**, 343–355.

Pissart, A. (1966). 'Expériences et observations à propos de la genèse des sols polygonaux triés', *Revue Belge de Géographie*, **90**, 55–73.

Pissart, A. (1969). 'Le mecanisme périglaciaire dressant les pierres dans le sol. Résultats d'expériences', *Comptes Rendus de L'Académie des Sciences*, Paris, **268**, 3015–3017.

Pissart, A. (1970). 'Les phénomènes physiques essentiels liés au gel, les structures périglaciaires qui en résultent et leur signification climatique', *Annales de la Société Géologique de Belgique*, **93**, 7–49.

Pissart, A. (1976). 'Sols à buttes, cercles non-triés et sols striés non-triés de l'Ile de Banks (Canada, NWT)', *Biuletyn Peryglacjalny*, **26**, 275–285.

Pissart, A. (1982). 'Déformations de cylindres de limons entourés de graviers sous l'action d'alternaces gel-dégel. Expériences sur l'origine des cryoturbations', *Biuletyn Peryglacjalny*, **29**, 219–229.

Research Institute of Glaciology, Cryopedology and Desert Research (1975). *Permafrost*, Acad. Sinica Lanchou, China, Traduction NRC, Ottawa 1981.

Romans, J. C. and Robertson, L. (1974). 'Some aspect of the genesis of alpine and upland soils in the British Isles', in *Soil Microscopy* (Ed. G. K. Rutherford), pp. 498–510, Limestone Press, Kingston, Ontario.

Romans, J. C., Stevens, J. and Robertson, L. (1966). 'Alpine soils in northeast Scotland', *Journal of Soil Science*, **17**, 184–199.

Romkens, M. J. and Miller, R. D. (1973). 'Migration of mineral particles in ice with a temperature gradient', *Journal of Colloid and Interface Science*, **42** (1), 103–111.

Rowell, D. and Dillon, P. (1972). 'Migration and aggregation of Na–Ca clays by freezing of dispersed and flocculated suspensions', *Journal of Soil Science*, **23**, 442–447.

Schenk, E. (1968). 'Fundamental process of freezing and thawing in relation to the development of permafrost', in *Alpine and Arctic Environment* (Eds E. Wright and W. Osburn), pp. 229–236.

Schunke, E. (1975). 'Die Periglazialerscheinungen Islands in Abhängigkeit von Klima und Substrat', *Akad. Wiss. Göttingen. Abh. Math. Phys. Kl. Folge,* **3**, 30.

Shumskii, P. A. (1964). *Principles of Structural Glaciology*, Taduction Krauss, Dover Publication Company.

Smalley, I. J. and Davin, J. E. (1982). 'Fragipan in soils: a bibliographic study and review of some of the hard layers in loess and other materials', *New Zealand Soil Bureau Bibliographic Report* 30, Department of Scientific and Industrial Research, New Zealand.

Smith, J. (1956). 'Some moving soils in Spitsbergen', *Journal of Soil Science*, **7**, 10–21.

Sharp, R. P. (1942). 'Periglacial involutions in Northeast Illinois', *Journal of Geology*, **50**, 113–133.

Sokolov, I. A. (1980). Variety of forms of non-gley hydromorphic soil formation', *Soviet Soil Science*, **12**, 41–55.

Sokolov, I. A., Naumov, Y. M., Gradusov, B. P., Tursina, T. V. and Tsyurupa, V. V. (1976). Ultracontinental taiga soil formation on calcareous loams in Central Yakutia', *Soviet Soil Science*, **8**, 144–160.

Svenson, H. (1964). 'Structural observations in the minerogenic core of a palsa', *Svensk Geografisk Arbok*, **40**, 138–142.

Taber, S. (1929). 'Frost heaving', *Journal of Geology*, **37**, 428–461.

Taber, S. (1943). 'Perennially frozen ground in Alaska: its origin and history', *Geological Society of America Bulletin*, **54**, 1433–1548.

Tessier, D. (1978). 'Étude de l'organisation des argiles calciques. Evolution au cours de la déssiccation', *Annales Agronomiques*, **29**, (4), 4319–4355.

Tessier, D. (1984). *Étude Experimentale de l'Organisation des Matériaux Argileux. Hydratation, Gonflement et Structuration au Cours de la Déssiccation et de la Réhumectation*, Thèse d'État, Physique de la Terre, Université de Paris VI.

Tyutyunov, I. A. (1964). *An Introduction to Theory of the Formation of Frozen Rocks* (Traduction I. Muhlhaus), Pergamon, Oxford.

Vandenberghe, J. and Van den Broek, P. (1982). 'Weichselian convolution phenomena and processes in fine sediments', *Boreas*, **11**, 299–315.

Van Vliet-Lanoë, B. (1976). 'Traces de ségrégation de glace en lentilles associées aux sols et phénomènes périglaciaires fossiles', *Biuletyn Peryglacjalny*, **26**, 41–54.

Van Vliet-Lanoë, B. (1980). 'Corrélations entre fragipan et permagel. Application aux sols lessivés glossiques', *Comptes-rendus du Groupe de travail 'Régionalisation du Périglaciaire, Strasbourg*, **5**, 9–22.

Van Vliet-Lanoë, B. (1982). 'Structures et microstructures associées à la formation de glace de ségrégation. Leurs conséquences', *Proc. 4th Canadian Permafrost Confer. Calgary, 1981*, 116–122.

Van Vliet-Lanoë, B. (1983). 'Études cryopédologiques au Sud du Kongsfjord, Svalbard. Rapport de la mission Spitzberg 1982', publication interne du Centre de Géomorphologie du CNRS, Caen.

Van Vliet-Lanoë, B. and Coutard, J. P. (1984). 'Observations à propos de la genèse de sols cryoturbé ou en gouttes par gonflement cryogénique différentiel', in 'A propos de deux thèmes de recherche abordés au Centre de Géomorphologie du CNRS, by J. P. Lautridou, J. C. Ozouf, B. Van Vliet-Lanoë and J. P. Coutard, *Biuletyn Peryglacjalny* 30 (in press).

Van Vliet-Lanoë, B., Coutard, J. P., and Pissart, A. (1984). 'Structures caused by repeated freezing and thawing in various loamy sediments. A comparison of active, fossil and experimental data', *Earth Surface Processes and Landforms*, (in press).

Van Vliet-Lanoë, B. and Flageollet, J. C. (1981). 'Traces d'activité périglaciaire dans les Vosges Moyennes', *Biuletyn Peryglacjalny*, **28**, 209–219.

Van Vliet-Lanoë, B. and Langohr, R. (1981). 'Correlation between fragipan and permafrost—with special reference to Weichsel silty deposits in Belgium and Northern France', *Catena*, **8**, 137–154.

Van Vliet-Lanoë, B. and Langohr, R. (1983). 'Evidence of disturbance of pore ferriargillans in silty soils of Belgium and Northern France', in *Soil Micromorphology*. (Eds P. Bullock and C. P. Murphy), pp. 511–518, AB Academic Publishers, Berkhamsted.

Van Vliet-Lanoë, B. and Valadas, B. (1983). 'A propos des formations déplacées de versants cristallins des massifs anciens: le rôle de la glace de ségrégation dans la

Soils and Quaternary landscape evolution

dynamique', *Bulletin de l'Association Francaise pour l'Étude du Quaternaire*, **4**, 153–160.

Velichko, A. A. and Morozova, T. D. (1976). 'Stages of development and paleogeographical inheritance of recent soil features in the Central Russian Plain', *Catena*, **3**, 169–189.

Volk, O. H. and Geyger, E. (1970). 'Schaümböden als Ursache der Vegetations Losigkeit in arieden Gebieten', *Zeitschrift für Geomorphologië*, **14**, (1), 79–95.

Washburn, A. L. (1956). 'Classification of patterned ground and review of suggested origins', *Geol. Soc. Amer. Bull.*, **67**, 823–865.

Washburn, A. L. (1973). *Periglacial Processes and Environments*, Arnold, London.

Whittow, J. B. (1968). 'A note on present day crypedological phenomena in Northern New South Wales, Australia', *Biuletyn Peryglacjalny*, **17**, 305–310.

Williams, P. J. (1959). 'An investigation into processes occurring in solifluction', *Amer. Jour. Science*, **257**, 481–490.

Williams, P. J. (1966). 'Downslope soil movement at a sub-arctic location with regard to variation with depth', *Canadian Geotechnical Journal*, **3**, 191–203.

Williams, P. J. and Wood, J. A. (1982). 'Investigations of moisture movements and stresses in frozen soils', Final Report of the Geotechnical Sciences Laboratories, Carleton University, Ottawa, Contract serie no. OSV81–OO119.

Woo, M. K. and Heron, R. (1981). 'Occurrence of ice layer at the base of High Arctic snowpacks', *Arctic and Alpine Research*, **13**, 225–230.

Zoltai, S. and Tarnocai, C. (1981). 'Some non-sorted patterned ground types in Northern Canada', *Arctic and Alpine Research*, **13**, 139–151.

APPLICATIONS:
BRITISH ISLES

Soils and Quaternary Landscape Evolution
Edited by J. Boardman
© 1985 John Wiley & Sons Ltd.

7

Soils and Quaternary Stratigraphy in the United Kingdom

J. A. Catt

ABSTRACT

The routine mapping of surface soils and studies of their genesis, especially by petrographic methods (granulometry, mineralogy and micromorphology), have greatly expanded knowledge of Quaternary events in Britain. They have shown that thin deposits of Late Devensian loess are widespread in southern and eastern England, and that patches of older loess also occur in situations favouring its preservation. Plateau deposits over Chalk in southern England (the Clay-with-flints of geological maps) were found to be derived mainly from a thin veneer of basal Palaeogene sediment. This shows that the sub-Palaeogene surface is more widespread than was previously thought, and was probably exhumed mainly by the Middle Pleistocene. Neogene peneplanation and Plio-Pleistocene marine transgressions had little influence in shaping the present Chalk landscape. The recognition of thin, weathered outliers of Late Devensian till has shown that the ice margin reconstructed from supposed end moraines or marginal drainage channels is often inaccurate. In eastern coastal areas the Late Devensian glacial sequence was originally thought to include three lithologically distinct tills, but the uppermost of these (the Hessle) proved to be a Holocene weathering profile, in which oxidation of pyrite and siderite has caused slight reddening to 5 m depth and some erratics have been destroyed near the surface.

Laterally extensive buried soils, such as the Valley Farm Rubified *Sol Lessivé*, can clarify stratigraphic relationships between Quaternary deposits, but they are less common in the UK than in many other countries. Relict soils with evidence of pre-Devensian pedogenesis are much more common, and often contain micromorphological evidence of development during several Quaternary stages. Characteristic features resulting from soil development during individual interglacials cannot yet be distinguished, but intensity of rubification, weathering of certain minerals and some micromorphological features are useful for dating within broad limits.

161

INTRODUCTION

The study of soils and their history is an indispensible aspect of stratigraphy and palaeoenvironmental reconstruction, especially in the Quaternary. Many branches of Quaternary studies involve knowledge of processes on land surfaces, such as subaerial erosion, glacial, fluvial, aeolian or slope deposition, and development of vegetation types and of animal or human environments. Soils provide evidence for episodes of land-surface stability, when there was little or no erosion or deposition and when upper layers of the lithosphere were modified by physical, chemical and biological processes dependent upon proximity to the atmosphere. As evidence for sequences of events in terrestrial (as opposed to marine) environments, they are consequently just as important as deposits and erosion surfaces.

Soils formed during past periods have often been removed by erosion. However, some are buried, either so deeply that they are beyond the influence of later soil-forming processes, or by thin deposits which do not completely seal them from later pedological modification. Other soils of past periods may be buried and subsequently exhumed, or simply remain beneath a stable land surface, surviving to the present day but often being modified by intervening climatic and related environmental changes. The last type, often called relict soils, cannot easily be distinguished from exhumed soils, and according to Morrison (1978) have less stratigraphic value than buried soils. The development of deeply buried soils must have commenced after the youngest parent material was deposited and ceased before or during deposition of the oldest overlying sediment, so there is a finite interval during which the soil-forming processes occurred. In contrast, in relict soils at the surface there is no means of dating the end of processes not attributable to present-day pedogenesis, though as in buried soils pedogenesis must have begun sometime after deposition of the youngest parent material. Nevertheless, because they occur at the present land surface and are more accessible, they can be studied more easily and their distribution can be determined quite precisely; consequently they do have some advantages over their buried equivalents.

With a few exceptions, notably the Valley Farm Soil of Rose and Allen (1977), buried soils have contributed little to our present knowledge of the Quaternary of Britain. This contrasts with the position in USA and many other countries. Catt (1979) discussed possible reasons for this, and drew attention to information concerning Quaternary events that can be obtained from profile studies and mapping of the contemporary soil mantle. As the contributions of buried soils to British Quaternary stratigraphy are covered by other papers in this volume, evidence from them will be discussed only where it relates to the main point — that knowledge of the origin and distribution of surface soils is important evidence for Quaternary stratigraphy and landscape evolution.

The system of soil classification currently used by the Soil Survey of England

and Wales has been outlined by Avery (1980). The primary units (soil profiles) are distinguished according to characteristics described in the field (Hodgson, 1976) and evaluated in part by supporting laboratory analyses. Profiles are divided into horizons differentiated by colour, extent of mottling, textural class, amount and type of stones, organic matter and carbonate content, pH, the type and extent of development of soil structure (peds and voids), consistence, type and abundance of roots, coatings, concretions and concentrations of secondary minerals. For mapping purposes, profiles in which the composition of the soil materials is within prescribed limits, and which have the same diagnostic horizons, are grouped as soil series. Units shown on medium- and large-scale soil maps are identified by the chief soil series they contain.

Some horizon features result from soil development processes, such as incorporation of humus, decalcification, gleying, illuviation of clay, humus and sesquioxides, and development of structure by desiccation, root penetration and formation of clay–humus bonds. Other characteristics are inherited from the parent material or materials of the soil. In temperate regions mineral weathering during the Quaternary has usually been limited to dissolution of carbonates, loss or alteration of the small amounts of some heavy minerals in sand and silt fractions, and the loss of interlayer potassium from clay micas. The particle-size distribution and mineralogical composition of deposits (expressed on a carbonate-free basis and allowing for changes to the weatherable minerals) therefore can often be used to identify soil parent materials, even in horizons where other important lithological characteristics, such as colour, sedimentary structures, carbonate or fossil content, have been modified or lost by pedogenesis. This is usually done by comparing detailed particle-size distributions (e.g. at whole or half phi intervals), the amounts of resistant minerals in selected size fractions (e.g. clay, coarse silt or fine sand) and the lithological types of stones (if present) in acetic-acid treated samples of successive soil horizons with the same characteristics of fresh, unaltered deposits beneath or nearby.

The extent to which the mineralogy and particle-size distribution of parent materials have been modified in a soil profile can usually be judged by examination of thin sections prepared from resin-impregnated blocks of soil (Bascomb and Bullock, 1974, pp. 7–13). For example, these may show weathered mineral grains, bodies of clay illuviated from higher horizons (argillans), or accumulations of sesquioxides translocated from elsewhere in the profile. They also reveal the small-scale heterogeneity that arises, especially in upper parts of profiles, when two or more parent materials or soil horizons are mixed by soil faunal activity, cryoturbation, tree-fall or deep cultivation. Consequently, micromorphological studies have proved a useful accompaniment to bulk granulometric and mineralogical analyses in distinguishing the parent materials of British soils and the ways in which they have been modified by pedogenesis.

The combined petrographic studies of selected soil profiles often show that successive horizons are formed from parent materials of different composition, mode of formation and age. Some soils are derived from the Quaternary or bedrock formation that, according to geological maps, they overlie. However, many contain evidence for thin, previously unrecognized, superficial deposits, which are often partly mixed with underlying materials that may or may not bear the effects of intervening episodes of pedogenesis.

DISTRIBUTION OF LOESS IN ENGLAND AND WALES

The commonest deposits identified by petrographic studies of soil profiles in England and Wales are thin aeolian sediments such as coversand and loess. Although some loess deposits of limited lateral extent were identified by Trechmann (1920), Palmer and Cooke (1923) and Pitcher *et al.* (1954), and many of the superficial deposits shown on geological maps of south-east England as 'Brickearth', 'Head Brickearth', 'River Brickearth' or 'Loam' were sometimes interpreted as loess (e.g. by Dines *et al.*, 1954), the extensive occurrence of thin, weathered, loess deposits in various parts of England and Wales (Catt, 1978, 1984a) became known only through routine soil profile descriptions and mineralogical studies at selected sites.

Soils in thick (> 1 m) loess deposits are shown on Soil Survey maps as Hamble, Hook and Park Gate series, depending on drainage status and the resulting depth and intensity of greyish and ochreous mottling (Hodgson, 1967). They occur mainly in north Kent, south-east Essex, parts of the Thames and Lea Valleys and the West Sussex Coastal Plain. The upper horizons of these soils differ from the fresh, unaltered loess in being decalcified, in showing weak clay illuviation from A and E into underlying Bt horizons, and in the alteration or loss of a few weatherable fine sand and coarse silt minerals (mainly apatite, collophane, augite, biotite and glauconite). A, E and C horizons are typically silt loams with < 18 per cent clay (< 2 μm) and < 10 per cent sand (> 63 μm), whereas Bt horizons are silty clay loams with 20–25 per cent clay. The silt in all horizons is dominantly coarse (16–63 μm).

Many other soils in south and east England and in parts of south Wales, the Midlands, Pennines and north-west England, show an upward increase in coarse silt, which is mineralogically similar to the same size fraction from the Hamble, Hook and Park Gate soils. These occur on all landscape facets except the lowest parts of some river valleys occupied by Holocene alluvium, indicating that the silt can only have been deposited by the wind. They overlie almost all substrata in southern England, but in Wales and northern England are commonest on limestone surfaces, such as the Chalk of Lincolnshire and Humberside (Catt *et al.*, 1974) and the Carboniferous Limestone of the Pennines (Pigott, 1962; Bullock, 1971). In eastern parts of the country the loess-containing soils occur west of the feather edge of the weathered Devensian till (the Skipsea Till of Madgett and Catt, 1978), but thin head deposits composed of loess and frost-shattered chalk extend eastwards beneath it. The mineralogical similarity of the loess to the coarse silt fraction of the Skipsea Till suggests that it was derived from Devensian glacial outwash sediments, which would have accumulated in lower parts of the North Sea basin, though the loess must have been deposited before the ice reached its outermost limit soon after 18 250 BP (Penny *et al.*, 1969). Other indirect evidence for a Late Devensian age was summarized by Catt (1978), and Wintle (1981) obtained thermoluminescence dates on loess from various parts of southern England ranging from 14 500 to 18 800 BP.

The countrywide distribution of soils derived wholly or partly from loess is best shown at present by the 1:250 000 soil map of England and Wales (Soil Survey of England and Wales, 1983). The relevant map units are listed by Catt (1984a). Most of the silt corresponds granulometrically and mineralogically with the Late Devensian loess of eastern areas, though there is a progressive westward decrease in modal size and increase in amounts of flaky minerals because of winnowing by easterly periglacial winds. However, the Devensian loess of the Lizard (Coombe *et al.*, 1956) and other parts of Cornwall and the Scillies is coarser and mineralogically distinct, possibly because it was derived from the Irish Sea basin instead of the North Sea area (Catt and Staines, 1982).

In parts of eastern England some small loess deposits are evidently older than Devensian. At Barham, Suffolk (Rose and Allen, 1977) and Northfleet, Kent (Burchell, 1933), their stratigraphic relationships in buried situations indicate Anglian and Wolstonian ages respectively. On the Chilterns, loess which has been mixed with weathered material from Plateau Drift overlying the Chalk and washed into dolines penetrating the Plateau Drift, is distinguished from the Devensian loess cover by reddish colours and slightly different coarse silt mineral assemblages (Avery *et al.*, 1982). Acheulian artefacts in these deposits at Caddington (Smith, 1894; Sampson, 1978) and Gaddesden Row (Smith, 1916) show that the dolines were partly filled before the Devensian, and mineralogical comparisons suggest that the loess was originally deposited on surrounding land surfaces during both the Anglian and Wolstonian cold stages.

Although Britain is on the western margin of the Eurasian loess belt, and has much less impressive aeolian deposits even than Belgium and northern France on the opposite side of the English Channel, soil mapping and petrographic profile studies have nevertheless shown that loess is much more common in England and Wales than was previously suspected. The evidence from Chiltern dolines suggests that loess veneers similar to the present one were deposited in the Anglian and Wolstonian, but were almost entirely removed by subsequent surface washing, in much the same way as the Devensian loess has been reworked and locally removed by Late Devensian gelifluction and Holocene fluvial and slope erosion.

THE NATURE OF PLATEAU DRIFTS

The superficial deposits capping many interfluves and plateau surfaces south of the glacial limits in England are typically reddish and clayey. The most extensive are those on Chalk surfaces south of the Anglian glacial limit, which crosses the Chilterns from Luton to Welwyn; they are shown on modern geological maps as Clay-with-flints, with some patches of Pebbly Clay and Sand or Glacial Gravel with Bunter Pebbles on the Chilterns, and of Sand in Clay-with-flints on the North Downs. Similar deposits on the Upper Greensand plateau of east Devon are mapped as Clay-with-flints-and-cherts, and others on the Hythe Beds in west Kent and east Surrey as Angular Chert Drift.

Because these deposits all contain siliceous nodules, or broken fragments thereof, derived from subjacent limestones (flint from the Upper Chalk, chert from the Upper Greensand or Hythe Beds), they were originally interpreted as residual accumulations resulting from prolonged weathering and dissolution of those rocks. However, the Upper Chalk is an extremely pure limestone and the small amounts of acid-insoluble residue it contains have a much smaller clay/flints ratio than does Clay-with-flints (Jukes-Browne, 1906). Mineralogical studies of the clay and other fine fractions (<2 mm) in soils with horizons composed of Clay-with-flints on the South Downs (Hodgson *et al.*, 1967), North Downs (John, 1980) and Chilterns (Avery *et al.*, 1959, 1969, 1972; Loveday, 1962) have shown that Chalk residues form a very small proportion of the deposit; its fine sand, coarse silt and clay fractions resemble mineralogically those of the basal Palaeogene deposit, which is usually the Reading Beds. This suggests that most of the Clay-with-flints was formed by weathering of a thin veneer of basal Palaeogene sediment left on the Sub-Tertiary Surface (a multi-facetted marine erosion surface cut in the Chalk by repeated Palaeogene transgressions) when it was exhumed by subaerial erosion (Catt and Hodgson, 1976; Catt, 1984b). Other Plateau Drifts have not been studied in the same detail, but probably originated in similar ways.

The main soil with horizons of Clay-with-flints or Clay-with-flints-and-cherts is the Batcombe series (Robinson, 1948; Avery, 1964; Harrod, 1971), a fine silty over clayey stagnogleyic paleo-argillic brown earth (Avery, 1980). A typical profile consists of dark greyish brown surface (A) and yellowish brown subsurface (E) horizons, both of silt loam or silty clay loam texture, overlying brightly mottled (strong brown, yellowish red, red, pale brown and grey) paleo-argillic B horizons, which are clayey in texture but also contain moderate amounts of sand. The silt component of the A and E horizons is Devensian loess. The E horizon usually merges gradually into the paleo-argillic B, though this boundary is sometimes sharp and irregular, with the pale E horizon often forming downward-pointing tongues, which are probably Devensian frost structures. The flints are angular, frost-shattered fragments throughout the profile, though they are smaller and more abundant in A and E horizons than in the Bt. The red mottles and micromorphological evidence for clay illuviation often decrease downwards and Chalk is usually encountered at 2–5 m depth. Associated soils in Plateau Drifts are the Carstens series (fine silty over clayey typical paleo-argillic brown earth), Marlow series (fine loamy over clayey typical paleo-argillic brown earth), Berkhamsted series (coarse loamy over clayey stagnogleyic paleo-argillic brown earth) and Hornbeam series (fine loamy over clayey stagnogleyic paleo-argillic brown earth).

A yellowish red, reddish brown or black layer composed almost entirely of illuvial clay and unworn flint nodules often occurs at the base of the Plateau Drift above the irregular Chalk surface, lining solution cavities and covering residual Chalk pinnacles. This deposit, termed Clay-with-flints *sensu stricto* by Loveday (1962), contains very little sand and silt, the mineralogical composition of which shows that, like the unworn flint nodules, they are derived from the Chalk beneath. As the Chalk surface beneath the Plateau Drift was irregularly lowered by dissolution, many of the cavities so formed were filled with illuvial clay, which thus surrounded the flints and small amounts of other insoluble residue from the Chalk. On upper valley slopes and spurs the Clay-with-flints *sensu stricto* often emerges from beneath the Plateau Drift, to form a narrow zone of clayey paleo-argillic brown earths (Winchester series), which have extremely flinty and clay-rich, unmottled Bt horizons.

Berkhamsted and Hornbeam series occur on the Chiltern plateau in the Pebbly Clay and Sand, which is usually derived from Reading Beds sand and pebble beds. Soils on lower valley and scarp slopes locally contain horizons resembling Plateau Drift or Clay-with-flints *sensu stricto*, which are probably masses that were gelifluncted from the plateau during cold stages of the Quaternary. Horizons resembling Clay-with-flints *sensu stricto*, but often with shattered flints, also occur beneath loamy head or fluvial deposits on younger chalk surfaces representing successive stages of valley incision below the Sub-Tertiary Surface; the youngest of these horizons, probably associated with Devensian deposits, are thin and do not qualify as paleo-argillic (Chartres and Whalley, 1975).

Fossiliferous deposits of Miocene or early Pliocene (Diestian) and late Pliocene or early Pleistocene (Red Crag) age occur at a few high-level sites on the Chilterns and North Downs, namely Rothamsted Farm (Dines and Chatwin, 1930), Netley Heath (Chatwin, 1927) and Lenham (Preswich, 1858; Worssam, 1963). Wooldridge (1927) correlated many unfossiliferous sands and gravels with these, but apart from those on the North Downs grouped as the Headley Formation of John (1980) and the gravels at Little Heath, originally described by Gilbert (1920) and recently reinvestigated by Moffat and Catt (1983), most of the unfossiliferous deposits are probably weathered outliers of coarse Palaeogene sediments or remnants of old fluvial terrace gravels. Wooldridge and Linton (1939) suggested that the marine transgression (or transgressions) cut a widespread platform and cliff through the Palaeogene deposits into Chalk, thus bevelling the dipslope throughout south-east England. However, this concept conflicts with the distribution of Plateau Drift derived from basal Palaeogene beds (Catt and Hodgson, 1976), and the platform and cliff are often better explained as landscape features by exhumation of a folded Sub-Tertiary Surface (John, 1980; Moffat *et al.*, in press). The few genuine Plio-Pleistocene deposits either rest unconformably on thin Palaeogene beds that are scarcely weathered and disturbed, or form small disturbed patches within the Plateau Drift. This suggests that surrounding parts of the Sub-Tertiary Surface were exhumed, and their remnant veneer of Palaeogene beds converted to Plateau Drift, after the marine transgressions.

The late Cenozoic history of the southern chalk plateaux, inferred from the study of soils, is therefore:

1   During the Neogene, temperate subaerial denudation partly exhumed the Sub-Tertiary Surface, which had previously been deeply buried by Palaeogene sediments and gently folded by various Tertiary tectonic stresses (Jones, 1980).

2   A veneer of basal Palaeogene sediment (usually Reading Beds clay) was left on many parts of the exhumed surface, because stream erosion ceased when the Paleogene cover was thin enough to be permeable (Catt and Hodgson, 1976).

3   The veneer was then gradually converted to Plateau Drift by weathering, clay translocation, disturbance and incorporation of flints released by dissolution of the subjacent Chalk; the voids formed by dissolution of Chalk were often filled with mixed accumulations of insoluble Chalk residue and illuvial clay (= Clay-with-flints *sensu stricto*).

4   During the earliest Pleistocene (?Waltonian) in Hertfordshire and Surrey, the sea invaded a landscape largely of undisturbed Palaeogene deposits above what is now the Chalk dipslope; in places where marine erosion penetrated almost to the Sub-Tertiary Surface, leaving only a thin veneer of Palaeogene beds, the marine deposits were preserved by soon being incorporated into Plateau Drift.

5   The marine episode in east Kent occurred much earlier, but probably under similar circumstances, as the undoubted (i.e. fossiliferous) Lenham Beds occur only in deep solution pipes (Prestwich, 1858).

6   During later Quaternary temperate stages, subaerial denudation continued to exhume lower parts of the Chalk dipslope, Clay-with-flints *sensu stricto* continued to accumulate beneath the Plateau Drift, and upper layers of the Plateau Drift were rubified.

7   During cold Quaternary stages, Clay-with-flints *sensu stricto* was mixed with the Plateau Drift by deep-reaching cryoturbation, the flints were frost-shattered, and in the Devensian and at least two earlier cold stages thin layers of loess were deposited on (and partly mixed with) the Plateau Drift; the loess covers were subsequently washed or geliflucted into dolines (Avery *et al.*, 1982), which were probably formed at the margins of unaltered Palaeogene outliers that also have since been thinned and converted to Plateau Drift.

Some parts of the Sub-Tertiary Surface adjacent to undisturbed Palaeogene outliers or the main Palaeogene outcrop, for example in the Marlow and Reading areas and close to the Thanet Beds outcrop in Surrey and Kent (John, 1980), are occupied by soils with ordinary argillic horizons formed in remnants of the Bullhead Bed or Woolwich and Reading Bottom Bed (Yattendon series) or a thin veneer of Thanet Beds or other loamy basal Palaeogene sediment (Frilsham series). The relatively simple profiles on these parts suggests that they have been exhumed since the Ipswichian.

The Batcombe profiles studied by Avery *et al.* (1959, 1969, 1972) and Loveday (1962) have microfabrics approximately as complex (in terms of the evidence for multiple episodes of clay illuviation, reorganization by cryoturbation and shrink–swell stresses, rubification and hydromorphism) as those of relict soils thought to have been affected by two or more episodes of interglacial pedogenesis (e.g. Bullock and Murphy, 1979). The parts of the Sub-Tertiary Surface on which these profiles occur were therefore exhumed since the marine episode, probably during or just before the Middle Pleistocene. Parts of the Sub-Tertiary Surface exhumed before the Red Crag transgression have still to be identified. They will probably have Plateau Drift in which the flints are more intensely weathered (softened) and the subsoil (Bt) microfabric is at least locally more complex even than that of the Batcombe profiles already studied.

THE STRATIGRAPHY AND LIMITS OF LATE DEVENSIAN GLACIAL DEPOSITS

Wood and Rome (1868) originally proposed a three-fold subdivision of the tills exposed in Holderness, which persisted in little modified form until quite recently. Their lowest division, the Basement Clay, is now known to include tills of two separate glaciations, the Wolstonian Basement Till (Catt and Penny, 1966) and the Late Devensian Skipsea Till (Madgett and Catt, 1978). Their middle division, the Purple Clay, was renamed the Withernsea Till by Madgett and Catt (1978); it is exposed only in the coastal cliffs of south Holderness from Hornsea to Easington. Their uppermost division, named the Hessle Clay, was said to occur at the surface throughout the area, and to be the only division present where the glacial deposits are thin on the lower eastern slopes of the Yorkshire Wolds. As Catt and Penny (1966) noted, the main field characteristics Wood and Rome used to distinguish the Hessle from lower divisions (more earthy, less tenacious, foxy-red colour variegated by cinereous vertical partings, irregular shapes of many chalk stones) could result from soil development processes such as desiccation, gleying and decalcification. Despite this, all publications about the glacial deposits of Holderness and adjacent coastal areas of eastern England (including Catt and Penny, 1966) preserved the Hessle as a separate lithostratigraphic unit. This was mainly because its matrix colour (even at depths of 2–5 m), its erratics and heavy sand mineral suites all differ from those of lower tills.

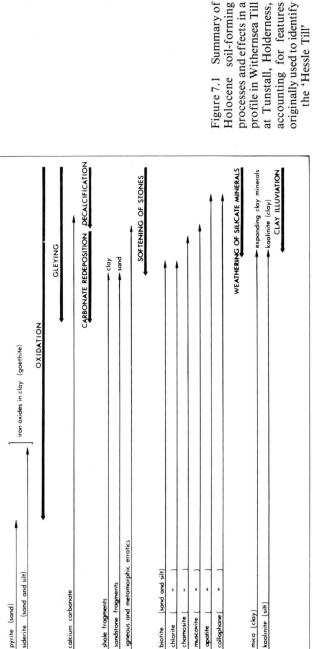

Figure 7.1 Summary of Holocene soil-forming processes and effects in a profile in Withernsea Till at Tunstall, Holderness, accounting for features originally used to identify the 'Hessle Till'

However, particle size and mineralogical analyses of scattered subsoil samples within the Late Devensian glacial limit in Holderness, eastern Lincolnshire and north Norfolk showed that, when allowances were made for complete or partial removal of some heavy minerals by weathering, the composition of Hessle Till in an arcuate area adjacent to the south Holderness coast is similar to the Withernsea (= Purple) Till, whereas elsewhere it resembles the Skipsea Till (Madgett, 1975). Detailed petrographic study of deep profiles in the two areas, at Tunstall on the Holderness coast and at Holkham in north Norfolk, showed that some weathering changes affect the till at 1–5 m below the ground surface (Madgett and Catt, 1978). At Tunstall the profile included 5 m of Hessle Clay overlying the Purple. The dark, reddish brown (5 YR 3/4–4/4), 'foxy-red' colour was attributed to formation of hydrated iron oxides by deep oxidation of the iron-containing heavy minerals pyrite, siderite and chamosite. Other weathering processes had removed further heavy minerals from upper (decalcified) parts of the profile and had softened or destroyed many of the larger erratic clasts, such as basic igneous and metamorphic rocks, sandstones, shales and limestones. Figure 7.1 summarizes these and other weathering changes at Tunstall.

The main conclusion of Madgett and Catt (1978) was that all horizons of the Tunstall profile, including those equivalent to Hessle Clay, could have been derived from Withernsea Till. The unusual thickness of the Holderness soil mantle, which must have formed over the last 13 000 years only and in climatic conditions no different from much shallower Holocene soils elsewhere in Britain, reflects the weatherability of the original till in terms of its mineralogical composition and tendency to develop deep vertical fissures on exposure to the atmosphere. The main soil in the Withernsea and Skipsea Tills is mapped as Holderness series, a fine loamy typical stagnogley soil; less strongly gleyed stagnogleyic argillic brown earths (Salwick series) occur in small, randomly distributed patches.

The effects of weathering in the profile at Holkham were similar to those at Tunstall, except that the till was only 1.25 m thick and oxidized throughout, so that evidence for weatherable iron-containing heavy minerals in the original till was absent. However, in other aspects of its composition (particle-size distribution, stone content and resistant sand and silt minerals), it resembled the Skipsea Till of Holderness, rather than any other till in eastern England.

The detailed study of two deep soil profiles, and the supplementary analyses of many subsoil samples taken throughout eastern Yorkshire, Lincolnshire and north Norfolk, and of unweathered tills from deeper exposures mainly on the Holderness coast, thus allowed a major revision of the Devensian glacial stratigraphy to be made (Madgett, 1975; Madgett and Catt, 1978). The sequence was decreased from three to two lithostratigraphic units by elimination of the Hessle (now a soil stratigraphic unit), and the approximate geographic extent of the upper unit (the Purple or Withernsea Till) inland from the south

Holderness coast was delimited. The lower (Drab or Skipsea Till) unit in Holderness was correlated with the single known Devensian till further south, which had previously been termed the Hessle in Lincolnshire and the Hunstanton Till in Norfolk. Straw (1969) divided the Lincolnshire Hessle into two (the Upper and Lower Marsh Tills), but on the basis of geomorphological evidence only; his units were morphostratigraphic and not lithostratigraphic. Finally, the relationship between successive Devensian tills in eastern England was shown to be offlapping and not overlapping, as proposed by Catt and Penny (1966).

Throughout eastern England and in many other parts of Britain the Late Devensian till thins progressively to a 'feather edge'. For example, on the dipslope of the Yorkshire Wolds, there is often a narrow zone in which the soil comprises surface and subsurface horizons of reddish brown clay loam with erratics over a subsoil usually < 30 cm thick composed of loess mixed with frost-shattered chalk and flint (Catt *et al.*, 1974). This zone occurs between the areas of soils in thick (> 1 m) Skipsea Till and those at slightly higher levels in thin flinty loess over Chalk (mainly Andover series). The till is easily recognizable in the field until it is about 40 cm thick, but thinner layers usually have been mixed with the underlying loessial horizon by cryoturbation or cultivation. Because the ice was here moving from north-east to south-west, effectively uphill, the highest remnants of till or other glacial sediment cannot have been subsequently geliflucted beyond the ice margin, and therefore indicate the minimum westward extent of the glacier. Other features normally accepted as indicating the ice margin, such as supposed end moraines or marginal drainage channels, often lie some distance to the east. For example, thin, weathered, Skipsea Till occurs at least 1 km west of the ridge at Kirmington (north Lincolnshire) that has been interpreted as part of the Devensian end moraine (Madgett and Catt, 1978, pp. 99 and 105). Similarly, the 1:250 000 soil map (Soil Survey of England and Wales, 1983) shows that soils derived from till extend up to 2 km south of the large moraine at Reighton on the northern side of Flamborough Head.

Both ridges and channels can originate subglacially or may be formed proglacially during minor readvances within the outermost ice limit, so soil evidence for a till margin, especially in an area where the ice was forced uphill, is a better indication of the extent of a glacier. However, the feather edge of till may also lie some way inside the original ice limit if, for example, there was erosion of the feather edge after the ice reached its maximum position. Exact delimitation of former glaciers in lowland areas, even ones as recent as Late Devensian, is therefore more problematical than is generally assumed. Other evidence from soil mapping and petrographic studies for the position of the main Devensian glacial limit in England and Wales is given by Crampton (1959, 1961), Hodgson and Palmer (1971), Hollis (1975, 1978) and Heaven (1978).

THE NATURE OF PEBBLY CLAY DRIFT AND EVIDENCE FOR
A PRE-ANGLIAN TILL IN HERTFORDSHIRE

Another example of confusion between lithostratigraphic and soil stratigraphic units is provided by the Pebbly Clay Drift of south-east Hertfordshire. Thomasson (1961) recognized this during a soil survey of part of the area, and suggested that it is a deeply weathered till older than the (Anglian) Chalky Boulder Clay, but younger than the early Pleistocene Pebble Gravel. Using a 3-foot (90-cm) soil auger, he distinguished profiles with characteristic red-mottled pebbly clay subsoil horizons (the Pebbly Clay Drift) from those in which chalky till occurs at various depths within 90 cm, and from others (usually podzols) in sandy and gravelly materials (the Pebble Gravel). In places these soils occurred in sequence on slopes from interfluve cappings of Chalky Boulder Clay down to the outcrop of the London Clay, which is the main bedrock formation of the area, and thus gave the impression of a layered tripartite Quaternary succession. Elsewhere the brightly mottled pebbly clay subsoil was mapped extensively on some of the broader interfluves, and temporary exposures showed that locally it is more than 3.3 m thick.

However, Sturdy *et al.* (1979) described thick, red-mottled paleo-argillic horizons derived from Chalky Boulder Clay in Essex, and Avery and Catt (1983) showed that brightly mottled soils in Northaw Great Wood, mapped as Pebbly Clay Drift by Thomasson (1961), are also derived from chalky till and often overlie it at depths > 90 cm. Moffat and Catt (1982) tried to show that the Pebbly Clay Drift is a distinct lithostratigraphic unit between the Chalky Boulder Clay and Pebble Gravel by drilling through the Chalky Boulder Clay at several plateau sites in south-east Hertfordshire. They found a thin layer resembling the Pebbly Clay Drift at only one site, but petrographic studies suggested it was a buried paleo-argillic horizon formed in the Pebble Gravel, and was probably equivalent to the Valley Farm Soil of Rose and Allen (1977).

The materials that Thomasson (1961) called Pebbly Clay Drift therefore seem to be paleo-argillic horizons formed during two different periods, one before and one after deposition of the Chalky Boulder Clay. Most of the Pebbly Clay Drift on interfluves is probably deeply weathered Chalky Boulder Clay, and many of the occurrences on valley sides could be transported masses of the same soil material, the recent removal of which has brought unweathered chalky till closer to the surface on parts of the plateau. This casts considerable doubt on the value of the Pebbly Clay Drift as evidence for a pre-Anglian glaciation of Hertfordshire.

DISCUSSION AND CONCLUSIONS

The three examples described show how mapping and profile genesis studies of soils on the present land surface have provided considerable evidence for the extent and relative age of Quaternary deposits, and for a history of erosion and other processes in extraglacial areas. In terms of lithostratigraphy, this evidence is quite as valuable, and often complementary to that obtained from buried soils. It also has considerable geomorphological significance, in that soils with relict interglacial features (red mottles, red or yellow argillans disrupted and incorporated into the soil matrix), periglacial structures (wedges or cryoturbations) or horizons containing loess of one or more ages, such as those on Chalk plateaux, provide evidence for considerable periods of land surface stability, with little erosion or deposition (Catt, 1983, 1984a). In contrast, the absence of these features over large areas usually implies widespread erosion during or after the episodes of interglacial pedogenesis, periglacial disturbance or loess deposition.

Although some soil features clearly indicate that parts of the land surface are older than others, many surfaces cannot yet be dated very precisely from soil characteristics. Paleo-argillic horizons recognized by the Soil Survey of England and Wales (Avery, 1980) are quite variable, but this is only partly because they represent different periods of pedogenesis in terms of length of time and environmental conditions. Differences in particle size and mineralogical composition of parent materials and in hydrologic conditions also contribute to their variability. Other types of soil horizons recognized by the Soil Survey are also quite variable for the same reasons. These are usually assumed to result from contemporary pedogenesis, but some of their features may be inherited from an early Holocene period when soil-forming conditions were slightly different from the present (Catt, 1979); others could even be relict from interstadials or interglacials, though there is no direct evidence for this. They may well have some potential for dating the surfaces beneath which they occur, but this remains to be investigated.

No factor or combination of factors is a universally reliable indicator of soil age, because too little is known at present about the range of pedogenetic effects in successive interglacials or other parts of the Quaternary. The age distinction between paleo-argillic (interglacial) and ordinary argillic (Holocene) horizons seems consistent in south and east England, but may not be quite so clear-cut in wetter western areas such as Wales, where Clayden (1977) reported reddish horizons with paleo-argillic features in Devensian gravels. Micromorphological differences between paleo-argillic horizons in some Plateau Drifts and those in somewhat younger Quaternary sediments suggest that some of these horizons have undergone longer periods of development than others, though precise dating is still uncertain. This is partly supported by the relative extents of sand and clay mineral weathering in different paleo-argillic brown earths, though the pattern is complicated by the original variability of the parent materials.

Better dating of paleo-argillic or other relict soils mapped by the Soil Survey of England and Wales will probably come from detailed comparisons with buried soils, which usually result from shorter or more easily dated episodes of soil formation. Comparisons are especially useful if a relict soil at the surface can be traced laterally into a buried soil, because the effects of pedogenesis before and after burial can then be distinguished, providing other soil-forming factors are the same. However, examples of this situation are not common; the most promising is the Valley Farm Soil, described elsewhere in this volume.

Although the value of buried soils in clarifying stratigraphic relationships has been recognized later in Britain than in the USA and many other countries, the contribution that genetic studies of unburied soils can make to reconstructions of Quaternary events is perhaps as fully appreciated in Britain as anywhere. What is now required for the benefit of both Quaternary geology and soil science is an effective integration of both approaches to palaeopedology.

REFERENCES

Avery, B. W. (1964). 'The soils and land use of the district around Aylesbury and Hemel Hempstead', *Mem. Soil Surv. Gt. Brit.*

Avery, B. W. (1980). 'Soil classification for England and Wales (Higher Categories)', *Soil Surv. Tech. Mono.* **14**.

Avery, B. W. and Catt, J. A. (1983). 'Northaw Great Wood', in *Quaternary Research Association Field Guide: Diversion of the Thames* (Ed. J. Rose), pp. 96–101, Birkbeck College, London.

Avery, B. W., Stephen, I., Brown, G. and Yaalon, D. H. (1959). 'The origin and development of brown earths on Clay-with-flints and coombe deposits', *J. Soil Sci.* **10**, 177–195.

Avery, B. W., Bullock, P., Weir, A. H., Catt, J. A., Ormerod, E. C. and Johnston, A. E. (1969). 'The soils of Broadbalk', *Roth. Exp. Stn Rept for 1968*, Part 2, 63–115.

Avery, B. W., Bullock, P., Catt, J. A., Newman, A. C. D., Rayner, J. H. and Weir, A. H. (1972). 'The soil of Barnfield', *Roth. Exp. Stn Rept for 1971*, Part 2, 5–37.

Avery, B. W., Bullock, P., Catt, J. A., Rayner, J. H. and Weir, A. H. (1982). 'Composition and origin of some brickearths on the Chiltern Hills, England', *Catena*, **9**, 153–174.

Bascomb, C. L. and Bullock, P. (1974). 'Sample preparation and stone content', in *Soil Survey Laboratory Methods* (Eds B. W. Avery and C. L. Bascomb) pp. 5–13, *Soil Surv. Tech. Mon.* **6**.

Bullock, P. (1971). 'The soils of the Malham Tarn area', *Field Stud.* **3**, 381–408.

Bullock, P. and Murphy, C. P. (1979). 'Evolution of a paleo-argillic brown earth (Paleudalf) from Oxfordshire, England', *Geoderma*, **22**, 225–252.

Burchell, J. P. T. (1933). 'The Northfleet 50-foot submergence later than the coombe rock of post-Early Mousterian times', *Archaeologia*, **83**, 67–92.

Catt, J. A. (1978). 'The contribution of loess to soils in lowland Britain', in *The Effect of Man on the Landscape: The Lowland Zone* (Eds S. Limbrey and J. G. Evans) pp. 12–20, *Council Brit. Archaeol. Res. Rept*, **21**.

Catt, J. A. (1979). 'Soils and Quaternary geology in Britain', *J. Soil Sci.* **30**, 607–642.

Catt, J. A. (1983). 'Cenozoic pedogenesis and landform development in south-east England', in *Residual Deposits: Surface Related Weathering Processes and Materials* (Ed. R. C. L. Wilson), pp. 251–258, Blackwell, Oxford.

Catt, J. A. (1984a). 'Soil particle size distribution and mineralogy as indicators of pedogenic and geomorphic history: examples from the loessial soils of England and Wales,' in *Geomorphology and Soils* (Eds K. S. Richards, R. Arnett and S. Ellis), Allen and Unwin (in press).

Catt, J. A. (1984b). 'The nature, origin and geomorphological significance of Clay-with-flints', in *The Scientific Study of Flint and Chert: Papers from the Fourth International Flint Symposium* (Eds G. de G. Sieveking and M. B. Hart), Cambridge University Press (in press).

Catt, J. A. and Hodgson, J. M. (1976). 'Soils and geomorphology of the chalk in south-east England', *Earth Surf. Processes*, **1**, 181–193.

Catt, J. A. and Penny, L. F. (1966). 'The Pleistocene deposits of Holderness, East Yorkshire', *Proc. Yorks. Geol. Soc.* **35**, 375–420.

Catt, J. A. and Staines, S. J. (1982). 'Loess in Cornwall', *Proc. Ussher Soc.* **5**, 368–375.

Catt, J. A., Weir, A. H. and Madgett, P. A. (1974). 'The loess of eastern Yorkshire and Lincolnshire', *Proc. Yorks. Geol. Soc.* **40**, 23–39.

Chartres, C. J. and Whalley, W. B. (1975). 'Evidence for Late Quaternary solution of Chalk at Basingstoke, Hampshire', *Proc. Geol. Assoc.* **86**, 365–372.

Chatwin, C. P. (1927). 'Fossils from the ironsands on Netley Heath', *Geol. Surv. Summ. Prog. for 1926*, 154–157.

Clayden, B. (1977). 'Paleosols', *Cambria*, **4**, 84–97.

Coombe, D. E., Frost, L. C., LeBas, M. and Watters, W. (1956). 'The nature and origin of the soils over the Cornish serpentine', *J. Ecol.* **44**, 605–615.

Crampton, C. B. (1959). 'Analysis of heavy minerals in certain drift soils of Yorkshire', *Proc. Yorks. Geol. Soc.* **32**, 69–82.

Crampton, C. B. (1961). 'An interpretation of the micro-mineralogy of certain Glamorgan soils: the influence of ice and wind', *J. Soil Sci.* **12**, 158–171.

Dines, H. G. and Chatwin, C. P. (1930). 'Pliocene sandstone from Rothamsted (Hertfordshire)', *Geol. Surv. Summ. Prog. for 1929*, 1–7.

Dines, H. G., Holmes, S. C. A. and Robbie, J. A. (1954). 'Geology of the country around Chatham', *Mem. Geol. Surv. Gt. Brit.*

Gilbert, C. J. (1920). 'On the occurrence of extensive deposits of high-level sands and gravels resting upon the Chalk at Little Heath near Berkhamsted', *Q. J. Geol. Soc. Lond.* **75**, 32–43.

Harrod, T. R. (1971). 'Soils in Devon I sheet ST10 (Honiton)', *Soil Surv. Record*, **9**.

Heaven, F. (1978). 'Soils in Lincolnshire III sheet TF 28 (Donington-on-Bain)', *Soil Surv. Record*, **55**.

Hodgson, J. M. (1967). 'Soils of the west Sussex coastal plain', *Soil Surv. Gt. Brit. Bull.* **3**.

Hodgson, J. M. (1976). *Soil Survey Field Handbook. Soil Surv. Techn. Mon.* **5**.

Hodgson, J. M., Catt, J. A. and Weir, A. H. (1967). 'The origin and development of Clay-with-flints and associated soils horizons on the South Downs', *J. Soil Sci.* **18**, 85–102.

Hodgson, J. M. and Palmer, R. C. (1971). 'Soils in Herefordshire I sheet SO53 (Hereford South)', *Soil Surv. Record*, **2**.

Hollis, J. M. (1975). 'Soils in Staffordshire I sheet SK05 (Onecote)', *Soil Surv. Record*, **29**.

Hollis, J. M. (1978). 'Soils in Salop I sheet SO79E/89W (Claverley)', *Soil Surv. Record*, **49**.

John, D. T. (1980). 'The soils and superficial deposits on the North Downs of Surrey', in *The Shaping of Southern England* (Ed. D. K. C. Jones), *Inst. Brit. Geogr. Spec. Publ.* **11**, 101–130.

Jones, D. K. C. (1980). 'The Tertiary evolution of south-east England, with particular reference to the Weald', in *The Shaping of Southern England* (Ed. D. K. C. Jones), *Inst. Brit. Geogr. Spec. Publ.* **11**, 13–47.

Jukes-Browne, A. J. (1906). 'The Clay-with-flints; its origin and distribution', *Q. J. Geol. Soc. Lond.* **62**, 132–164.

Soils and Quaternary stratigraphy in the United Kingdom 177

Loveday, J. (1962). 'Plateau deposits of the southern Chiltern Hills', *Proc. Geol. Assoc.* **73**, 83–102.

Madgett, P. A. (1975). 'Re-interpretation of Devensian till stratigraphy of eastern England', *Nature* **253**, 105–107.

Madgett, P. A. and Catt, J. A. (1978). 'Petrography, stratigraphy and weathering of late Pleistocene tills in East Yorkshire, Lincolnshire and north Norfolk', *Proc. Yorks. Geol. Soc.* **42**, 55–108.

Moffat, A. J. and Catt, J. A. (1982). 'The nature of the Pebbly Clay Drift at Epping Green, south-east Hertfordshire', *Trans. Herts. Nat. Hist. Soc.* **28**, 5, 16–24.

Moffat, A. J. and Catt, J. A. (1983). 'A new excavation in Plio-Pleistocene deposits at Little Heath', *Trans. Herts. Nat. Hist. Soc.* **29**, 1, 5–10.

Moffat, A. J., Catt, J. A., Webster, R. and Brown, E. H. (in press). 'A re-examination of the evidence for a Plio-Pleistocene marine transgression on the Chiltern Hills. III Structures and surfaces', *Earth Surf. Processes and Landf.*

Morrison, R. B. (1978). 'Quaternary soil stratigraphy—concepts, methods and problems', in *Quaternary soils* (Ed. W. C. Mahaney), pp. 77–108, GeoAbstracts, Norwich.

Palmer, L. S. and Cooke, J. H. (1923). 'The Pleistocene deposits of the Portsmouth district and their relation to man', *Proc. Geol. Assoc.* **34**, 253–282.

Penny, L. F., Coope, G. R. and Catt, J. A. (1969). 'Age and insect fauna of the Dimlington Silts, East Yorkshire', *Nature*, **224**, 65–67.

Pigott, C. D. (1962). 'Soil formation and development on the Carboniferous Limestone of Derbyshire. I. Parent materials', *J. Ecol.* **50**, 145–156.

Pitcher, W. S., Shearman, D. J. and Pugh, D. C. (1954). 'The loess of Pegwell Bay and its associated frost soils', *Geol. Mag.* **91**, 308–314.

Prestwich, J. (1858). 'On the age of some sands and iron-sandstones on the North Downs', *Q. J. Geol. Soc. Lond.* **14**, 322–335.

Robinson, K. L. (1948). 'The soils of Dorset', in *A Geographical Handbook of the Dorset Flora* (Ed. R. L. Good), pp. 14–28, Dorset Natural History and Archaeological Society, Dorchester.

Rose, J. and Allen, P. (1977). 'Middle Pleistocene stratigraphy in south-east Suffolk', *J. Geol. Soc.* **133**, 83–102.

Sampson, C. G. (Ed.) (1978). *Paleoecology and Archeology of an Acheulian site at Caddington, England*, Department of Anthropology, Southern Methodist University, Dallas.

Smith, W. G. (1894). *Man the Primeval Savage*, Stanford, London.

Smith, W. G. (1916). 'Notes on the Palaeolithic floor near Caddington', *Archaeologia*, **67**, 49–74.

Soil Survey of England and Wales (1983). *Soil Map of England and Wales Scale 1:250,000*. Soil Survey of England and Wales, Harpenden.

Straw, A. (1969). 'Pleistocene events in Lincolnshire: a survey and revised nomenclature', *Trans. Lincs. Nat. Union*, **18**, 85–98.

Sturdy, R. J., Allen, R. H., Bullock, P., Catt, J. A. and Greenfield, S. (1979). 'Paleosols developed on Chalky Boulder Clay in Essex', *J. Soil Sci.* **30**, 117–137.

Thomasson, A. J. (1961). 'Some aspects of the drift deposits and geomorphology of south-east Hertfordshire', *Proc. Geol. Assoc.* **72**, 287–302.

Trechmann, C. T. (1920). 'On a deposit of interglacial loess, and some transported preglacial freshwater clays on the Durham coast', *Q. J. Geol. Soc. Lond.* **75**, 173–201.

Wintle, A. G. (1981). 'Thermoluminescence dating of late Devensian loesses in southern England', *Nature*, **289**, 479–480.

Soils and Quaternary landscape evolution

Wood, S. V., and Rome, J. L. (1868). 'On the glacial and postglacial structure of Lincolnshire and south-east Yorkshire', *Q. J. Geol. Soc. Lond.* **24**, 146–184.

Wooldridge, S. W. (1927). 'The Pliocene history of the London basin', *Proc. Geol. Assoc.* **38**, 49–132.

Wooldridge, S. W. and Linton, D. L. (1939). 'Structure, surface and drainage in south-east England', *Trans. Inst. Brit. Geogr.* **10**.

Worssam, B. C. (1963). 'Geology of the country around Maidstone', *Mem. Geol. Surv. Gt. Brit.*

Soils and Quaternary Landscape Evolution
Edited by J. Boardman
© 1985 John Wiley & Sons Ltd.

8

The Valley Farm Soil in southern East Anglia

R. A. KEMP

ABSTRACT

The Valley Farm Soil has been recognized as an important stratigraphical unit extending over large parts of southern East Anglia. It clearly separates Anglian deposits from sediments attributed to earlier cold stages. Rubification and clay illuviation were the major pedogenic processes responsible for the development of the soil, with surface water gleying playing an important subsidiary role. The limited micromorphological evidence for weathering is not sufficient to account for the accumulation of large amounts of clay and iron oxides (hematite and goethite), which must have been derived from eluvial horizons formed in overbank sediments and/or loess-enriched materials. These eluvial horizons have been removed by erosion and the truncated Bt has been disturbed by cryogenic processes.

The Valley Farm Soil may be more complex than originally perceived with soils of more than one age represented within the unit. For the present, however, it is proposed that it should be considered as a single soil stratigraphic unit having a Cromerian age as a minimum, and a maximum age range equivalent to at least the Cromerian, Beestonian and Pastonian stages.

INTRODUCTION

In revising the Middle Pleistocene stratigraphy of southern East Anglia (Table 8.1), Rose et al. (1976) introduced for the first time in Britain the concept of soils as stratigraphic units. Recognition of the Valley Farm and Barham Soils was found to be important in the separation of sediments previously considered as equivalent in age and origin. This paper reviews and reappraises the areal extent, characteristics, genesis and stratigraphical significance of the Valley Farm Soil.

The Valley Farm Soil, named after its type site in south-east Suffolk (Rose and Allen, 1977), is developed in the Kesgrave Sands and Gravels, a fluvial sediment deposited by a north-eastwards flowing proto-Thames in a periglacial environment (Rose et al., 1976). The soil, which formed on several low-relief

179

Soils and Quaternary landscape evolution

Table 8.1   Middle Pleistocene stratigraphy and paleoenvironments in southern East
Anglia (Rose *et al.*, 1976)

Stage name	Formation name in south-east Suffolk	Environment	Sediments and soils in Essex and West Suffolk
Anglian	Lowestoft Till	Glacial	Till
	Barham Sands and Gravels	Glacifluvial	Sands and Gravels
	Barham Loess	Periglacial	Loess, coversand and head
	Barham Arctic Structure Soil		Arctic structure soil
Cromerian	Valley Farm Rubified Sol Lessivé	Humid, warm temperate	Rubified sol lessivé
Beestonian	Kesgrave Sands and Gravels	Periglacial	Sands and gravels

terrace surfaces of different altitudes, is normally truncated and disturbed by
involutions and wedges that often contain coversand. These features are
characteristic of the superimposed Anglian Barham Arctic Structure Soil (Rose
et al., 1976).

The Valley Farm and Barham Soils are buried beneath loess, coversand, head,
glacifluvial Barham Sands and Gravels or Lowestoft Till, all of which have been
assigned to the Anglian Stage (Rose *et al.*, 1976). Recognition of the buried
soils is important as they indicate a previous landsurface representing a break
in deposition between these sediments and the underlying Kesgrave Sands and
Gravels. Mineralogical and sedimentological analyses confirmed the distinction
between the two sand and gravel units (Rose *et al.*, 1976; Rose and Allen, 1977)
which had previously been considered in the Chelmsford and Harlow regions
as a single unit (Clayton, 1957; Bristow and Cox, 1973).

The Middle Pleistocene stratigraphy for southern East Anglia (Table 8.1)
proposed by Rose *et al.* (1976) places considerable emphasis on the Valley
Farm Soil as a stratigraphic unit. They assigned a Cromerian age to the
soil on the basis of its stratigraphic position and assumed temperate stage
status. This dating was justified at the time in view of the Anglian age of the
overlying sediments and the assumed Beestonian age of the Kesgrave Sands and
Gravels (Hey, 1980).

The Valley Farm Soil has subsequently been identified and used in research
studies (Hey, 1980; Allen, 1982, 1983; Kemp, 1983. Whiteman, 1983) and recent
mapping projects (Auton, 1982; Hobson, 1982; Thomas, 1982). However, fundamental objections to it have been voiced by Lake *et al.* (1977) and Wilson and
Lake (1983) who disputed not only its paleoenvironmental significance but also
its pedogenic origin. Baker and Jones (1980) proposed that it is an Anglian interstadial rather than pre-Anglian temperate stage soil, whilst Rose (1983) suggested
that it is more complex with the soil developing over a number of temperate

and cold stages on the upper terrace surfaces and only during the C₁ on the lowest levels. The relative merits of each of these proposal considered later in the paper.

CLASSIFICATION AND AREAL EXTENT OF THE VALLEY FARM SOIL

The Valley Farm Soil was described as a truncated rubified *sol lessivé modaux* by Rose *et al.* (1976). Only the illuvial (Bt) horizon survives beneath the Anglian sediments, the overlying eluvial (E) having been removed by erosion. According to the system used in England and Wales (Avery, 1980), this truncated and buried soil, modified by Anglian periglacial surface processes, would be classified as a stagnogleyic or typical paleo-argillic brown earth. The distinction between stagnogleyic and typical depends on the presence or absence of greyish mottles attributed to gleying.

The Valley Farm Soil has been reported beneath Anglian sediments at many sites in southern East Anglia extending from Burgate to Chelmsford and Kesgrave to Furneux Pelham (Figure 8.1). Moffat and Catt (1982) have also suggested that it is laterally equivalent to a buried soil formed in Pebble Gravel beneath the Anglian till, near Epping Green in south-east Hertfordshire. Direct correlation is not possible, however, because the possible differences in age of the Pebble Gravel and Kesgrave Sands and Gravels parent materials. At some sites, such as Stebbing in Essex, the Valley Farm Soil has been exhumed in places and has had post-Anglian pedological features superimposed upon it. In the south-east, beyond the limits of the Anglian ice margins, the Valley Farm Soil has been covered by only thin aeolian deposits during cold stages and never completely isolated from post-Anglian pedogenic processes by deep burial (Figure 8.1). Consequently, the soils of this area, mapped as stagnogleyic paleo-argillic brown earths by the Soil Survey of England and Wales (Hodge *et al.*, 1983), contain some relict features derived from the Valley Farm Soil.

More detail on the extent of the Valley Farm Soil has been recently provided by the staff of the Minerals Assessment Unit of the Institute of Geological Sciences, who have identified it in boreholes (Auton, 1982; Hobson, 1982; Thomas, 1982). Previously it was either included within the Barham Sands and Gravels (Marks and Merritt, 1981; Marks and Murray, 1981; Marks, 1982) or not recognized at all.

CHARACTERISTICS AND GENESIS OF THE VALLEY FARM SOIL

The Valley Farm Soil is recognized in the field by its red, reddish brown and reddish yellow colours, the evidence of clay illuviation and its lateral continuity across sedimentary structures (Rose *et al.*, 1976; Rose and Allen, 1977). Munsell colour hues vary from 10 R to 7.5 YR with 2.5 YR and 5 YR being the most common. This contrasts strongly with the brownish yellow, white or grey colours

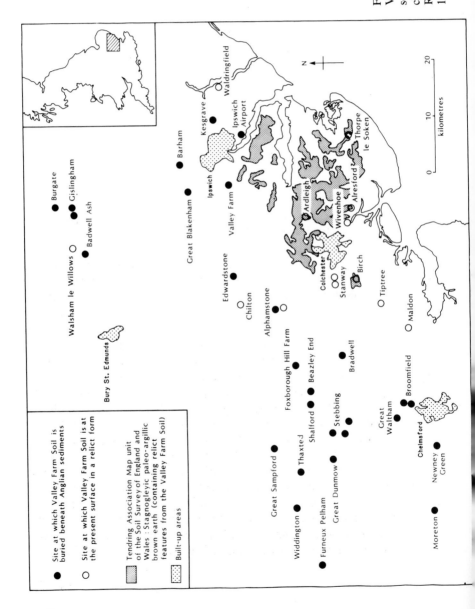

Figure 8.1 Extent of the Valley Farm Soil in southern East Anglia (including sites referred to in Rose *et al.*, 1976; Auton, 1982; Thomas, 1982; Hodge *et al.*, 1983)

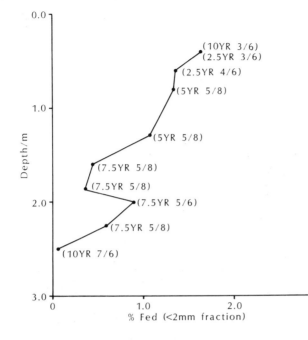

Figure 8.2 Dithionite iron content and Munsell colour variation with depth for the exhumed Valley Farm Soil at Stebbing, Essex

(7.5 YR, 10 YR or 2.5 Y hues) of the Kesgrave Sands and Gravels parent material with which the 0.4–3.0 m thick soil has a sharp to diffuse boundary.

These variations in colour reflect the dominating influence of iron oxides in the soil. Over 4 per cent dithionite-extractable iron (Fed) occurs in the <2 mm fraction of the redder parts of the soil, although values more typically lie between 0.3 and 1.5 per cent. These represent up to twenty times the amounts present in the white sand and gravel parent material. However, the direct relationship between colour and iron oxide content apparently indicated by some Fed depth functions (Figure 8.2) can be rather misleading. Large Fed values (2–4 per cent) have also been recorded from ferruginous lenses within the Kesgrave Sands and Gravels, though these discontinuous oxidation zones, which are frequently associated with sedimentary structures, are rarely as red as the Valley Farm Soil.

The colour of the soil is determined not so much by the quantity of iron oxides as by the form in which they occur. Sensitive differential x-ray diffraction techniques recently developed at Rothamsted Experimental Station have revealed that the iron oxides in the buried Valley Farm Soil at Great Blakenham and Stebbing are predominantly clay-size crystalline goethite and hematite (G. Brown and I. Wood, 1983, pers. comm.). The latter mineral, absent in the parent material, was formed by rubification which is a pedogenic process involving

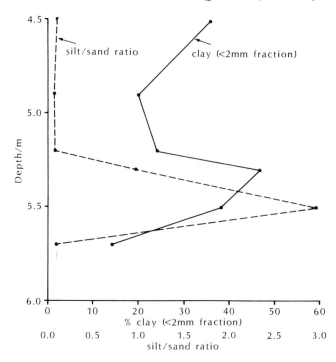

Figure 8.3   Gravimetric clay content and silt/sand ratio variation with depth for the
buried Valley Farm Soil at Stebbing, Essex

the internal dehydration of amorphous iron oxides and subsequent crystallization
to hematite (Fischer and Schwertmann, 1975; Duchaufour, 1982, p. 385).
Although the exact formative conditions have not been unequivocally
established, it appears that a long soil-forming interval with relatively
high temperatures and distinct dry periods is most conducive to the
process (Folk, 1976). Evidence from Germany, however, indicates that a
cooler climate may be suitable providing a warm pedoclimate, associated
with a coarse-textured parent material, is maintained (Schwertmann *et al.*,
1982). In Britain rubification has traditionally been considered as a soil
process mainly active during the temperate stages prior to the Devensian
(Avery, 1980).

The rubified Valley Farm Soil commonly has between 15 and 30 per cent more
clay than the underlying sandy parent material. Usually it has bimodal sand
and clay particle-size distribution with little silt (0–5 per cent). Enrichment of
clay in the soil relative to the parent material is interpreted as evidence for the
physical translocation of clay-size particles from overlying eluvial horizons, and
clay coatings on the upper surfaces of pebbles support this assertion (Rose and
Allen, 1977).

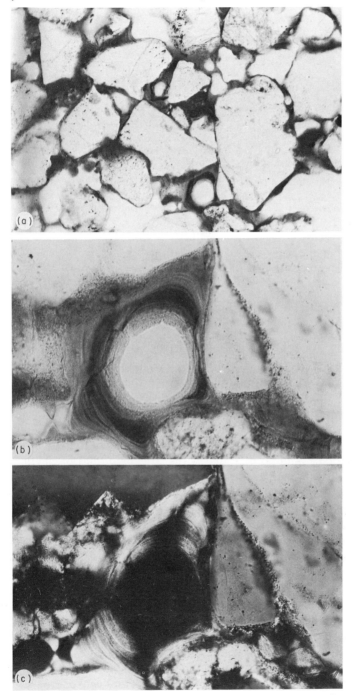

Figure 8.4   caption overleaf

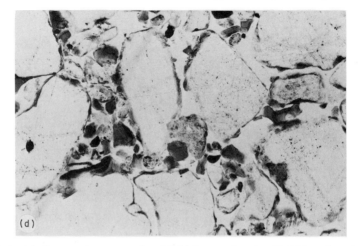

Figure 8.4   (a) Clay coatings around and bridging sand grains (Stebbing, Essex). Plane-polarized light. Frame length: 1.348 mm. (b) Well-oriented clay coating from Figure 8.4a at a larger scale. Plane-polarized light. Frame length: 340 μm. (c) Well-oriented clay coating from Figure 8.4a at a larger scale. Cross-polarized light. Frame length: 340 μm. (d) Loose, fragmented, clay coatings in intergranular voids (Beazley End, Essex). Plane-polarized light. Frame length: 1.348 mm

In some profiles the general trend of increasing clay content from the sandy parent material to the top of the truncated soil is not so distinct because of initial textural discontinuities within the Kesgrave Unit itself. A particularly striking example is shown in Figure 8.3 where the inflection in the curve of the clay content depth function is matched by an increase in the silt/sand ratio, both resulting from a poorly sorted silty clay lens in the parent material.

The periodic hold-up of water by the large quantities of clay in the Valley Farm Soil has resulted in surface-water gleying playing a significant secondary role in the genesis of the soil. Some of the medium to coarse grey mottles, however, may be more correctly associated with the impedence of water by frozen layers or ice lenses within the superimposed Barham Soil.

Micromorphology provides possibly the most conclusive evidence of the pedogenic origin of the Valley Farm Unit. In thin sections taken from horizons which have the typical sand/clay bimodal particle-size distribution, virtually all the clay is illuvial in origin. It is continuously oriented and occurs as either coatings to voids and grains (Figures 8.4a, b and c) or as fragmented (Figure 8.4d) and deformed coatings. In more poorly-sorted horizons the illuvial clay has similar characteristics, although its clear differentiation from stress-orientated, non-illuvial clay is often difficult.

Point counted or visually estimated areas of illuvial clay in thin sections from the Valley Farm Soil are commonly between 20–40 per cent (on a gravel- and

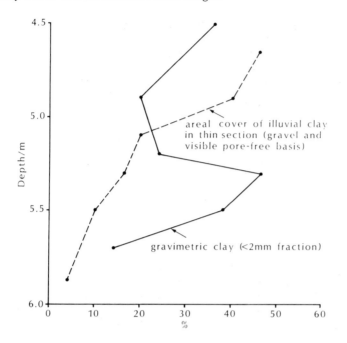

Figure 8.5   Variation in gravimetric clay content and illuvial clay cover (as derived from thin-section analyses) with depth for the buried Valley Farm Soil at Stebbing, Essex. At the top of the profile the illuvial clay values are greater than the gravimetric clay contents. This discrepancy reflects the differing bases upon which the respective amounts were calculated. The relationship between the two measurements is discussed in detail by Murphy and Kemp (1984)

visible-pore-free basis (Murphy and Kemp, 1984)), values far in excess of those normally obtained from Flandrian soils (< 10 per cent) in Britain. Illuvial clay depth functions confirm the original pedological interpretation of the Valley Farm Unit by Rose *et al.* (1976), and also allow the relative effects of parent material and pedogenesis on particle-size characteristics to be evaluated (Figure 8.5).

In mottled zones, iron oxides occur partly as nodules and coatings or as segregations and the contrasted grey-red mottles (Figure 8.6b) imply that the movement and redistribution of iron oxides occurred separately or after the translocation of the clay. In contrast, the illuvial clay in the non-mottled zones is often colour laminated (Figure 8.6c), its reddish-brown, orange or yellow-brown pigments indicating the linkage of iron oxides to clay minerals prior to their translocation. This observation confirms the close association between rubification and clay illuviation, the two processes responsible for the major morphological features of the soil. Thin sections from soil horizons developed

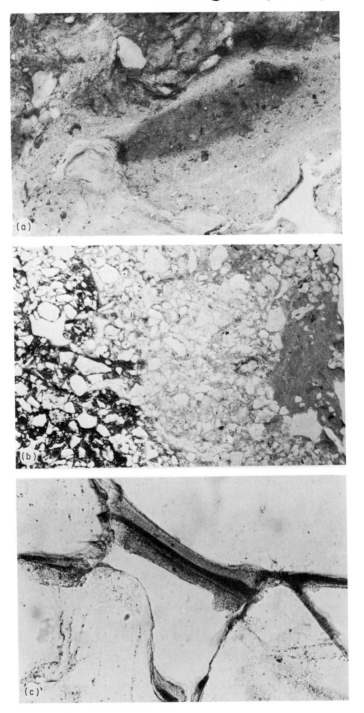

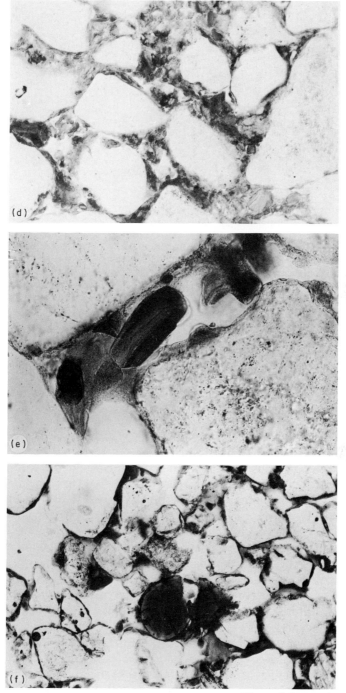

Figure 8.6 caption overleaf

Figure 8.6   (a) Sesquioxide segregation superimposed across a clay infilling (Stebbing, Essex). Plane-polarized light. Frame length: 1.348 mm. (b) Prominent colour contrast across mottle boundary (Stebbing, Essex). Plane-polarized light. Frame length: 5.225 mm. (c) Colour microlaminations within a bridge clay coating (Ipswich Airport, Suffolk). Plane-polarized light. Frame length: 340 μm. (d) Fragmented clay coatings embedded within deformed clay coatings (Great Blakenham, Suffolk). Plane-polarized light. Frame length: 1.348 mm. (e) Fragmented clay coatings surrounded by a later clay coating (Stebbing, Essex). Plane-polarized light. Frame length: 340 μm. (f) A weathered glauconite grain (Great Blakenham, Suffolk). Plane-polarized light. Frame length: 1.348 mm. (g) A weathered feldspar grain (Stebbing, Essex). Plane-polarized light. Frame length: 1.348 mm

in the relatively rare fine-textured lenses of the Kesgrave Unit, however, also contain reddened non-illuvial clay, presumably indicating that this material has been rubified *in situ*.

The complex associations of textural pedofeatures (Bullock *et al.*, 1984) in some horizons of the Valley Farm Soil are difficult to interpret in terms of soil-forming environments or number and order of processes. The rubification and considerable clay illuviation must reflect the stable environment of a temperate stage, during which conditions were at least as warm as during the Flandrian in this part of the country. Except at the base of the profiles, most of the illuvial clay is disturbed, occurring as fragmented or deformed coatings (Figure 8.4d). The assumption that this disruption of coatings has been caused by cryoturbation (Bullock and Murphy, 1979) during a cold stage is justifiable in view of the scale of the disturbance and the evidence of associated cold-climate features such as involutions, ice-wedge casts, sand wedges and banded fabrics (Rose *et al.*, 1985). These features are characteristic of the superimposed Barham Soil and are usually related to the single Anglian cold stage. The frequent occurrence of fragmented coatings within or surrounded by later deformed coatings in thin sections from some profiles (Figures 8.6d and 8.6e), however, suggests a more complex sequence of processes. In some profiles it may indicate a much longer pre-Anglian soil-forming interval than originally perceived, perhaps covering two or more temperate stages, separated by cold stage(s).

There is considerable micromorphological evidence of mineral weathering in the Valley Farm Soil. Surface pitting of stable minerals such as tourmaline could be inherited from earlier weathering cycles, but the alteration of glauconite (Figure 8.6f) and feldspar (Figure 8.6g) must have occurred *in situ*, as mineral grains weathered to such an extent would not have survived fluvial transport. However, neither heavy sand mineral nor clay mineral analyses reveal any vertical weathering trend within the Valley Farm Soil. The Kesgrave Sands and Gravels have a heavy mineral suite deficient in easily weatherable components (Kemp, 1983), and the clay minerals in both the soil and the parent material are predominantly interstratified smectites within subsidiary amounts of mica and kaolinite.

The *in situ* weathering of feldspar and glauconite is unlikely to have provided sufficient iron oxides and clay to account for the large amounts of both components present in the soil. Their origin is difficult to establish in view of the absence of eluvial horizons from which they would have been translocated. The most plausible explanation is that the clay and iron oxides were derived from eluvial horizons developed in fine-textured overbank sediments or sands and gravels mixed with loess or coversand, which have since been removed by erosion. The introduction of more weatherable minerals from aeolian sources would provide the opportunity for neoformation of clay minerals and crystalline iron oxides which were subsequently translocated down the profile. Much of the clay and iron oxides eluviated from horizons formed in overbank sediments, however, could presumably be primary minerals, unaltered except for minor interspecies transformations and crystallization of particular iron oxides.

At Broomfield in Essex and other sites in the region (Rose *et al.*, 1985), the soil buried beneath Anglian till is much greyer (10 YR 4/1) in its upper 60 cm than other exposures of the Valley Farm Soil. It may represent either a poorly drained component of a Valley Farm paleocatena or an Anglian humic gley soil superimposed upon the rubified illuvial horizon (Rose *et al.*, 1978).

REAPPRAISAL OF THE VALLEY FARM SOIL CHRONOLOGY

The Valley Farm Soil was originally assigned to the Cromerian on the basis of its stratigraphical position and assumed temperate stage status (Rose *et al.*, 1976; Rose and Allen, 1977). The temperate environment is consistent with the extent of rubification and clay illuviation, both of which exceed what has been recorded for soils formed in similar materials in this country during the Flandrian or any interstadial. This discounts the possible interstadial age for the soil suggested by Baker and Jones (1980). The soil must have formed prior to the

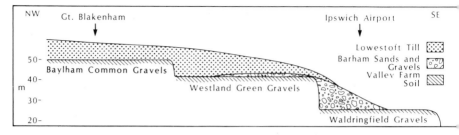

Figure 8.7   Schematic cross section through the Quaternary sediments in South-east
Suffolk showing location of Great Blakenham and Ipswich Airport sites

Anglian, as it is buried by sediments deposited during that cold stage,
yet it must post-date deposition of its parent material. The Kesgrave Sands
and Gravels were initially considered by Rose *et al.* (1976) to be the time
equivalent of the Beestonian gravels on the Norfolk coast (Hey, 1980).
Altitudinal and lithological data collected over a large region have confirmed
their suggestion that the unit is composite in origin with surface height
ranges indicating a series of river terrace levels (Rose and Allen, 1977;
Hey, 1980; Allen, 1982, 1983; Green *et al.*, 1982; Rose, 1983). Following
the establishment of the Bramertonian and Pre-Pastonian stages within the
British Quaternary stratigraphy (Funnell *et al.*, 1979), the high-level Westland
Green Member of the Kesgrave Formation has been tentatively reassigned
to the earlier Pre-Pastonian (Hey, 1980). The Valley Farm Soil on these
upper terrace surfaces could therefore have developed over at least three
stages, two temperate (Pastonian and Cromerian) and one cold (Beestonian),
prior to burial by Anglian sediments. This could account for the
micromorphological evidence of disturbance between periods of clay and iron-
oxide illuviation (Kemp, 1983; Rose, 1983). As the low-level Kesgrave Sands
and Gravels are considered to be Beestonian in age (Hey, 1980), the soil formed
on this terrace surface should be less developed than the one on the upper terrace,
reflecting the effects of pedogenesis during only one temperate (Cromerian) stage
(Rose, 1983).

   This hypothesis is being investigated by the author in south-east Suffolk,
where Allen (1983) has separated the Kesgrave Formation into three
constituent members. The upper Baylham Common and Westland Green
Gravels, separated mainly on altitudinal differences, are both lithologically
distinct from the lowest Waldringfield Member. Preliminary data has been
obtained from buried soils developed on the highest and lowest terrace levels
at Great Blakenham and Ipswich Airport respectively (Figures 8.1 and 8.7).
Undoubtedly an age factor could account for the appreciable differences
between these two soils, notably the redder colours and greater degree of
mineral grain weathering observed at Great Blakenham. However, more

Table 8.2   The stratigraphic position of the Valley Farm Soil

Stage	Formation name	Soils	Environments
Anglian	Lowestoft Till		Glacial
	Barham Sands and Gravels		Glacifluvial
	Barham Loess	Barham Soil	Periglacial
Cromerian		↑	Temperate
Beestonian	Kesgrave Sands and Gravels (Lower levels; Hey, 1980)	VALLEY FARM SOIL ↓	Periglacial
Pastonian		(?)	Temperate
Pre-Pastonian	Kesgrave Sands and Gravels (Upper levels; Hey, 1980)		Periglacial
Bramertonian			Temperate

detailed work on these and other buried soils is required before any differences can be considered significant enough to allow the separation of the Valley Farm Soil into more than one stratigraphic unit. In particular, studies should be made of contrasting buried soils developed not only on these terraces in other parts of Suffolk, but also on equivalent surfaces further west in Essex.

A soil stratigraphic unit is normally considered older than the youngest sediment burying it and younger than the youngest parent material in which it has formed. However, the possibility that the Valley Farm Soil is a complex soil stratigraphic unit, which may in the future be separated into two or more units, necessitates the use of a more flexible approach to its dating. Consequently, accepting the respective ages assigned to the upper and lower levels of the Kesgrave Formation by Hey (1980), it is proposed that the Valley Farm Soil should for the present be considered as a single soil stratigraphic unit having a minimum Cromerian age and a maximum age range equivalent to the Pastonian, Beestonian and Cromerian stages (Table 8.2). Future revision of the regional stratigraphy involving the recognition of additional cold or temperate stages may alter this chronology.

CONCLUSIONS

The suggestion by Lake *et al.* (1977) and Wilson and Lake (1983) that the Valley Farm Soil is an Anglian flow till is not consistent with the field and laboratory data which undoubtedly indicate a pedogenic origin. The soil developed on terrace surfaces in sands and gravels deposited by a proto-Thames. It extends over large parts of the region, truncated and modified by Anglian cold-climate processes, and is either buried beneath Anglian sediments or is at the present surface where post-Anglian pedological features are superimposed on it. Where buried, the soil fulfills an important stratigraphic function marking a distinctive depositional and environmental break between the Kesgrave Sands and Gravels and the overlying Anglian sediments.

The red, reddish brown and reddish yellow colours associated with measurable hematite concentrations, and the high illuvial clay content reflect the two major processes active in the development of the soil; rubification and clay translocation. Thin sections from some profiles contain fragmented clay coatings embedded within later coatings, complex pedofeatures indicating phases of disruption have occurred during the soil development. This disturbance may be associated with cryoturbation dating from a pre-Anglian cold stage. Surface-water gleying was a subsidiary pedogenic process largely dependent on the illuvial clay accumulation. Evidence of *in situ* weathering is limited to micro-morphological indications of feldspar and glauconite disintegration.

It is difficult to deduce the origin of the clay and iron oxide components within the Valley Farm Soil, as any overlying eluvial horizons have been removed by erosion. These eluvial horizons probably developed in overbank sediments and/or loess- and coversand-enriched materials containing either the components themselves or minerals that would provide them by weathering.

The recognition that the Kesgrave Sands and Gravels include sediments of more than one age has led to reconsideration of the Valley Farm Soil and stimulated further work into the comparison of buried profiles developed on the various terrace levels. Differences between the buried soils on the highest and lowest terrace levels in Suffolk may reflect a longer soil-forming interval on the upper surface. However, further work is required to confirm the significance of these preliminary results before the status of the Valley Farm Soil can be revised. Thus, the Valley Farm Soil should, for the present, still be considered as a single (albeit possibly complex) soil stratigraphic unit having a Cromerian age as a minimum, and a maximum age range equivalent to at least the Cromerian, Beestonian and Pastonian stages.

ACKNOWLEDGEMENTS

The author thanks Dr P. Bullock, Dr J. A. Catt and Mr J. Rose for their advice and encouragement during this study and acknowledges the support given by a NERC Case Studentship. Thanks are also given to Miss Loraine Rutt for drawing the diagrams.

REFERENCES

Allen, P. (1982). *Quaternary Research Association Field Meeting Guide, Suffolk, May 7–9, 1982.*
Allen, P. (1983). *Middle Pleistocene Stratigraphy and Landform Development of South-east Suffolk,* Unpublished PhD Thesis, University of London.

Auton, C. A. (1982). 'The sand and gravel resources of the country around Redgrave, Suffolk. Description of 1:25 000 sheet TM07 and part of TM08', *Miner. Assess. Rept. Inst. Geol. Sci.* **117**.

Avery, B. W. (1980). 'Soil Classification for England and Wales (higher categories)', *Soil Survey Technical Monograph*, **14**, Harpenden.

Baker, C. A. and Jones, D. K. C. (1980). 'Glaciation of the London Basin and its influence on the drainage pattern: a review and appraisal', in *The Shaping of Southern England* (Ed. D. K. C. Jones), pp. 131–175, Academic Press, London.

Bristow, C. R. and Cox, F. C. (1973). 'The Gipping Till: a reappraisal of East Anglian glacial stratigraphy', *J. Geol. Soc. Lond.* **129**, 1–37.

Bullock, P. Federoff, N., Jongerius, A., Stoops, G. and Tursina, T. (1984). *Handbook for Soil Thin Section Description*, Wayne Research Ltd, Wolverhampton.

Bullock, P. and Murphy, C. P. (1979). 'Evolution of a paleoargillic brown earth (Paleudalf) from Oxfordshire, England', *Geoderma*, **22**, 225–253.

Clayton, K. M. (1957). 'Some aspects of the glacial deposits of Essex', *Proc. Geol. Ass.* **68**, 1–21.

Duchaufour, P. (1982). *Pedology* (translation by T. R. Paton), George Allen and Unwin, London.

Fischer, W. R. and Schwertmann, U. (1975). 'The formation of hematite from amorphous iron (III) hydroxide', *Clay. Clay Min.* **23**, 33–37.

Folk, R. L. (1976). 'Reddening of desert sands: Simpson Desert, N. T., Australia', *J. Sed. Pet.* **46**, 604–615.

Funnell, B. M., Norton, P. E. P. and West, R. G. (1979). 'The Crag at Bramerton, near Norfolk', *Phil. Trans. Roy. Soc.* **287B**, 490–534.

Green, C. P., McGregor, D. F. M. and Evans, A. H. (1982). 'Development of the Thames drainage system in Early and Middle Pleistocene times', *Geol. Mag.* **119**, 281–290.

Hey, R. W. (1980). 'Equivalents of the Westland Green Gravels in Essex and East Anglia', *Proc. Geol. Ass.* **91**, 279–290.

Hobson, P. M. (1982). 'The sand and gravel resources of the country around Sudbury, Suffolk: description of 1:25 000 resource sheet TL84', *Miner. Assess. Rep. Inst. Geol. Sci.* **118**.

Hodge, C. A. H., Burton, R. G. O., Corbett, W. M., Evans, R., George, H., Heaven, F. W., Robson, J. D. and Seale, R. S. (1983). *Soils of England and Wales. Sheet 4. Soils of Eastern England*, Ordnance Survey, Southampton.

Kemp, R. A. (1983). 'Stebbing: the Valley Farm Palaeosols layer', in *Quaternary Research Association Field Guide: Diversion of the Thames* (Ed. J. Rose), pp. 154–158, Birkbeck College, London.

Lake, R. D., Ellison, R. A. and Moorlock, B. S. P. (1977). 'Middle Pleistocene stratigraphy in southern East Anglia', *Nature*, **265**, 663.

Marks, R. J. (1982). 'The sand and gravel resources of the country around Clare, Suffolk: description of 1:25 000 resource sheet TL74', *Miner. Assess. Rep. Inst. Geol. Sci.* **97**.

Marks, R. J. and Merrit, J. W. (1981). 'The sand and gravel resources of the country north-east of Halstead, Essex: description of 1:25 000 sheet TL83', *Miner. Assess. Rep. Inst. Geol. Sci.* **68**.

Marks, R. J. and Murray, D. W. (1981). 'The sand and gravel resources of the country around Sible Hedingham, Essex: description of 1:25 000 resource sheet TL73', *Miner. Assess. Rep. Inst. Geol. Sci.* **82**.

Moffat, A. J. and Catt, J. A. (1982). 'The nature of the Pebbly Clay Drift at Epping Green, south-east Hertfordshire', *Herts. Nat. Hist. Soc.* **28**, 16–24.

Murphy, C. P. and Kemp, R. A. (1984). 'Overestimation of clay and underestimation of pores in soil thin sections', *J. Soil Sci.* **35**, 481–495.

Rose, J. (1983). 'Early Middle Pleistocene sediments and palaeosols in west and central Essex', in *Quaternary Research Association Field Guide: Diversion of the Thames* (Ed. J. Rose), pp. 135–139, Birkbeck College, London.

Rose, J. and Allen, P. (1977). 'Middle Pleistocene stratigraphy in south-east Suffolk', *J. Geol. Soc. Lond.* **133**, 83–102.

Rose, J., Allen, P. and Hey, R. W. (1976). 'Middle Pleistocene stratigraphy in southern East Anglia', *Nature*, **263**, 492–494.

Rose, J., Kemp, R. A., Whiteman, C. A., Allen, P. and Owen, N. (1985). 'The early Anglian Barham Soil of Eastern England', in *Soils and Quaternary Landscape Evolution* (Ed. J. Boardman), Wiley, Chichester (in press).

Rose, J., Sturdy, R. G., Allen, P. and Whiteman, C. A. (1978). 'Middle Pleistocene sediments and palaeosols near Chelmsford, Essex', *Proc. Geol. Ass.* **89**, 91–96.

Schwertmann, U., Murad, E. and Schulze, D. G. (1982). 'Is there Holocene reddening (hematite formation) in soils of axeric temperate areas?', *Geoderma*, **27**, 209–223.

Soil Survey of England and Wales (1983). *'Soils of south-east England, 1:250 000 soil map'*, Harpenden.

Thomas, C. W. (1982). 'The sand and gravel resources of the country around Great Dunmow, Essex: description of the 1:25 000 resource sheet TL62', *Miner. Assess. Rept. Inst. Geol. Sci.* **109**.

Whiteman, C. A. (1983). 'Great Waltham', in *Quaternary Research Association Field Guide: Diversion of the Thames* (Ed. J. Rose), pp. 163–169, Birkbeck College, London.

Wilson, D. and Lake, R. D. (1983). 'Field Meeting to north Essex and west Suffolk, 20–22 June 1980', *Proc. Geol. Ass.* **94**, 75–79.

Soils and Quaternary Landscape Evolution
Edited by J. Boardman
© 1985 John Wiley & Sons Ltd.

9

The Early Anglian Barham Soil of Eastern England

J. ROSE, P. ALLEN, R. A. KEMP, C. A. WHITEMAN AND N. OWEN

ABSTRACT

The Barham Soil is a soil stratigraphic unit formed in a periglacial environment in eastern England during the early Anglian Stage of the Middle Pleistocene. It is developed in soil material and freshwater sediments of Cromerian age and buried beneath glacial and glacifluvial deposits of the Anglian Glaciation. Typical large-scale soil structures include involutions, ice-wedge casts, frost cracks and sand wedges. Smaller-scale soil features include platy aggregates, banded fabrics, silt cappings, disrupted clay skins and fractured gravel clasts. Occasionally, organic material in the soil causes dark grey colour horizonation. Throughout the region the soil is associated with glacially-derived wind-blown sediments in the form of sand-wedge infills, wind-polished flint pebbles, and beds of coversand and loess. The soil characteristics vary with parent material and drainage, with involutions best developed in the moisture-retentive argillic horizon of the Cromerian soil, and sand wedges and colour horizonation generally associated with better-drained localities. The relationships of soil properties one with each other, with the wind-blown sediments, and with the vegetational evidence from Norfolk and north Suffolk reveals sequences of soil development and permits the reconstruction of the environmental history of the region. The Barham Soil first developed under birch woodland during Zone IVc of the Cromerian Interglacial. During the pre-glacial part of the Anglian Stage the landscape was characterized by disrupted grass and herb tundra vegetation, sorted circles and non-sorted polygonal patterns, with patches of polar desert and accumulations of wind-blown sands and silt. Many features of the Barham Soil developed when the ice sheets had reached the vicinity of eastern England, and sand wedges were the last structure to form before the land surface was overridden by Anglian ice or buried by till or outwash.

The Barham Soil as described here is recognized because of its buried position. Beyond the limits of Anglian glacigenic deposits it may form a component of relict palaeosols. However, in these localities recognition is difficult because of the effects of post-Anglian pedogenesis.

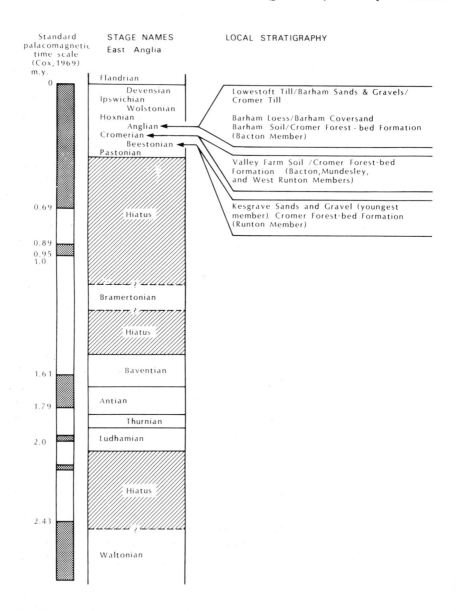

Figure 9.1   Quaternary stratigraphy of the British Isles and the local stratigraphy associated with the Barham Soil. The British stratigraphic column is based on Zagwijn (1975) with the Bramertonian added from Funnell *et al.* (1979). The positioning of the Bramertonian within the hiatus is largely speculative being correlated tentatively with the main climatic amelioration of the early Waalian. It is probable that this scheme will change in the near future

INTRODUCTION

The Barham Arctic Structure Soil was defined formally as a buried palaeosol formed by periglacial processes during the early part of the Anglian Glacial Stage of the Middle Pleistocene (Rose and Allen, 1977) (Figure 9.1). It was identified on the basis of periglacial structures, such as involutions, ice-wedge casts and sand wedges, preserved in fragments of a former land surface across parts of Essex and Suffolk in eastern England (Rose *et al.*, 1976, 1977, 1978a and b; Allen, 1982, 1983; Whiteman, 1983). Throughout the region it is associated with wind-blown sand and silt defined stratigraphically as the Barham Coversand (Allen, 1983) and the Barham Loess (Rose and Allen, 1977). The term 'arctic structure soil', introduced by Mückenhausen (1962), was used initially because evidence for a major episode of subaerial activity was reflected in the form of physical structures without noticeable soil horizonation. The environment deduced from this evidence was considered to be that of polar desert with permafrost and active-layer processes.

Since the initial study, further data relating to the Barham Arctic Structure Soil and the associated wind-blown sediments have been collected from field observations, laboratory analyses and other published work. The aims of this paper are threefold. First, to review the geographical distribution of the soil and associated sediments, their stratigraphical positions and their significance within the Quaternary history of eastern England. Secondly, the paper gives detailed descriptions of the properties of the Barham Soil and associated wind-blown sediments, and discusses their processes of formation and palaeoenvironmental significance. Finally, the Barham Soil is discussed in terms of its status as a soil stratigraphic unit and its relevance to the understanding of Quaternary soils and stratigraphy in adjacent areas.

TERMINOLOGY

The Barham Arctic Structure Soil fulfils the requirements of a soil stratigraphic unit (American Commission on Stratigraphic Nomenclature, 1961) in that it consists of a series of soil profiles that have a similar stratigraphic relationship, and it can be traced regionally on the basis of its pedological properties, although these may vary from place to place according to local site conditions. Since the original definition of the arctic structure soil was published, stratigraphic evidence for the early Anglian age has been reinforced and a wider range of soil properties have been recognized as additional sites have been discovered at locations with different parent material, topography, or drainage. The result is that the original term 'arctic structure soil' is now too restrictive. Therefore, throughout the remainder of this paper and elsewhere (Rose *et al.*, 1985) the soil developed during the early Anglian Stage of the British Pleistocene will be known as the Barham Soil and will have the status of a soil stratigraphic unit.

LOCATION AND PUBLISHED RECORDS OF THE BARHAM SOIL

All locations at which the Barham Soil has been recorded are shown on Figure 9.2, ranging from Epping Green in south-east Hertfordshire, through Essex and Suffolk to West Runton and Bacton in north Norfolk. The majority of the sites have been described in Rose *et al.* (1976, 1978a and b), Rose and Allen (1977), Allen (1982, 1983) and Whiteman (1983) and are developed on terrace surfaces formed when the Thames flowed through East Anglia, and are buried by Anglian wind-blown and glacigenic sediments. Additionally, other publications give evidence of a periglacial soil of early Anglian age, although

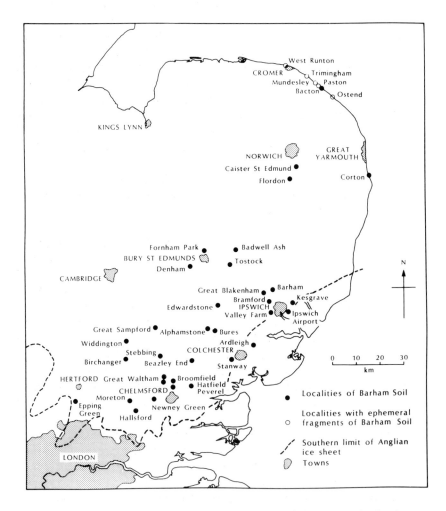

Figure 9.2 Localities showing the Barham Soil in eastern England

with one exception (Moffat and Catt, 1982) they do not discuss its wider significance.

In south-east Hertfordshire Moffat and Catt (1982) have described a 'sand-filled fissure' (p.17) developed in a 'red-mottled, variable stoney, sandy clay' (p.18) and buried by chalky till, reflecting an 'interglacial soil (presumably Cromerian in age) [which] was deformed by the development of an Arctic Structure Soil and invaded by Coversand' (p.23). Its position beneath Lowestoft Till indicates an early Anglian age.

At Ardleigh, north-east of Colchester, Spencer (1966) described 'a zone of disturbed deposit caused when conditions of freezing and thawing caused the soil to flow, crumpling the strata' (p.352), and represented its form and position with a diagrammatic section (p.349 Figure 2). The position between 'glacial outwash gravel', and 'lower gravel' (p.352) that would now be classified as Kesgrave Formation, suggests equivalence of these freeze-thaw structures with the Barham Soil.

Further north-east, at Corton, 'ice-wedge polygons' have been carefully mapped and described by Gardner and West (1975) and West (1980). From their position beneath Cromer Till and across a succession of tidal silts, peat and Rootlet Bed of Cromerian age the authors were able to say that 'a true large-scale arctic polygon system developed in early Anglian times before the arrival of Anglian ice' (Gardner and West, 1975, p.52), thus recognizing the Barham Soil before its formal designation and clearly stating its stratigraphical position. The authors note that this succession and the periglacial structures had been drawn by Blake in 1884 although at that time their palaeoenvironmental significance was not appreciated.

The first specific reference to periglacial structures of early Anglian age was by West and Donner in 1958, when they described a section at Bacton on the north-east Norfolk coast showing an horizon with involutions and the caste of an ice wedge disturbing clays and sands at the top of the Cromer Forest Bed Formation, and buried beneath the First Cromer Till of the North Sea Drift. They conclude that 'the cold climate causing the involutions and the wedge . . . antedates the advance of the North Sea ice and this advance may be the result of the same climatic conditions' (p.9). In reviews of the stratigraphic significance of periglacial structures in Britain, West (1968, 1969) related this periglacial activity at Bacton to the early Lowestoftian (Anglian) glacial stage.

At other sites along the north Norfolk coast: Ostend, Paston, Mundesley, Trimingham and West Runton (Figure 9.2), West and Wilson (1966) and West (1980) have described ice-wedge casts and involutions in a succession of river and lake sands, silts and muds (Table 9.1). Precise written description and section drawing, combined with palaeobotanical evidence, allows careful analysis of these features, and shows that although they clearly represent the effects of ground-ice activity in early Anglian times, they developed on ephemeral land surfaces composed of fluvial and lacustrine sediments in a region of aggrading sedimentation. Thus, although they correspond with the same climatic event as

Table 9.1   Stratigraphic position of periglacial structures formed in the Barham Soil in Norfolk and north Suffolk (after West, 1980)

Stage	Sediments	Pollen assemblage biozones	Vegetation	Environment	Ice-wedge casts	Involutions	Sites
An.	Till	—	—	Glacial	+	+	West Runton, Corton, Trimingham, Bacton/Ostend, Paston/Mundesley
	Freshwater laminated silty silty clay	—				+	Paston/Mundesley
	Fluviatile sand	—			+	+	West Runton, Corton, Paston/Mundesley
eAn.	Fluviatile silty mud and sand	Cyperaceae	Herb	Cold	+	+	Trimingham
	Freshwater laminated silty clay	Gramineae-Cyperaceae	Grassland	Cold	+		Paston/Mundesley
	Freshwater laminated silt and stony detritus mud	Gramineae-Cyperaceae-*Betula-Pinus*	Grassland	Cold	+		Paston/Mundesley
	Fluviatile cross-bedded sand	—	—		+	+	Paston/Mundesley
	Freshwater laminated clay, sand and silty mud	Gramineae-*Betula-Pinus-Alnus*	Grassland with some woodland	Cool			Bacton/Ostend
Cr IVc	Freshwater silt and clay mud	*Betula*	Woodland	Cool	+	+	West Runton, Paston/Mundesley
Cr IVa	Tidal laminated sand with silty clay and organic laminae	*Pinus-Alnus-Picea-Betula*	Woodland	Cool			Bacton/Ostend
Cr IIIb	Tidal laminated silty clay	*Abies-Carpinus*	Woodland	Temperate			Trimingham, Corton

the Barham Soil they represent elements of a land surface with far less spatial and temporal significance than the Barham Soil as formally defined. Their limited development and continuity suggests that they are short duration sub-divisions of the Barham Soil each with a restricted range of pedological features. However, because of their association with palaeobotanical evidence, it is possible to associate their development with a detailed sequence of climatic and environmental change from the end of the Cromerian temperate Stage to the onset of glacierization in the region during the Anglian. In particular, they show that ice-wedge forms first developed in Cromerian IVc while the region still experienced at least patches of birch woodland, and continued to develop throughout the early Anglian as the vegetation types changed from grassland with some trees, through a grassland phase and herb phase to the conditions of polar desert (Table 9.1).

Finally, the Barham Soil can be seen in a photograph in the British Regional Geology of East Anglia (Chatwin, 1961, Plate III). It is located between 'glacial gravels' and 'chalky boulder clay' at Bramford near Ipswich, but no reference to its existence is made in the text.

BACKGROUND INFORMATION

Soil parent material

The parent material of the Barham Soil is the argillic horizon of the Valley Farm Soil throughout most of the region (Kemp, 1985) and freshwater sediments at Corton and along the north Norfolk coast (West, 1980) (Figure 9.3). The Valley Farm Soil is Cromerian in age at its youngest (Rose, 1983; Kemp, 1985) and is developed in fluviatile silts, sands and gravels of the Kesgrave and 'Pebble Gravel' formations (Rose and Allen, 1977; Moffat and Catt, 1982). The presence of the argillic horizon in the river deposits means that the periglacial soil-forming processes operated in material with between 30 and 15 per cent more clay than would be expected from unmodified river deposits, which in an area like south-east Suffolk have between 1.4 and 17.8 per cent silt and clay (Allen, 1983). The freshwater sediments at Corton and in Norfolk are of late Cromerian and early Anglian age (West, 1980) (Fig. 9.1, Table 9.1). Beds within these deposits show abrupt size changes with sand and clay laminations as at Bacton (West and Donner, 1958) and clay, silt, sand, detritus mud and peat as at Corton (Gardner and West, 1975). Only rarely, as at West Runton, is the Barham Soil developed on relatively thick and uniform beds of unmodified sand and gravel (West, 1980).

Burial material

Throughout the whole region the Barham Soil is buried by Anglian glacigenic sediments. These vary from facies of Lowestoft Till in Hertfordshire, Essex

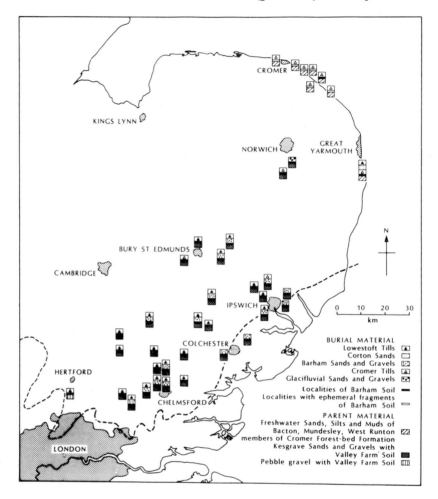

Figure 9.3   Parent material and burial material at localities where the Barham Soil has
been observed

and south and east Suffolk, to facies of Cromer Till in north Suffolk and
Norfolk, and glacifluvial sands and gravels at scattered sites throughout the
region (Figure 9.3). Occasionally slope deposits derived from glacial material,
as at Broomfield (Rose *et al.*, 1978b), or soil material, as at Ipswich Airport,
occur between the soil and the glacigenic overburden. The thickness of the
overlying material inevitably varies between sites and along exposures depending
upon the present-day topography, but typical situations can be represented by
Great Waltham near Chelmsford, where there is up to 8 m of till (Whiteman,
1983) and Barham near Ipswich where there is up to 4 m of sands and gravels

and 2.5 m of till (Allen, 1973). At Ipswich Airport, Kesgrave and Caistor St Edmunds till is absent but the overlying glacifluvial gravels reach thicknesses of between 5 and 10 m.

Perhaps it is worth noting at this point that there are many localities throughout the region where the palaeosols are absent. This can be seen at localities like Twitch Cross and Amigers (Rose *et al.*, 1976) where the soils have been destroyed by glacial and glacifluvial erosion. At some localities such as Newney Green and Barham destruction of the palaeosols is restricted to places where ice or meltwater has cut channels into the Kesgrave Sands and Gravels (Rose *et al.*, 1978a and b). At Broomfield, Great Blakenham and other sites, the palaeosols have been sheared and laterally displaced. In the region of the Fen basin and the area of the degraded Chalk escarpment the palaeosols are continuously absent (Rose *et al.*, 1985, Figure 5). This is attributed to extensive erosion by the Anglian ice which formed and deposited the Lowestoft Till (Perrin *et al.*, 1979).

Quaternary significance

Within the Quaternary history of the British Isles the Barham Soil is significant because it formed on the landscape that was overridden by Anglian ice-sheets (Perrin *et al.*, 1979) and is therefore a key to the reconstruction of the topography and environment of eastern England before major changes occurred as a result of the Anglian glacial erosion and deposition. Within East Anglia the landscape that existed while the Barham Soil formed was dominated by extensive terraces formed when the Thames flowed through East Anglia (Hey, 1980; Rose *et al.*, 1985). Anglian glaciers altered this landscape by eroding the Wash and Fen basin and reducing the relief of the Chalk scarp, and by extensive deposition with a radial drainage pattern throughout the rest of East Anglia. Of the original rivers only the Thames survived, albeit along a more southerly route (Rose, 1983).

DESCRIPTION AND PROCESSES

Soil colour

At most of the sites at which the Barham Soil is recorded soil colours are determined by the lithology of the parent material. However, at Newney Green, Great Waltham, Broomfield, Great Blakenham and Florden (Figure 9.2), the soils show weak colour horizonation parallel with the land surface, with an upper dark grey band which changes downwards through grey-brown to the original colours of the parent material. Generally, these colour changes are independent of the soil texture, as at Broomfield which is a gravelly clay loam overlain by some 20 cm of coversand and about 5 m of chalky till. Here, the uppermost

60 cm is dark grey (N 4/0) to dark greyish brown (10 YR 4/2) with dark yellowish brown (10 YR 4/4) mottles. This is separated by a relatively clear boundary from a strong brown (7.5 YR 5/6) to yellowish brown (10 YR 5/8) band with grey mottles (5 Y 6/1) which, in turn, merges with yellowish red (5 YR 5/8) parent material.

Micromorphological analysis of the soil material from Broomfield shows that the dark colour is due to dark, clay coatings around sand grains, and chemical analysis of the dark grey soil material from Newney Green, Broomfield and Flordon gives organic carbon contents of between 1.58 and 0.17 per cent.

In many respects this colour horizonation and organic carbon content is similar to that described from well-drained soils forming beneath tundra vegetation in arctic regions at the present time (Hill and Tedrow, 1961; O'Brien *et al.*, 1979), suggesting an affinity with arctic brown soils (Tedrow, 1978; Bockheim, 1978). Where this property is developed, therefore, horizonation processes are more important in the formation of the Barham Soil than physical disruptive processes that are typical of the Barham Soil elsewhere throughout the region.

However, some caution is required when considering soil colour as sedimentary and micromorphological evidence from Great Blakenham, Newney Green and Broomfield indicates that, in places, the Barham Soil, has been eroded with the transportation and subsequent redeposition of soil material as sedimentary units.

Involutions

The localities at which involutions have been observed are shown on Figure 9.4. These are the most commonly developed macrostructure and occur as regular patterns. Examples are illustrated on Figure 9.5 and dimensions are given on Table 9.2. Individual features range from simple lobe-shaped forms with a symmetrical shape (Rose and Allen, 1977, Figure 3) to ball and pillow structures at Badwell Ash (Allen, 1983, Figure 13), and complex lobe-shaped forms with extreme elongation and irregular protrusions as at Newney Green (Figure 9.5). In some cases different size involutions are developed together, either conformably with a small lobe forming the centre of a large one, or discordantly with the smaller truncating part of the larger structure. Typical spacing between lobes is about 1 m with a typical depth of about 0.4 m. The maximum depth of sediment affected by involutions is about 1.8 m.

The material in which the involutions are developed includes all sizes and associations of sizes, ranging from clayey gravel to well-sorted sand, and its arrangement varies from a fine-textured core surrounded by coarse material to a coarse-textured core surrounded by fine material. Where involutions can be observed along a single section such as at Barham, they are formed of both coarse and fine material, although those with gravel cores are generally wider and deeper (mean width 0.98 m, mean depth 0.53 m) than those with sand cores

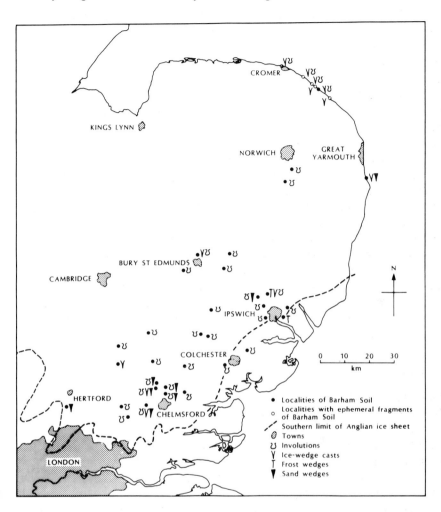

Figure 9.4   Localities where involutions, ice-wedge casts, frost cracks, and sand wedges have been observed as part of the Barham Soil

(mean width 0.58 m, mean depth 0.35 m) (Allen, 1983). However, in all cases their occurrence is independent of the primary sedimentary structures, and in places such as Great Blakenham and Newney Green they are visually impressive, consisting of red (10 R 4/8) and light grey (5 Y 7/2) Valley Farm Soil mixed with yellow (10 YR 7/6) and strong brown (7.5 YR 5/6–5/8) Barham Coversand.

In plan the involutions show either a polygonal pattern with a diameter of about 1.5 m as at Newney Green (Figure 9.9) or a series of circles as at Barham.

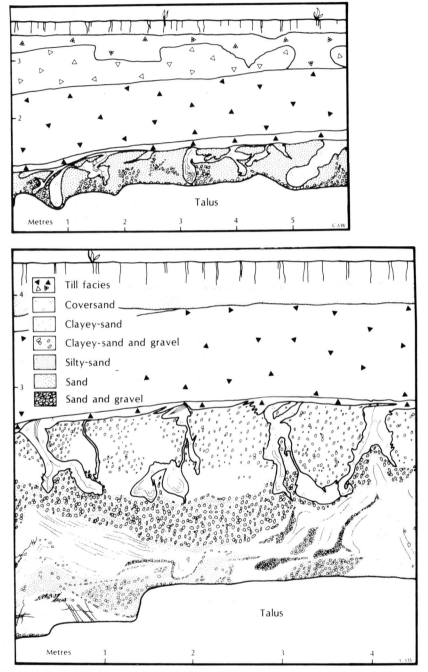

Figure 9.5 Involution as part of the Barham Soil at Newney Green, near Chelmsford

Table 9.2   Dimensions of involutions formed in the Barham Soil

Site	Spacing (metres)			Depth (metres)			
	n	$\bar{x}$	σ^{n-1}	n	$\bar{x}$	σ^{n-1}	max.
Newney Green	21	0.85	±0.28	25	0.75	±0.30	1.79
Broomfield	21	1.83	±1.00	22	0.61	±0.29	1.51
Valley Farm	10	0.85	±0.42	10	0.24	±0.10	0.40
Barham Face 1	52	1.45	±0.80	56	0.41	±0.20	0.80
Barham Face 5	5	1.34	±0.64	6	0.51	±0.25	0.80
Kesgrave	23	0.70	±0.67	24	0.16	±0.11	0.45
Badwell Ash	17	0.63	±0.37	18	0.45	±0.26	0.85

In terms of periglacial soil formation the involutions represent non-sorted patterns (Washburn, 1973). Although there is controversy about their mechanism of formation (French, 1976) it is generally accepted that they represent active-layer processes in areas of seasonal freezing with cryostatic forces operating during the autumn freeze (Washburn, 1973) and gravitational forces operating on soil with low cohesion during the spring and summer thaw (Shilts, 1978). Although these processes may operate in conditions of seasonally frozen ground without permafrost, involutions clearly indicate cold climate conditions with a depth of disturbance reflecting either the depth of the active-layer or the depth of the ground-water table (Van Vliet-Lanoë, 1983).

Throughout this study care has been taken to differentiate periglacial involutions from load or dewatering structures caused by shock or overburden pressures operating on sediments with high pore-water pressures and inverted density gradients (Butrym *et al.*, 1964; Lowe, 1975). Dewatering structures do exist at localities such as Barham and Valley Farm (Rose and Allen, 1977; Allen, 1983) but in all cases they are formed in fine-grained sediments arranged with the finest at the base. A periglacial origin is invoked in cases where original density gradients are normal or disturbed materials are of the size that would require improbably high stresses. For instance at Badwell Ash, the forces required to produce an upward movement of gravel-size clasts (8 mm) in a fluidized state would need velocities of at least 8 cm sec^{-1}, which would in turn require an unlikely thickness of overburden (Allen, 1983).

Ice-wedge casts and frost cracks

The localities at which ice-wedge casts and frost cracks have been observed are shown on Figure 9.4. Occurrences are not common, the best development being at Corton where ice-wedge casts show a spacing of about 12 m and can be identified as part of a non-sorted polygonal pattern (Gardner and West, 1975). Typically the frost cracks are about 30 cm deep and 1–10 cm wide as at Ipswich Airport and Barham. In Norfolk and at Corton the ice-wedge casts are

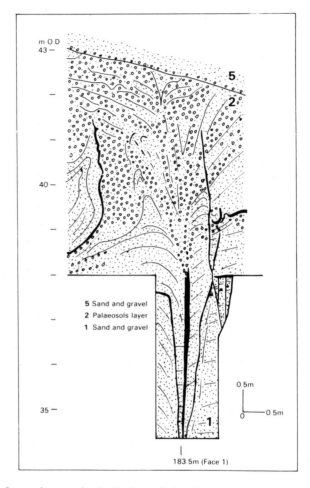

Figure 9.6   Ice-wedge cast in the Barham Soil at Barham, near Ipswich. Symbols are the same as those shown in Figure 9.5

1–2 m deep with maximum widths of about 10 cm–1 m (Gardner and West, 1975; West and Donner, 1958; West, 1980). Over the remainder of the region, at localities like Newney Green, Great Waltham, Broomfield and Fornham Park they are larger, with typical depths of about 4 m and a maximum depth of more than 8 m at Barham (Figure 9.6).

Frost cracks and ice-wedge casts are developed in the full range of parent materials from organic muds at Corton to clay-rich sands and gravels at the majority of sites, to reasonably well-sorted silts, sands and gravels in north Norfolk. Their shape is an elongate wedge, and they are composed primarily of allogenic material, which in some cases like Newney Green, can be shown to be derived from the adjacent soil. In cases like the wedges at Great Waltham, Barham and Corton, normal faulting parallel with the wedge sides can be observed in the adjacent sediments (Figure 9.6), and at Barham the adjacent sediments show evidence of upturning (Figure 9.6).

These features are considered to represent the effects of vein-ice growth in fissures caused by thermal contraction of the permafrost with the parent material. The likelihood of initiation and growth is determined by the rate of cooling and the ground-ice content of the soil, with fine-grained sediments more susceptible than well-drained coarse-grained material (Leffingwell, 1919; Lachenbruch, 1962). By analogy with ice-wedge formation in Alaska at the present time, Péwé *et al.* (1969) has suggested that initiation requires a mean annual temperature of at least $-6\,°C$ and Price (1972) has suggested that formation in well-sorted gravels, similar to those in which the deepest wedges of the Barham Soil are developed, is only likely when the mean annual temperature is as low as $-12\,°C$. Butrym *et al.* (1964) and Black (1976) have emphasized the problem of distinguishing between wedge-shaped structures developed due to thermal contraction and those due to the effects of loading, shock or dessication.

Within the area of the Barham Soil readjustment structures or clastic dykes are present, but they are restricted to lower levels of the Kesgrave Sands and Gravels (as at Valley Farm (Rose and Allen, 1977, Figure 3)) and merge with the overlying sediments. In all the cases referred to the Barham Soil the wedges develop below an unconformity that represents the land surface in vertical section. The cryostatic origin of the wedges is demonstrated by the upturned structures adjacent to the wedge casts, where buckling has occurred due to lateral stresses caused by the growth of ground-ice or the expansion of the ground on thawing. Evidence for subsequent ice melting is demonstrated by the parallel normal faults. Therefore, ice wedges developed in the Barham Soil indicate continuous permafrost, thermal contraction of the ground, vein-ice growth, and mean annual temperatures as low as $-12\,°C$.

Sand wedges

The occurrence of sand wedges is shown in Figure 9.4 with typical examples given in Figure 9.7. They occur throughout the region but are more common in Essex and south-east Suffolk. In form they vary from a simple elongate wedge shape, to complex forms with bulbous protrusions at the base (Figure 9.8). Sizes vary from about 0.5–2 m deep to 5–75 cm wide at the top. In some cases (i.e. Figure 9.7) adjacent beds show upturning, but in most there is no discernible disturbance. The plan-form of these wedges was revealed at Newney Green (Figure 9.9) and shows well-defined, if complex, polygonal patterns with a hierarchy of sizes ranging from about 10 m diameter for the largest to about 1 m diameter for the smallest. These small polygons also include coversand-filled involutions.

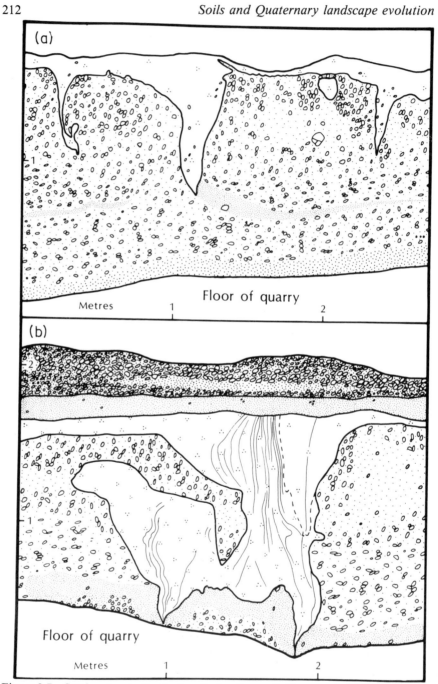

Figure 9.7   Sand wedges in the Barham Soil at Broomfield, near Chelmsford. Symbols
are the same as those shown in Figure 9.5

Figure 9.8   Sand wedge with a bulbous base at Great Blakenham, near Ipswich. Symbols are the same as those shown in Figure 9.5

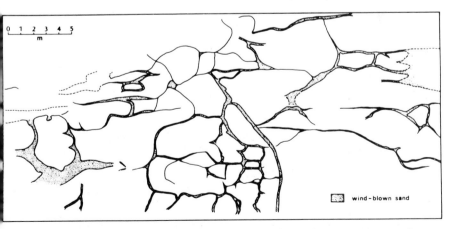

Figure 9.9   Plan of sand-wedge polygons and involution polygons at Newney Green, near Chelmsford

The sand wedges are filled with well-sorted medium and fine sand with small quantities of silt and clay (Figure 9.11). Some wedges contain sand with occasional wind-polished pebbles (Figure 9.7a). These materials are similar in size range and mineralogy to the Barham Coversand. The wedges filled solely with sand show vertical laminations which are arranged, roughly, with a mirror-image pattern either side of the central laminae which generally extends down to the deepest point of the wedge. Each lamina is in the order of 2–5 mm thick and is differentiated by slight variations in colour. In particular cases, such as that at Broomfield (Figure 9.7b), complex assemblages of laminae are developed with one group cutting through another, indicating sequences of wedge development.

Sand wedges are attributed to thermal contraction in dry permafrost where fissures are not filled by vein-ice but remain open to trap wind-blown sediment (Péwé, 1959). Vertical laminations reflect individual increments of sediment, with the thickness of each pair of laminae reflecting the width of the fissure, and the number of laminations reflecting the number of cooling events leading to thermal contraction. Upturning of the adjacent material is due to lateral stresses caused by the net addition of sediment to the soil, and the absence of collapse structures is due to the dry permafrost and the absence of a void during the melt season. The occurrence of pebbles among the wind-blown sediment in some of the wedges is attributed to possible changes in permafrost conditions leading to the development of composite wedges (Berg and Black, 1966). Although sand wedges were originally considered to indicate aridity (Péwé, 1959; Berg and Black, 1966), emphasis is now placed on local controls of soil moisture, such as variations in relief, and observations record sand wedges developed on well-drained areas in close proximity with ice wedges developed in wetter localities (P. Worsley, pers. comm.).

With regard to the Barham Soil, sand wedges provide further evidence for periglacial conditions, and indicate the occurrence of arid permafrost environments with contemporary, wind-blown sand. However, not all the sand wedges may indicate thermal contraction, as many of the smaller features penetrate into the ground only as far as the base of the clay-rich Cromerian soil material. In these cases fissuring may be due to drying of the fine-grained sediments rather than due to intense cooling, but as the larger wedges extend well below this material thermal contraction must also have applied. Using estimates of growth rates of between 0.3 mm year^{-1} and about 1.0 mm year^{-1} for sand wedges in Antarctica (Berg and Black, 1966) the largest sequence of sand wedges from the Barham Soil (Broomfield, Figure 9.7b) indicate a duration of between about 1250 and 4200 years. Finally, the derivation of the wind-blown sand from material entrained and transported by the Anglian ice sheets indicates that the development of sand wedges in the Barham Soil coincided with the presence of ice in the vicinity of eastern England.

Vertically orientated stones

These are frequently observed in the Barham Soil, usually in association with involutions, as at Stebbing, Broomfield (Figure 9.7) and Newney Green (Figure 9.5). They are considered to reflect differential particle movement in response to freezing and thawing in the active-layer (Watson and Watson, 1971; French, 1976).

Platy aggregates

These features consist of platy peds, 1–3 cm thick, and 10–30 cm apart. They are typical of sandy loam soil materials, but tend to ignore minor textural variations and occasional clasts. The material between the discontinuities is densely packed. The whole soil has a roughly rectilinear brecciated appearance and is strong when in a confined state but brittle when unconfined. This soil property is also known as indurated horizon (Fitzpatrick, 1956) and fragipan (Van Vliet and Langohr, 1981). Platy aggregates have been observed at several sites in the region such as Great Waltham, Stebbing, Beazley End, Barham, Fornham Park and Caistor St Edmunds.

Although different explanations have been proposed to account for the development of platy aggregates (Van Vliet and Langohr, 1981) it is generally accepted that they are due to ground-ice segregation and accompanying desiccation with the growth of ice lenses. Dense packing is attributed to the high stresses between the ice lenses. Platy aggregates in the Barham Soil are therefore considered to be additional evidence for ground-ice processes, and specifically reflect small, ice segregations in moderately moisture-retentive soil materials.

Banded fabric

Linear, sub-horizontal concentrations of silt and clay-size material beneath bands of uncoated or partially coated sand grains are common micromorphological features at Stebbing, Great Blakenham and Ipswich Airport (Figure 9.10c). Their upper boundary is relatively even with planar or convex forms, and the length of individual bands varies from about 1 to 3 mm (Figure 9.10c). Similar features have been called sorted lamellae structures by Van Vliet-Lanoë (1976), banded fabric by McKeague *et al.* (1974) and silt droplet fabric by Romans *et al.* (1980). Although detailed differences in the mechanisms of formation have been proposed, all invoke the growth and melt of ground-ice. The small-scale vertical pattern of relative depletion and concentration of fine material reflects the translocation of clay and silt following melt of ground ice during spring and summer thaw periods. Their occurrence within the Barham Soil, therefore provides further evidence for periglacial soil formation.

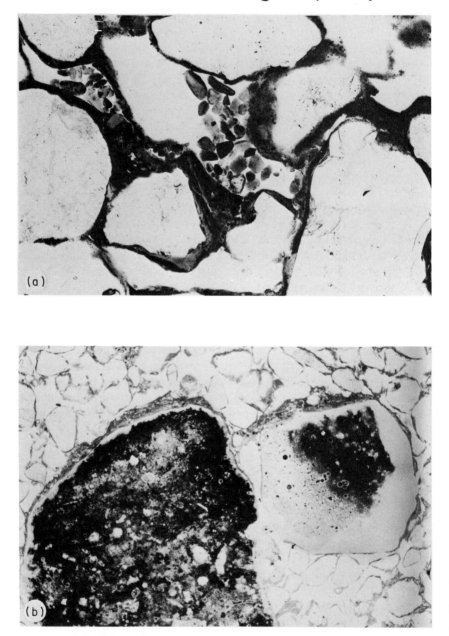

Figure 9.10 Micromorphological characteristics of the Barham Soil. (a) Rounded silt-size papules in a void between coated sand grains at Ipswich Airport. Frame length: 1.35 mm. (b) Cappings of silty-clay size material on sand grains at Ipswich Airport

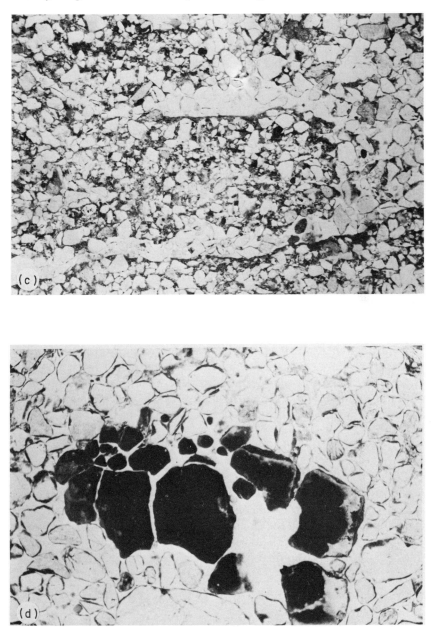

Frame length: 1.35 mm. (c) Banded fabric in a sand matrix at Stebbing. Frame length: 5.35 mm. (d) Fractured fine gravel particle at Ipswich Airport. Frame length: 5.35 mm. All are in plane-polarized light

Silty-clay cappings

Further evidence for the translocation of silt and clay-size material is provided by the development of silt and clay cappings on the upper surfaces of sand-size grains (Figure 9.10b). As with banded fabrics these features reflect the migration of relatively coarse material in soils disrupted by the growth and melt of ground-ice (Boulton and Dent, 1974: Romans *et al.*, 1980).

Disrupted argillans

Disrupted argillans are a common micromorphological feature of the Valley Farm and Barham palaeosols. Those surrounded by later clay skins clearly relate to disruptive processes prior to clay translocation during the temperate Cromerian Stage, and do not relate to the formation of the Barham Soil. Those which can be related specifically to the Barham Soil are either locally fractured, angular, or conjoined, or are more intensely deformed and fragmented with sub-rounded to rounded edges, showing no relation to their original position of formation. These latter type, also known as papules, are shown in Figure 9.10a where they reflect disruption and movement by cryoturbation or gelifluction, although the rare incomplete fillings of voids by rounded papules are due to translocation processes (Figure 9.10a).

The disruption process reflects high stresses in the soil material, and is sufficiently common, with a wide enough vertical range, to suggest a ubiquitous process such as the growth of ground-ice rather than fracture caused by root-growth or tree-fall (Van-Vliet and Langhor, 1983). The infillings of rounded papules are further evidence for translocation during the formation of the Barham Soil.

Fractured soil particles

Finally, a thin section from Ipswich Airport shows a fractured, fresh, fine-gravel particle (Figure 9.10d). The components of this fractured grain remain juxtaposed, and edges can be matched, indicating little transport since breakage. This suggests that parts of the Barham Soil experienced fracturing by ground-ice stresses as one of the final processes before burial and fossilization.

WIND-BLOWN SEDIMENTS ASSOCIATED WITH THE BARHAM SOIL

Barham Loess

The Barham Loess exists as a sedimentary body at Barham (Rose and Allen, 1977; Allen, 1983) and as a constituent of the Barham Soil at Edwardstone (Figure 9.2) (Rose *et al.*, 1976). At Barham it is composed of 85 per cent silt,

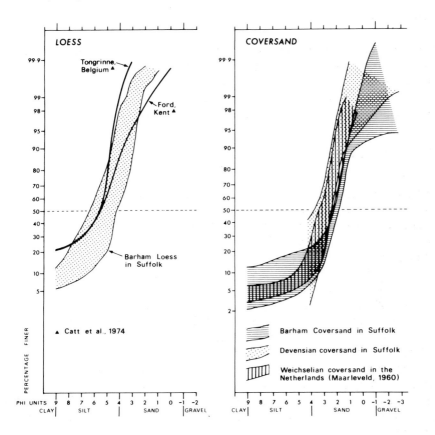

Figure 9.11 Particle-size distribution of the Barham Loess, Barham Coversand, and Devensian and Weichselian loess and coversand from southeast England and the Low Countries

10 per cent clay and 5 per cent sand with a mode in the coarse and medium silt range ($+5\phi$ to $+7\phi$) (Figure 9.11), and has a similar size distribution to Devensian loess in eastern England and Weichselian loess in Belgium (Catt *et al.*, 1974) (Figure 9.11). Mineralogically it is dominated by quartz and feldspar, with a wide range of heavy minerals including weatherable materials such as apatite and collophane (Table 9.3). The mineral assemblage is qualitatively and quantitatively similar to the same size fraction from the Anglian glacigenic deposits (Table 9.3), and includes many minerals not present in the Kesgrave Sands and Gravels, suggesting that it was derived from glacigenic material when Anglian ice was in the vicinity of eastern England.

Table 9.3 Mineralogical composition of Barham Loess, Barham Coversand, Barham/Sandy Lane Gravels, and Kesgrave Sands and Gravels. All these determinations were done by Dr J. A. Catt and they are reproduced with his kind permission

Lithology and mineralogy (percentage values)

Mineral Composition	16–23 μm fraction				63–250 μm fraction		
	Barham Loess	Barham Coversand	Barham Coversand	Sandy Lane Gravels (silt bed)	Barham Coversand	Barham Sands and Gravels	Kesgrave Sands and Gravels
	Barham	Newney Green	Newney Green	Great Blakenham	Newney Green	Newney Green	Newney Green
Light fraction							
Quartz	83.0	80.3	84.6	82.3	88.1	90.5	83.5
Alkali felspar	12.0	14.8	11.8	7.7	10.2	7.2	11.8
Flint	1.0	1.4	1.3	2.0	1.4	1.4	4.7
Mica	4.0	2.0	1.6	5.4	—	—	—
Glauconite	<1.0	1.5	0.7	2.6	0.3	0.9	—
Heavy fraction							
Epidote	35.4	21.5	23.5	27.6	6.6	7.3	3.8
Zoisite	1.2	3.1	2.6	0.5	0.9	1.6	—
Zircon	20.6	26.3	30.4	15.7	15.1	40.8	26.4
Tourmaline	0.9	7.9	7.5	6.0	19.8	13.0	50.3
Chlorite	9.5	5.5	7.1	19.5	14.9	0.6	—
Biotite	2.2	0.7	0.8	1.9	3.4	—	—
Green Hornblende	2.2	1.7	3.2	1.5	0.4	0.9	—
Blue Hornblende	—	—	—	—	—	0.1	—

Table 9.3 *(continued)*

Tremolite	—	1.0	1.2	1.1	—	0.2	—
Garnet	6.9	11.5	8.4	2.6	6.9	6.8	1.4
Yellow Rutile	11.5	8.0	3.7	13.2	1.1	1.8	0.9
Blue Rutile	—	3.2	4.4	0.8	4.3	9.2	5.0
Red Rutile	—	0.2	—	0.1	0.7	1.2	0.2
Anatase	6.9	1.2	1.8	6.8	0.9	0.8	0.5
Brookite	2.1	0.3	0.2	0.2	—	0.1	—
Staurolite	—	1.4	0.6	0.3	4.4	5.8	8.2
Kyanite	—	1.5	0.9	0.3	3.9	5.5	3.1
Apatite	0.1	2.0	1.7	0.1	2.4	2.1	—
Collophane	0.5	2.0	2.0	1.7	14.3	2.2	—
Brown Spinel	—	0.8	—	—	—	—	—
Andalusite	—	0.2	—	—	—	—	—

Barham Coversand

The Barham Coversand exists as thin beds up to about 30 cm thick, and sand wedge and involution infills. It occurs at the majority of sites south of, and including, Barham. It may also exist at Corton (Gardner and West, 1975). It has a distinctive yellow (10 YR 7/6) to strong brown (7.5 YR 5/6–5/8) colour, and contains between 75 and 92 per cent sand with a mode in the fine and medium sand range ($+3\phi$ to $+1\phi$) (Figure 9.11), similar to Devensian coversand in Suffolk and Weichselian coversand in the Netherlands (Maarleveld, 1960) (Figure 9.11). Additionially it contains between 8 and 23 per cent silt and clay, and up to 7 per cent gravel. Some of the included flint pebbles show the effects of wind polishing. Like the Barham Loess its mineralogy is dominated by quartz and feldspar with a wide range of heavy minerals (Table 9.3) unlike equivalent size fractions of the Kesgrave Sands and Gravels, but similar to the Anglian glacigenic deposits (Table 9.3). The Barham Coversand is attributed to wind transport and deposition at a time when Anglian ice was in the vicinity of eastern England and the Barham Soil was in process of formation.

Significance of wind-blown sediments

The Barham Loess and Coversand are important sediments, because they are a component of the early Anglian periglacial landscape in eastern England and indicate the role of aeolian processes at that time. The fact that they are derived from glacigenic material and mixed with the Barham Soil means that the elements of the soil with which they are associated can be ascribed to the period of time when Anglian ice was in the vicinity of eastern England. Finally, the distinctive mineralogy of these wind-blown sediments means that they provide complementary evidence for the use of the Barham Soil as a stratigraphic indicator.

POST-FORMATIONAL MODIFICATION

At many of the exposures, sections of the Barham Soil show evidence of post-formational modification. Reference has already been made (Section on soil colour) to transported dark-grey soil material at Broomfield and Great Blakenham. Glacitectonic deformation by the overriding Anglian ice exists in the form of low-angle shears, deformed ice-wedge casts and sand wedges, and nappe-like structures at Newney Green, Great Waltham, Broomfield, Great Blakenham and Barham. Disturbance of the soil by London Clay diapirs has been observed at Broomfield and Stanway, and large gull-like structures, attributed to contemporaneous or later periglacial activity have been described at Stebbing (Whiteman, 1983). Translocation of dissolved calcium carbonate from the overlying chalky till has resulted in the development of carbonate nodules at Newney Green where the buried soil is relatively close to the present-day ground surface.

However, in terms of the total exposure of Barham Soil these examples of post-formational modification are rare. Typically it shows no discernible evidence of modification since burial, and as far as can be determined at the current state of knowledge there is no evidence for diagenesis.

CONCLUSIONS

History of development of the Barham Soil

At many sites the structures and features of the Barham Soil show relationships, with each other and with the Barham Loess and Coversand, that enable the history of soil development to be reconstructed. For instance, at Newney Green the upper parts of ice-wedge casts and some sand wedges are disturbed by involutions indicating that thermal contraction in both icy and dry permafrost was succeeded by cryoturbation. At the majority of sites, represented by Moreton, Badwell Ash and Caistor St Edmunds, involutions are formed in soil material, whereas at Newney Green, Barham and Valley Farm, they involve Barham Coversand showing that cryoturbation occurred both prior to, and following, aeolian deposition. However, undisturbed sheets of coversand at sites such as Broomfield, Stebbing and parts of Newney Green, show also that in certain localities aeolian sedimentation persisted after cryoturbation has ceased. At Barham an undisturbed sand wedge penetrates an ice-wedge cast, and at Great Blakenham an undisturbed sand wedge penetrates involutions, indicating that thermal contraction and wind transport were the final processes. At a different scale, micromorphological evidence also shows sequences of soil development. In general, it shows cold climate soil properties superimposed on the temperate climate properties of the Valley Farm Soil (Kemp, 1985). At the local level, for instance, at Ipswich Airport, clusters of rounded papules indicate translocation after argillan fragmentation.

Interpretation is far from simple as the evidence exists in a variety of combinations due to the variations in site conditions, and some structures, like the platy aggregates at Stebbing which are below the zone of involutions, cannot be placed in a wider context. Nevertheless, the occurrence of the various processes associated with the formation of the Barham Soil is illustrated in Figure 9.12, which has been constructed on the basis of all the evidence available (shown by solid lines), and the likelihood of development although there is no direct evidence available (dashed lines). The relationship of the soil with ice in the vicinity of eastern England is established through the glacigenic provenance of the Barham Loess and Coversand, and its position in the British pollen stratigraphy is derived from the associated vegetational evidence in Norfolk and north Suffolk (West, 1980).

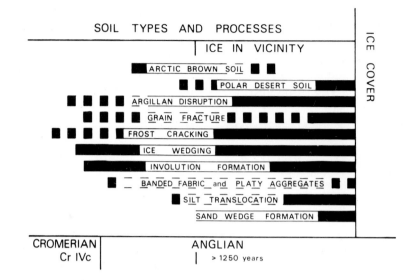

Figure 9.12   Development of the processes associated with the formation of the Barham Soil in relation to the British Pleistocene pollen stratigraphy, ice in the vicinity of eastern England, and ice cover

The first evidence for the development of the Barham Soil is during Zone IVc of the Cromerian Interglacial when ice-wedge casts and involutions formed in association with birch woodland in Norfolk. It is likely that at the same time, frost cracks were formed, susceptible clasts were fractured, and clay skins, formed during the preceding temperate climate, were disrupted by the growth of ground-ice. With the replacement of birch woodland by grass and herb tundra well-drained sites developed profiles similar to that of arctic brown soils, while at less well-drained localities active-layer processes, in particular cryoturbation and silt translocation, became more important leading to the disruption of the vegetation and the extension of bare ground. By this time many ice-wedges had already become casts and were deformed by involutions. This association of processes persisted while ice reached the vicinity of eastern England and wind-blown sediments contributed to the landscape. Local, and perhaps regional, desiccation became more important, leading to the extension of dry permafrost, the persistence of thermal contraction, and the diminution of active layer processes. At this time much of the region had the character of polar desert and using minimum estimates for the growth of sand wedges, it is possible to suggest that such conditions may have existed for at least 1250 years. Finally, undisturbed sand wedges, sorted lamellae, silt and clay cappings, disrupted argillans and a fractured clast were in the process of formation when the Barham Soil was overridden and fossilized beneath Anglian ice, outwash, and till.

The burial of the Barham Soil was a time-transgressive event. It was overridden first by Scandinavian ice moving from the north-east into north-east Norfolk and north Suffolk, and buried beneath Cromer Till and outwash. Scottish ice then moved into the region from the north and through the Fen Basin, before extending south and eastward as far south as Hertfordshire, Essex and south-east Suffolk, burying the Barham Soil beneath Lowestoft Till and outwash (Perrin *et al.*, 1979). This pattern most probably accounts for the restriction of the wind-blown sediments to sites south of, and including, Barham (although coversand may also exist at Corton (Gardner and West, 1975)), and helps explain the more complex development of the soil in the southern part of the region.

Status of the Barham Soil as a palaeosol

Although the structures that characterize the Barham Soil are commonly studied as part of periglacial geomorphology all those considered here relate to surface processes and therefore testify to the existence of a land surface. Because of the importance of physical processes the term 'arctic structure soil' is a convenient label, but detailed analysis shows that smaller-scale soil structures such as platy aggregates, micromorphology and organic content also provide evidence of soil formation and indicate a wider range of soil-forming processes. All these properties represent the products of soil formation in a periglacial environment and should be given equal status to those properties representing different processes operating under other environmental conditions.

Different degrees of soil development can be recognized across the region. In Norfolk and north Suffolk where the Barham Soil is developed in ephemeral land surfaces formed by aggrading fluvial and lacustrine sediments it consists of relatively simple ice-wedge casts and involutions. In south Suffolk and Essex where it developed through much of early Anglian time it shows a complex association of soil structures and features with particularly large ice-wedge casts, complete patterns of sand wedges and well-developed micromorphology.

Relationship of the Barham Soil with other palaeosols

At all the sites south of Corton the Valley Farm Soil forms the parent material of the Barham Soil. The Valley Farm Soil is a complex palaeosol which last developed in the temperate climate of the Cromerian Interglacial (Kemp, 1985). The boundary between the two palaeosols is essentially arbitrary, with soil properties attributed to temperate processes, such as clay translocation and chemical weathering, being ascribed to the Valley Farm Soil and soil processes attributed to cold climate, such as thermal contraction, cryoturbation, clast and clay-skin disruption, and silt translocation being ascribed to the Barham Soil. Consequently, the actual boundary between the two soils will vary from place to place depending on site conditions and the susceptibility to particular processes. This is typical of many stratigraphic sub-divisions, particularly where the evidence is based on individual sites rather than aggregate evidence, such as pollen assemblages, which usually represent a wider spectrum of environmental conditions.

The problem is most clearly illustrated by the evidence from Norfolk and north Suffolk where ice-wedge casts and involutions at West Runton and in the Paston/Mundesley area indicate that the Barham Soil first began to form in Zone IVc of the Cromerian Interglacial, whereas at other sites in the region, where land surfaces are also known to have existed, there is no evidence of periglacial soil formation until the early part of the Anglian Stage.

Beyond the limit of Anglian glacigenic deposits the Barham Soil may form a component of relict palaeosols (Rose and Allen, 1977; Kemp, 1985), although, in these localities recognition is difficult because of the effects of post-Anglian pedogenesis. However, recognition may be possible because of the exceptionally deep ice-wedge casts, the distinctive sand wedges and the association with mineralogically diagnostic wind-blown sediments.

Palaeoenvironmental significance

The Barham Soil formed in a periglacial climate which during at least one phase experienced annual temperatures at least as low as $-12\,^\circ$C. It includes the most complete development of sand wedges and is associated with the most extensive spread of wind-blown sand in Britain. It formed at a time just prior to glaciers reaching their most southerly extent in eastern England. It is concluded, therefore, that the Barham Soil provides evidence for the period of most severe cold and regional aridity yet recognized in the British Pleistocene.

ACKNOWLEDGEMENTS

The authors wish to thank Dr J. A. Catt for allowing us to use the results of his mineralogical analyses on the Barham Loess, Barham Coversand, Kesgrave Sands and Gravels and Barham/Sandy Land Gravels. Discussion with Dr P. Bullock about micromorphology is also gratefully acknowledged. We also wish to thank Professor P. Worsley and Professor R. G. West for constructive comments on an earlier draft of the first part of this paper. Cartographic assistance provided by Loraine Rutt is also gratefully acknowledged.

REFERENCES

Allen, P. (1973). 'Barham', in *Quaternary Research Association Field Guide, Clacton,* (Eds J. Rose and C. Turner), pp. 63–69, Birkbeck College, London.

West, R. G. (1969). 'Stratigraphy of periglacial features in East Anglia and adjacent areas', in *The Periglacial Environment* (Ed. T. L. Péwé), pp. 411–415, McGill-Queens University Press.

West, R. G. (1980). *The Pre-glacial Pleistocene of the Norfolk and Suffolk coasts.* Cambridge University Press, Cambridge.

West, R. G. and Donner, J. J. (1958). 'Pleistocene frost structures', *Trans. Norfolk and Norwich Nat. Soc.* **18**, 8–9.

West, R. G. and Wilson, D. G. (1966). 'Cromer Forest Bed Series', *Nature*, **209**, 497–498.

Whiteman, C. A. (1983). 'Great Waltham', in *Quaternary Research Association Field Guide: The Diversion of the Thames* (Ed. J. Rose), pp. 162–169, Birkbeck College, London.

Zagwijn, W. H. (1975). 'Variations in climate as shown by pollen analysis, especially in the Lower Pleistocene of Europe', in *Ice Ages: Ancient and Modern* (Eds A. E. Wright and F. Moseley), pp. 137–152, Seel House Press, Liverpool.

Soils and Quaternary Landscape Evolution
Edited by J. Boardman
© 1985 John Wiley & Sons Ltd.

10

The Troutbeck Paleosol, Cumbria, England

JOHN BOARDMAN

ABSTRACT

The Troutbeck Paleosol, developed in pre-Devensian glacigenic sediments, occurs at several sites in the Mosedale and Thornsgill valleys, in Cumbria, England. Irregular bedrock topography afforded protection for the sediments and soil from glacial erosion. Relatively unweathered Devensian sediments overlie the paleosol and generally form the land surface. The Troutbeck Paleosol consists of a zone of severe weathering with pedogenic features, which at one site extends to bedrock at a depth 15 m. The soil is truncated with sheared blocks incorporated into the Devensian till. The paleosol may be recognized on the basis of its colour: it is typically yellowish brown (10 YR 5/6) having developed from dark-grey (N 3/0) and dark-bluish-grey (10 BG 4/1) parent material. It contains rotted clasts of andesite and microgranite up to 1 m in length which are yellowish-brown throughout and break down to sand. The exteriors of many clasts are stained by iron and manganese. The clay mineral alteration products which have been identified are illite, vermiculite and kaolinite. The paleosol contains pedogenic features indicative of gleying and clay translocation. The scale and severity of weathering suggest that the paleosol developed over long periods of humid temperate conditions; perhaps of the order of 100 000–150 000 years. Paleosol evidence for pre-Devensian landscape evolution is extremely rare in northern England, particularly within the zone of Devensian glacial erosion.

INTRODUCTION

The north-eastern Lake District in Cumbria, northern England, is a deeply dissected upland area composed principally of lavas and tuffs of the Borrowdale Volcanic Group (Millward, *et al.*, 1978), and thinly-cleaved mudstones of the Skiddaw Group (Jackson, 1978), both of Ordovician age. The highest peaks reach almost 1000 m and deep glacial troughs, often occupied by lakes, radiate from the central highland (Figure 10.1). There is abundant evidence for both regional glaciation, when the landscape was submerged by ice sheets, and for phases of restricted corrie and valley glaciation (Manley, 1959; Sissons, 1980). Valley-side slopes are mantled with till or periglacial deposits (Boardman, 1978).

231

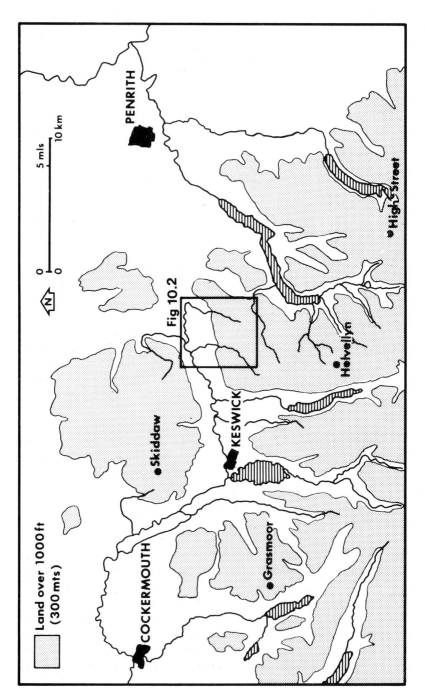

Figure 10.1   The northern Lake District, Cumbria

Table 10.1   Climatic data for the northern Lake District

| Station | Altitude (m) | Mean air temperature (°C) | | | Annual precipitation (mm) |
		Year	January	July	
Keswick	77	9.2	3.7	15.2	1470
Newton Rigg, near Penrith	168	8.3	2.4	14.7	896

Data from Manley (1973) and Matthews (1977).

The present climate of the northern Lake District may be represented by data from two sites (Table 10.1). Manley (1973) estimates that precipitation exceeds 2000 mm on the highest parts of Skiddaw. Sites on the lower ground between Keswick and Penrith, including the Mosedale basin, are likely to receive about 1700 mm (Matthews, 1977, Figure 3).

The natural vegetation of the Lake District during the Flandrian is mixed-oak forest which earlier reached to around 640 m on the Skiddaw massif. Since 5000 BP, clearance has greatly reduced the extent of forest, sheep grazing has prevented regeneration and the area is now dominated by open grassland (Pearsall and Pennington, 1973). Clearance of woodland has influenced soil development so that the natural process of acidification of originally base-rich soils has been hastened. At present the principal soil types in the area under consideration (Figure 10.2) are cambic stagnohumic gleys in the Mosedale basin and raw oligo-fibrous peat soils in the Thornsgill basin (Jarvis *et al.*, 1983).

QUATERNARY STRATIGRAPHY

The formal stratigraphic names proposed for the north-eastern Lake District (Boardman, 1981a and b) are given in Table 10.2. The two major litho-stratigraphic units in the area are the glacigenic deposits of the Threlkeld and Thornsgill Formations. The Threlkeld Formation consists of till, sands and gravels, and laminated beds which generally form the ground surface, and on which drumlins are the characteristic landform. It is the regional representative of the Late Devensian glaciation culminating about 18 000 BP. The Thornsgill Formation consists of a gravelly till and glacifluvial sands and gravels and occurs at several sites in the Mosedale and Thornsgill Beck valleys beneath the Threlkeld Formation (Figure 10.2). Soil development and weathering in the Thornsgill Formation resulted in the Troutbeck Paleosol. The evidence from two sites is considered in detail below and the relationship of the major units one to another and their position in the landscape is illustrated in Figure 10.3. Both till units contain mudstones from the underlying bedrock, transported Borrowdale Volcanic Group lithologies and Threlkeld Microgranite from outcrops 4 km west of Mosedale, reflecting ice movement from the west and south west.

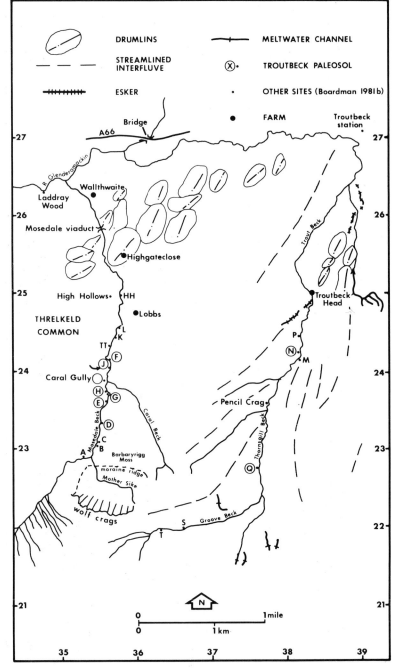

Figure 10.2   The valleys of Mosedale and Thornsgill Beck

Table 10.2 Formal stratigraphy proposed for the Quaternary of the north-eastern Lake District

Stage name	Sediments and soil properties	Lithostratigraphy		Soil stratigraphy	Environment
		Formation	Member		
Flandrian	Colluvium, alluvium, land-slip debris etc.		Not investigated		
	Rubification, silt and clay translocation, gleying			Laddray Wood Paleosol	Humid, warm temperate
	Scree	Millbeck	Dodd Wood Scree, Latrigg Grèzes Litées (stratified scree)		Periglacial
Devensian	Gravel, Till	Wolf Crags	Wolf Crags Gravel, Wolf Crags Till		Glacial/glacifluvial
	Till, Sand, gravel and laminated beds, Gravel	Threlkeld	Threlkeld Till, Lobbs Sand and Gravel, Mosedale Gravel		Glacial/glacifluvial
Ipswichian and earlier	Clast decomposition, rubification, solution, oxidation, clay flowage, clay translocation, gleying			Troutbeck Paleosol	Humid, warm temperate
Unknown	Till, sand and gravel	Thornsgill	Thornsgill Till including sand and gravel bed		Glacial/glacifluvial

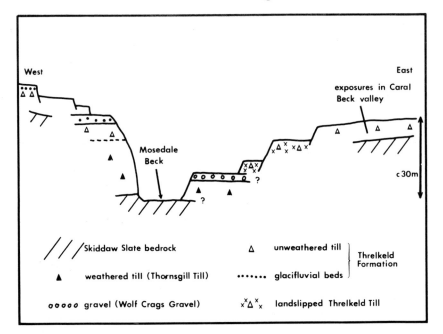

Figure 10.3   Mosedale Beck valley: diagrammatic cross-section in the Caral Beck area

METHODOLOGY

Particle-size analyses were by standard wet sieving and pipette techniques (British Standards Institution, 1975) and the following terms are used: gravel (>2 mm), sand (2 mm–63 μm), silt (2–63 μm) and clay ($<2\,\mu$m). Soil colours quoted are Standard Soil Colours (moist). Lithological analysis of clasts refers to samples from the 4–8 mm size range. Severely weathered clasts are defined as those non-mudstone clasts in the 4–8 mm size range which have totally lost their original colour and become yellow, brown or bleached. The term 'corestone' is applied to remnants of the interiors of Borrowdale Volcanic Group clasts. Micromorphological description of large thin sections (7.7×6.5 cm) uses the terminology of Brewer (1976), and Bullock and Murphy (1976) for voids.

DESCRIPTION

The Troutbeck Paleosol consists of severely weathered pre-Devensian sediments outcropping over an area of 8 km² in the Mosedale and Thornsgill valleys (Figure 10.2). At two sites the weathered zone gradually passes into relatively unweathered parent material. The greatest recorded depth of weathering is 15 m at Caral Gully (Figure 10.2). At one site, features indicative of gleying and clay translocation occur. Its main features are described at site N in Thornsgill, the type site, and D in Mosedale. At both of these sites the Thornsgill Formation overlies Skiddaw Group mudstones and is in turn overlain by deposits of the Threlkeld Formation.

Site N, Thornsgill (NY38142427)

Almost 5 m of sediments are here exposed in a river-cliff section on the west bank of Thornsgill Beck (Figure 10.4). Another 6 or 7 m are unexposed above the section before the top of the valley side is reached and these have been investigated by augering. The sequence is disturbed by glacitectonic structures. A raft of brecciated bedrock has been carried a short distance overriding sands and gravels. The latter are in places intensely folded, weathered, and contain manganese concretions. Major shear surfaces bound this unit and also occur within it.

At the base of the northern end of the section are dark-grey (N 3/0) till outcrops. This is rich in silt and clay and dominated by mudstone clasts with few severely weathered particles (Table 10.3). Above this unweathered till is a zone of partially weathered till. Lithologically this is similar to the underlying deposit except for more severely weathered clasts but is distinguished from adjacent units on the basis of olive, brown and grey colours with a texture (sample N3) intermediate between unweathered (N1) and severely weathered (N5) zones (Table 10.3, and sampling points on Figure 10.4).

The partially weathered unit is overlain by a layer of silty clay which is mainly orange (7.5 YR 6/8) with light-grey (5 Y 7/1) mottles to the south of a well-marked flexure. The main interest in this layer is its weathering features which were examined in thin section (Figure 10.4). The main features are as follows.

1   Mudstone particles, many of which are sericitized (a metamorphic effect), appear as ghosts and are difficult to distinguish from matrix material. Extensive, *in situ* weathering has occurred and the horizon has thereby acquired silt and clay.

2   Light-grey, iron-deficient areas and bright-brown areas of iron enrichment. Well-developed zones of iron depletion, or neoalbans (Veneman *et al.*, 1976) occur around natural voids. These are gley features; their composition is the same as other areas except for the removal of iron.

3   Voids occur in the form of chambers, channels, vughs and fissures. Many chambers are lined with silty argillans 0.08–0.25 mm thick which result from silt and clay being washed into voids and then settling out of suspension. Chambers and channels are generally of biological origin (Brewer, 1976).

4   Manganese is redeposited as quasicutans (Brewer, 1976) parallel to void walls but within the matrix; some voids are lined with manganese. Quasimangans and channel mangans are associated with wet conditions. Veneman *et al.*

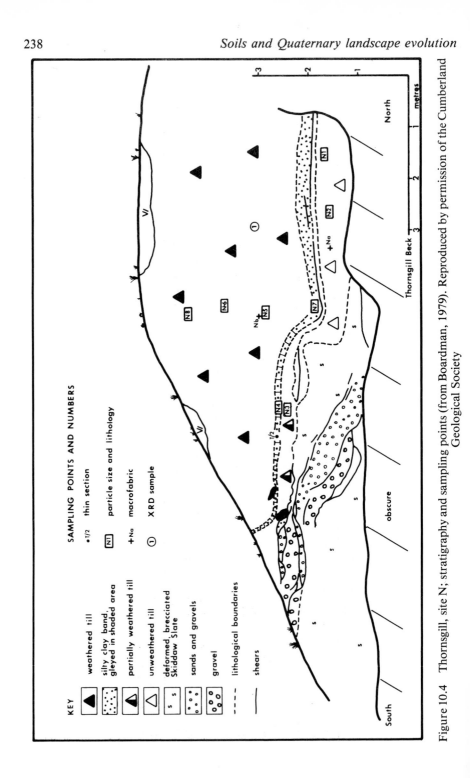

Figure 10.4   Thornsgill, site N; stratigraphy and sampling points (from Boardman, 1979). Reproduced by permission of the Cumberland Geological Society

Table 10.3 Site N, Thornsgill: particle size, lithological and weathering characteristics

Sample reference number	Percentage total sample				Percentage sample <2 mm			Percentage mudstone (n)	Percentage severely weathered (n)
	Gravel	Sand	Silt	Clay	Sand	Silt	Clay		
N9*	62.3	26.6	8.5	2.6	84.0	12.4	3.6		
N8	68.2	21.0	7.9	2.9	71.8	21.2	7.0		
N6	71.7	18.8	6.9	2.6	66.1	24.9	9.0		
N5					66.5	24.5	9.0	16.3 (1294)	100 (1074)
N4					18.9	52.2	28.9		
N7					18.7	46.0	35.3		
N3	35.3	25.4	28.2	11.1	39.2	43.7	17.1	91.1 (549)	83.6 (67)
N2	23.9	30.8	28.2	17.1	40.5	37.1	22.4		
N1	26.9	26.1	31.0	16.0	35.7	42.4	21.9	83.0 (1281)	23.1 (186)

*From laboratory crushing test: see text
(n), number in sample

(1976) found that low chromas within peds, and ped and channel neoalbans, are characteristic of horizons saturated for several months where removal of reduced iron and manganese has taken place.

Clay separations in the plasmic fabric are oriented parallel to two main directions of shearing and result from compression due to glacitectonic forces. As the voids are undisturbed they must have developed after the shearing.

The relative age of the gleying is not altogether clear. Some small-scale, iron-depleted areas as associated with voids, whereas, much larger areas of loss and enrichment of iron show no such relationship. There was probably more than one phase of reducing conditions, the earlier having been more intense or prolonged than the later ones. The small manganese deposits are associated with voids and therefore with the less-reduced conditions. Deposition of the silty argillans post-dates the formation of voids and is probably contemporaneous with the later phase of wet conditions. A tentative relative chronology is proposed.

1 Shearing and formation of sepic fabric.
2 Prolonged seasonal saturation, iron depletion and production of gley features.
3 Formation of voids by biological activity.
4 Small-scale iron depletion (neoalbans) and the development of gleying.
5 Deposition of silty argillans and manganese.

This is a horizon dominated by low chromas—a cambic horizon (Soil Survey Staff, 1975), formed at least 2.5 m below the original ground surface. Gley features are uncommon to the north of the flexure where the silty-clay layer is deeper.

About 3 m of bouldery, severely weathered till overlie the silty clay layer (Figure 10.4). Its colour is yellowish-brown (10 YR 5/8) with clasts often manganese stained. Particle-size analysis of three samples suggests that this part of the soil is relatively homogeneous with no vertical changes. It is differentiated from the partially weathered and silty-clay horizons by large gravel and small silt and clay content (Table 10.3). Sample N5 (Figure 10.4), has fewer mudstone clasts than the underlying beds; it is dominated by volcanic and microgranite clasts. In the severely weathered zone of the till, clasts are usually weathered throughout (grusified), capable of being broken in the hand and producing a yellow sand with little silt and clay (Table 10.3, sample N9). Boulders up to 0.3 m in length are seen in this condition. A sample of weathered granitic clasts obtained from the position shown in Figure 10.4 was submitted for XRD analysis which showed that the clay alteration products were illite and kaolinite (R. Merriman, pers. comm.).

Difficulties in assessing the relative importance of geological and pedogenic processes produced interpretational problems at this site. These may be summarized.

1   The character of the silty-clay layer results from deposition and flexing of a clay-rich band in the till and silt and clay enrichment by *in situ* rock-particle weathering. The relative importance of these two processes is difficult to evaluate. Micromorphological evidence suggests that weathering has picked-out and emphasized pre-existing textural and structural features.

2   The decrease in the percentage of mudstone down the profile is partly a weathering effect. However, the upper severely weathered till originally contained large granitic and volcanic boulders, for which there is no evidence in the basal unweathered mudstone-rich till unit, indicating primary lithological differences which could confuse the pedological interpretation.

With these reservations in mind it is suggested that the depth and character of the weathering at this site constitute the lower parts of a soil profile, and indicate pedogenesis under temperate conditions on a stable, vegetated land surface. Severe chemical weathering of clasts to considerable depths below a ground surface implies an advancing weathering front probably controlled by valley incision into the landscape. Gley features show that the deepening soil profiles were not always well drained but experienced periods of prolonged seasonal saturation. Pedogenesis was halted by deposition of the overlying till and glacifluvial beds which are of Late Devensian age.

Site D, Mosedale (NY35502332)

Site D is in the east bank of Mosedale Beck. The exposure does not reach the top of the valley side and the upper deposits are unexposed. The site is important because of a relatively complete sequence of deposits, representative of the area, which is exposed. The lithological succession consists of: (a) mudstone, Skiddaw Group; (b) weathered, reworked till, Thornsgill Till; (c) Troutbeck Paleosol; (d) unweathered gravel, Mosedale Gravel; and (e) unweathered till, Threlkeld Till. The relationships and sampling points for the sites are shown on Figures 10.5 and 10.6.

The weathered zone of the Thornsgill Till (the Troutbeck Paleosol) shows a downward decrease in the intensity of weathering to the extent that the basal zone appears to be little altered. Textural and lithological variability, plus considerable clast angularity, suggest that the till has suffered frost shattering and has been reworked under periglacial conditions. Many of the contrasts between units within the Thornsgill Till probably reflect deposition as a series of discrete flows. The resultant vector of the one statistically significant macrofabric is north-north-east to south-south-west, which does not conform with ice movement suggested by clast lithology and may imply movement down low-angle slopes into a pre-existing valley (Boardman, 1981b).

The basal unit of the Thornsgill Till has a maximum thickness of 0.3 m and variable, dark-greenish-grey (5 G 4/1) to dark-bluish-grey (10 BG 4/1) colour.

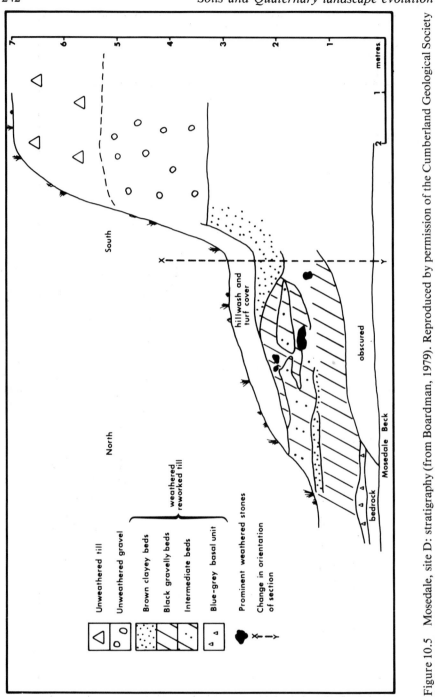

Figure 10.5   Mosedale, site D: stratigraphy (from Boardman, 1979). Reproduced by permission of the Cumberland Geological Society

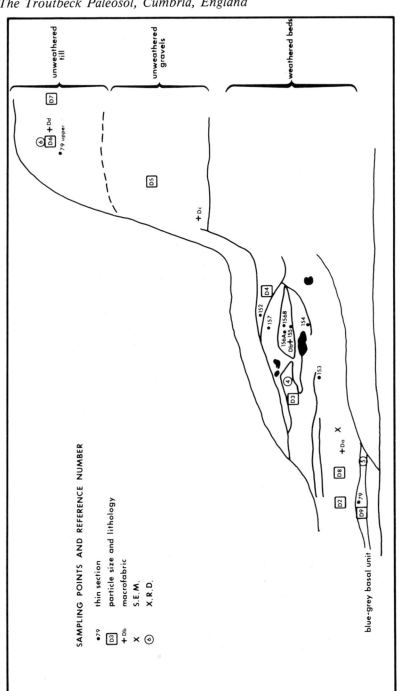

Figure 10.6   Mosedale, site D: sampling points

The unit is permanently wet with water draining from it and all clasts are soft and break easily in the hand. The particle-size data (D9, Table 10.4) reflect its weathered, friable condition: fines appear to result largely from *in situ* break down of mudstone, volcanic and granitic clasts. The volcanic clast population is severely weathered, composed of heavily pitted corestones which show evidence for a considerable change of colour, with up to 10 per cent being bleached (Table 10.4). XRD analysis of basic material shows the clay mineral alteration product to be vermiculite.

A thin section of the basal till unit shows many unweathered clasts, particularly chlorite-rich volcanic fragments which have retained their colour. Alteration of ferromagnesian minerals to clay is minimal. The basal unit has a relatively high calcium carbonate content (>4.75 per cent Ca CO_3 equivalent, by calcimeter), reflecting the condition of the original till and suggesting that the upper 2.5 m has been decalcified.

The blue-grey basal till unit grades, over about 10 cm, into a series of horizons which are differentiated on colour and texture (Figure 10.5). The black gravelly horizons are heavily manganese stained whereas horizons with greater percentages of fines are orange, yellow or brown due to iron oxide. The texture of the horizon reflects weathering of gravel-size clasts. Volcanic and granitic rocks which are dominant have been reduced to sand, and mudstones have been reduced to silt and clay. The clast population is lithologically variable, compare for example, D4 and D3 which have 57 per cent and 71 per cent mudstone respectively (Table 10.4).

Thin sections from the Troutbeck Paleosol at site D show the following features (Figure 10.6) (Boardman, 1981b, 1983).

1   Previously weathered material is incorporated into the Thornsgill Till. Some clasts have weathering rims which were broken during or after deposition. Some weathering rims have been reactivated *in situ.*
2   By comparison with the thin section from the blue-grey basal till unit, post-depositional weathering changes can be recognized as the development of yellow and brown colours and alteration of ferromagnesian minerals to clays. Most weathering must have occurred *in situ* as heavily weathered, rotted clasts could not have been transported by ice.
3   There are many examples of the flowage of clays due to hydration of mudstone clasts, producing trails of bright unweathered clays filling voids (Boardman, 1983, Figure 2). This feature is characteristic of areas where there are voids and signs of stress. The most likely cause of stress is overburden pressure of the Late Devensian ice sheet, which would account for the fresh unweathered condition of the hydrated clays, and accords with the fact that the hydrated clays disrupt iron oxide rims on weathered mudstone clasts (Boardman, 1981b, Plate 6).

Table 10.4 Site D, Mosedale: particle size, lithological and weathering characteristics

Sample reference number	Percentage total sample				Percentage sample <2 mm			Percentage mudstone (n)	Percentage severely weathered (n)
	Gravel	Sand	Silt	Clay	Sand	Silt	Clay		
D7	41.9	25.6	22.0	10.5	44.0	37.9	18.1	27.6 (500)	53.1 (346)
D6	30.5	29.4	26.3	13.8	42.3	37.9	19.8		
D5	72.5	22.8	3.1	1.6	82.9	11.2	5.9	26.1 (708)	79.3 (508)
Troutbeck Paleosol									
D4	33.7	42.5	18.5	5.3	64.1	27.9	8.0	57.3 (529)	98.2 (224)
D3	32.8	33.5	24.8	8.9	49.9	36.8	13.3	71.2 (755)	96.5 (200)
D2	68.8	27.5	3.0	0.7	88.1	9.6	2.3		
D8	68.9	25.6	4.6	0.9	82.3	14.7	3.0	42.1 (630)	99.7 (361)
D9	50.0	32.6	13.5	3.9	65.3	27.0	7.7	46.4 (760)	10.0 (401)

(*n*), number in sample

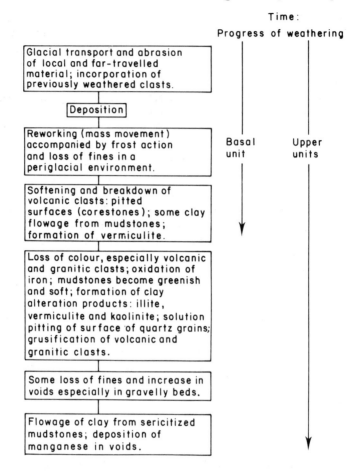

Figure 10.7   Tentative reconstruction of the erosional, depositional and weathering history of clasts in the Troutbeck Paleosol at site D (from Boardman, 1983). Reproduced by permission of AB Academic Publishers

Clay flowage is a mechanical weathering process which hitherto has not been recognized in thin section (Boardman, 1983). It demonstrates that *in situ* breakdown and physical dispersion of mudstone clasts occurred at a late stage within the Troutbeck Paleosol. The weakening of clasts which is associated with hydration helps to explain why clasts were deformed by increased overburden pressure. Stress-related features and clay flowage are particularly frequent in thin sections from the intermediate beds (Figures 10.5 and 10.6), which are sandy gravels with higher proportions of silt and clay than other samples from the paleosol at this site (Table 10.4).

XRD examination of a mixed sample from the intermediate beds of mainly granitic clasts with a little basic material, showed that the clay alteration products, in order of decreasing abundance, are illite, vermiculite and kaolinite (R. Merriman, pers. comm.). SEM examination of quartz grains from the black, gravelly beds shows a rough, flaking, surface texture similar to features described by Krinsley and Doornkamp (1973) as typical weathering in high-energy chemical environments (P. Wilson, pers. comm.).

Some tentative conclusions regarding the sequence of weathering within the Troutbeck Soil at site D are presented in Figure 10.7. It must be stressed that the impression of discrete events succeeding one another is an oversimplification. Changes would be gradual, overlap would occur and iron staining and other processes have probably been continuous.

Site D is the only site where a detailed sequence of weathering events within the paleosol can be deduced. This is because of the existence of a gradational weathering profile, with a less-weathered zone at the base, and also because of the variety of techniques and density of sampling adopted. It is likely that weathering proceeded in a similar manner at other sites.

The Troutbeck Paleosol at site D is overlain by outwash gravels and till, both of Late Devensian age and little weathered. The junction between the paleosol and the gravel is erosional and, at a higher level, blocks of the paleosol are incorporated into the Threlkeld Till. Both till and gravels contain occasional severely weathered clasts which have been eroded from the palaeosol (Table 10.4) suggesting that it may have been frozen and therefore relatively coherent. At sites outside the Mosedale and Thornsgill area, Threlkeld Till rarely contains any severely weathered clasts.

Other sites

In Mosedale, between sites G and F, there are almost continuous exposures of the paleosol in the western valley side, though some are inaccessible and landslipping complicates the interpretation of others. For example, at site J (NY35602401), almost 2 m of unweathered till underlies weathered Thornsgill Till. However, the whole sequence is part of a landslip and the Thornsgill Till may represent a sheared mass incorporated into Threlkeld Till.

Site E (NY35572356) is a section in a river cliff on the west bank of Mosedale Beck (Figure 10.8). The lowest unit is a weathered till of which neither the base nor the upper part can be seen as the base is concealed and the upper boundary is erosional. The till is cut into by a channel containing weathered fluvial beds. These deposits are overlain by unweathered fluvial gravels. The weathered till is yellowish brown (10 YR 5/6) near the base, becoming slightly yellower at higher levels. The proportion of rotted clasts appears to increase up the profile. Volcanic clasts are bleached angular corestones and some mudstones are also yellow throughout. Sample E1 (Figure 10.7), has the following particle-size

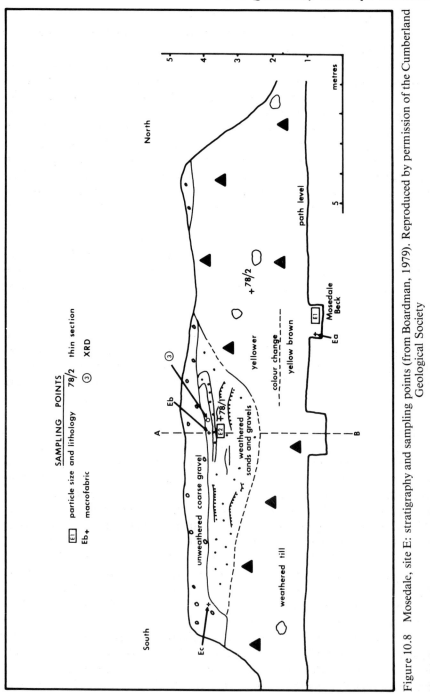

Figure 10.8   Mosedale, site E: stratigraphy and sampling points (from Boardman, 1979). Reproduced by permission of the Cumberland Geological Society

distribution: gravel 43.7 per cent, sand 30.6 per cent, silt 18.9 per cent and clay 6.7 per cent. These are typical features of the Troutbeck Paleosol developed in Thornsgill Till.

The weathered sands and gravels which overlie the till at this site are unique. A thin section of a well-sorted sand shows many signs of *in situ* fracturing of clasts. Weathering has caused oxidation, hydration with considerable clay flowage, and iron and manganese deposition. Clays which have flowed into voids can be shown, uniquely in this thin section, to have been subsequently weathered. Deformation and fracturing of clasts seems to have occurred after the weathering, probably as a result of ice overburden pressure during the Late Devensian glaciation. The unweathered gravels capping the section at site E are part of the terrace sequence within the Mosedale valley, formed by outwash from the Loch Lomond Stadial glacier at Wolf Crags (Boardman, 1981b; Rose and Boardman, 1983).

Site Q (NY37592277) is a west-bank river cliff in the Thornsgill valley. Severe weathering of Borrowdale Volcanic Group bedrock underlying Late Devensian till is considered to be part of the Troutbeck Paleosol. At this site the weathering is variable with sound areas adjacent to softened rock, which are dominated by either sandy material or by clay, presumably relating to lithological differences in the parent material.

The greatest thickness of weathering is at Caral Gully (NY35562388) in Mosedale where about 15 m of weathered Thornsgill Till overlies weathered bedrock. Typical colours of the paleosol are bright yellowish brown (10 YR 6/6) and yellowish brown (10 YR 5/8). Interiors of weathered volcanic and granitic clasts are also bright yellowish brown. Boulders, more than 1 m long, in the till and in the present river channel at the foot of the section, are also weathered throughout with large slabs flaking from their surfaces. The majority of the weathered boulders in this area are microgranite. A sample from the till is texturally similar to other weathered Thornsgill Till samples in that it has much very fine to medium sand; however, the abundant coarse silt is atypical. Weathering characteristics are similar to sites previously described in the severely weathered zone of the paleosol. A surprising observation is that despite its great thickness at Caral Gully, no down-profile changes have been recorded although the upper part of the profile has been removed by glacial erosion and the overlying Threlkeld Till incorporates sheared blocks of paleosol (Figure 10.9).

In the vicinity of Caral Gully several areas are of potential interest although landslipping complicates interpretation of much of this valley side. A section currently under investigation shows 11 m of the Troutbeck Paleosol with a reddish clayey horizon near the top.

A reddened horizon in the Troutbeck Paleosol was also recently discovered at site F (NY35622413). Reddish brown (2.5 YR 4/8) colours are recorded but a thin section shows very little illuvial clay. Lepidocrocite is the only iron oxide mineral detected by XRD and magnetic concentration, and the red pigmentation

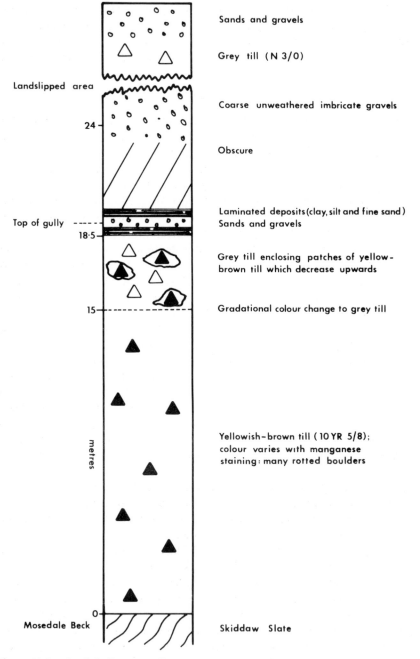

Figure 10.9   Caral Gully: generalized stratigraphy (from Boardman, 1979). Reproduced by permission of the Cumberland Geological Society

is readily reducible, consistent with the absence of haematite. The clay fraction of the sample is dominated by interstratified minerals (D. A. Jenkins, pers. comm.). The reason for reddening at this site is not known, nor is it clear if this represents typical B horizon material which has been eroded at other sites.

SYNTHESIS

The evidence available suggests that the Thornsgill Till, including till reworked after deposition, is a varied deposit, which causes difficulties in deducing the weathering histories of some sites. Unweathered parent material is, for example, dark bluish grey (10 BG 4/1) and dark grey (N 3/0). The paleosol is typically yellowish brown (10 YR 5/6) with manganese staining of clasts. Interiors of severely weathered clasts are also characteristically yellowish brown. Samples from the severely weathered horizons of the paleosol are of sandy loam, loamy sand and sand size range (Hodgson, 1974). Clasts are often rotted throughout.

Some of the more important micromorphological characteristics of the paleosol are listed in Table 10.5. These include two features which influence the character and development of the soil although they are not a product of pedogenic processes. First, signs of compressive stress are widespread, a high proportion of mudstone clasts being fractured. Fractures can be shown to have formed at different times at sites D, N and E: during deposition and reworking; while weathering was occurring; and after the main phase of weathering. Secondly, some clasts were in a partially weathered state prior to deposition and weathering features acquired at an earlier stage are sometimes reactivated during soil formation. It is tempting to regard glacial and periglacial sediments as providing fresh, unweathered parent material for soil formation but this is not the case here since some clasts show evidence of a previous cycle of weathering. In turn, weathered material from the Troutbeck Paleosol has been incorporated in the Threlkeld Till and becomes parent material for the soils of the present interglacial.

Apart from these two features, others listed in Table 10.5 are a result of *in situ* processes operating as the paleosol developed. It is worth emphasizing that clasts in the friable, decomposed condition, could not be transported without disintegration occurring. Comparison of the weathering profiles with the little weathered parent material at sites N and D illustrates the magnitude of *in situ* weathering which constitutes the Troutbeck Paleosol.

In addition to the features listed in Table 10.5, oxidation is ubiquitous except in the basal zones at sites N and D, as is the pitting of volcanic clasts. The latter process affects both the severely-weathered horizon where it has given rise to angular, bleached corestones, and has also affected some clasts in the basal unoxidized till at site D. It is assumed to result from solution by acidic groundwater although in that case it is difficult to explain the calcareous nature of the basal till at site D. Clasts decompose and are ultimately replaced by yellow

Table 10.5 Micromorphology of the Troutbeck Paleosol: selected features

Site, unit and thin-section ref	Clay flowage	Compression	Pre-transport clast weathering	Plasmic fabric	Manganese deposition	Chambers and channels	Cutans	Neo-albans
D, blue-grey basal unit	some	✓	✓	weak skel-insepic				
D, black gravel beds								
157	✓	✓	✓	nv skel-insepic	✓			
154					some			
D, brown clayey beds								
152	✓	✓	✓	lm insepic (vosepic) (skel-insepic)				
153					little			
D, intermediate beds								
156A	much	✓	✓	nv	✓			
156B	much	✓	✓	nv	✓			
155	much	much	✓	lm	✓			
E, sand and gravel								
78/1	✓	✓	✓	lm	✓			
E, Till								
78/2	nv	✓	✓	nv	✓			
N, silty clay band	✓	✓	✓	vo-masepic	✓	✓	silty argillans, mangans, quasimangans ferrans	✓

✓, present; nv, not visible; lm, little matrix.
From Boardman (1981b).

sand with a little silt and clay. Before that stage is reached the clasts are softened, the outer layer being affected first. Weathering of coarse, crystalline volcanic clasts produces angular, pitted corestones as mineral grains are softened and detached; removal of minerals in solution may also occur. The process would undoubtedly be aided by relatively high soil temperatures and long periods of time.

SEM examination of quartz grains from site D shows that chemical weathering has pitted the grain surfaces: this requires the removal of silica in solution although its chemical stability under all but the most extreme conditions is generally assumed, e.g. Folk (1974). However, to put this observation in perspective, most of the Borrowdale Volcanic Group lithologies are not quartz rich and therefore decomposition of clasts is probably achieved by chemical weathering of the much-less-stable felspars and ferromagnesian minerals. Alteration of the latter to clay minerals can be seen in thin section and is confirmed by XRD analysis.

Table 10.5 indicates that hydration of mudstones and flow of clay is seen in many thin sections. This represents late-stage weathering but the delicacy of the features is such that destruction of evidence of earlier hydration during soil formation is likely.

The weathering features so far reviewed are in varying degress present in most thin sections. Only in the silty clay layer at site N is a greater range displayed. Such features, representative of clast breakdown, gleying, and silty clay translocation, may have existed in other profiles and been lost during erosion of the upper parts of the soil profile elsewhere, or destroyed during subsequent weathering.

The plasmic fabric (Brewer, 1976) reflects the fact that clay orientation has occurred probably due to shrink-swell as wetting and drying took place. At site N, the preferred orientation of the clayey matrix results from glacitectonic stress prior to pedogenesis.

The character of the Troutbeck Paleosol is the result of pedological processes. At two sites a downward diminution of weathering is seen and relatively unweathered parent material occurs at the base of the profile. Hydrothermal activity—for which there is no evidence in the Quaternary deposits of the Lake District—would operate quite differently. Subsurface weathering beneath, in places, thick clayey unweathered Devensian till, is also inconceivable. The evidence strongly suggests that the Troutbeck Paleosol is the truncated remnant of a soil resulting from pedogenic processes operating at and below the contemporary land surface.

THE TROUTBECK PALEOSOL AND LANDSCAPE EVOLUTION

Field and micromorphological evidence show that a considerable thickness of the Thornsgill Formation was affected by oxidation, hydration, solution, the

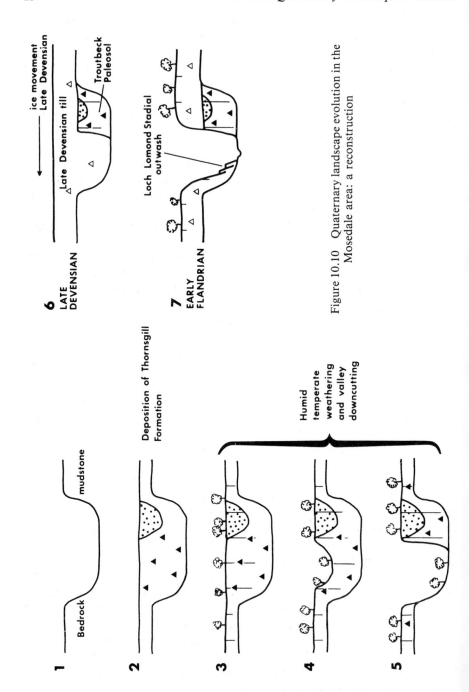

Figure 10.10 Quaternary landscape evolution in the Mosedale area: a reconstruction

throughflow of water and a fluctuating water table. The success of these processes in transforming till parent material suggests that soil development was aided by the incision of valleys into the land surface. Downcutting of valleys would be an important control on the level of the water table especially at sites near to valleys. The coarse texture of the till would facilitate the flow of water through the soil and would lead to accelerated weathering as the proportion of sand increased due to clast breakdown.

The degree of weathering of the paleosol, viewed both at macro and micro scale, is markedly greater than that which has affected the overlying deposits, those of the Devensian stage, during the Flandrian. The Troutbeck Paleosol must have formed during an interglacial, or interglacials, under a humid-temperate climate, with a stable, forested land surface.

Figure 10.10 is a reconstruction of the development of the paleosol in relation to the sedimentary and landform history of the area. The initial phase of valley formation is of unknown duration and could represent several discrete events. In Mosedale, the two later phases of valley incision have both exhumed the buried valley to a similar depth, that of the current stream channel. During the Loch Lomond Stadial (10 000–11 000 BP) a major phase of incision, followed by aggradation, occurred in Mosedale (Boardman, 1981b; Rose and Boardman, 1983). The non-cohesive character of the severely weathered deposits facilitated substantial erosion and the formation of a wide, deep valley wherever the stream incised into the paleosol (Figure 10.3). The deeply weathered condition of some of the large boulders in terraces and bars show that they were derived from the paleosol. The contrast between a Lateglacial phase of valley incision, and minimal fluvial modification of the valley during the Flandrian, suggests that in previous interglacials fluvial activity may also have been limited, especially as this area was presumably forested.

The survival of the Troutbeck Paleosol near to the base of the Glenderamackin valley which acted as a principal exit route for ice from the Lake District in the Late Devensian is surprising. It seems likely that, at most, a few metres of the upper horizons of the paleosol have been lost. The location of the paleosol within a buried valley would have afforded protection from erosion. However, despite an ice thickness of perhaps 1600 m in this area at the Late Devensian maximum (Boulton *et al.*, 1977), survival of the paleosol means very limited erosion occurred in the Mosedale and Thornsgill area. This may be because the ice sheet was cold-based and essentially non-erosive (Boardman, 1984). The reason for this must relate, in part, to temperature conditions beneath the ice sheet and may therefore reflect conditions which were regionally widespread. The assumption that the glacial landscape of the Lake District is essentially the product of the Late Devensian glacial episode (King, 1976; Moseley, 1978) is probably incorrect (Boardman, 1980).

AGE OF THE TROUTBECK PALEOSOL

Willow twigs from a peat overlying the paleosol at site H, Mosedale, have given a ^{14}C date of > 50 000 BP (Boardman *et al.*, in preparation). The Troutbeck Paleosol clearly pre-dates the Late Devensian glacial event, both stratigraphically, and by reason of the radiocarbon date. No other dateable material has been found in association with the paleosol. The stratigraphic position and character of the weathering imply formation in a pre-Devensian warm period which must include the Ipswichian Interglacial (Mitchell *et al.*, 1973). The scale and character of the weathering argue against formation during interstadial conditions and suggest a period of humid temperate conditions of greater length than that of the Flandrian.

Although there is evidence that the Last Interglacial in southern Britain was slightly warmer than the present (3 °C (Coope, 1977)), it is difficult to envisage the formation of the Troutbeck Paleosol during a period which may not have lasted more than 11 000 years (Woillard, 1979). Pre-Ipswichian interglacials, the Hoxnian and Cromerian, appear to have been similar climatically to the Flandrian (Coope, 1977).

The detailed relationship of soil development to climatic change in present-day, humid, temperate areas is still largely unknown (Boardman, 1984a). However, interglacials similar to the Flandrian, though perhaps longer, seem to have been the principal periods for pedogenesis of the type and scale of that observed in the Troutbeck Paleosol. In western Europe, glacial stages were characterized either by soil formation of a different type (e.g. Van Vliet-Lanoë, 1985), or by solifluction and erosion. Shifts of climate from glacial to interglacial modes appear to have been rapid. There is, however, no general agreement with respect to the British Isles as to the length of particular glacial and interglacial stages. Thus, in order to proceed with a brief discussion of the age of the Troutbeck Paleosol, a speculative note has to be introduced.

Harmon *et al.* (1977) define an interglacial as: 'those portions of the isotopic record which are more ^{18}O-depleted than the samples representing isotopic stage 3'. Using this criterion, recent interglacials may be identified (Figure 10.11). For western Europe this would seem to provide a reasonable approximation to reality: in the Grande Pile pollen diagram (Woillard, 1979), the fluctuating presence of mixed-oak forest trees closely reflects interglacial time as defined above.

The Troutbeck Paleosol includes profiles of 15 m depth showing no diminution in the intensity of weathering until bedrock is reached. Profiles are everywhere truncated reflecting a possible loss of 2–5 m of upper horizons. In contrast, Flandrian soil profiles in the same area show oxidation to about 2 m and contain fresh, unweathered clasts. These profiles represent about 10 000 years of pedogenesis. The Troutbeck Paleosol may represent 10–15 times the

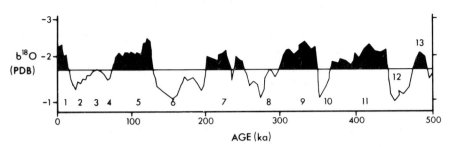

Figure 10.11 The isotopic record and interglacial periods (Harmon *et al.*, 1977). Interglacials are shown as shaded areas. Numbers below the curve are isotopic stages. Reproduced by permission of the National Research Council of Canada from the *Canadian Journal of Earth Science*, **14**, 1977

duration of weathering seen in Flandrian profiles, perhaps, 100 000 or 150 000 years of temperate conditions. Reference to Figure 10.11 therefore suggests that the formation of the paleosol is likely to have occurred in stages 9, 7 and 5.

REGIONAL SIGNIFICANCE OF THE TROUTBECK PALEOSOL

The Troutbeck Paleosol is significant because of the sparcity of evidence, both biogenic and pedogenic, for pre-Devensian conditions during the Quaternary in northern England. Interglacial deposits, probably of Ipswichian age, are described from Austerfield (Gaunt *et al.*, 1972) and Hutton Henry (Beaumont *et al.*, 1969) in the north-east. In the north-west there is the pollen record of the close of an interglacial from Scandal Beck (Carter *et al.*, 1978). Pre-Devensian pedogenesis in northern Britain is represented by paleosol features from a site outside the Devensian ice limit on the North Yorkshire Moors (Bullock *et al.*, 1973) and two paleosols beneath till at Kirkhill Quarry, north-east Scotland (Connell *et al.*, 1982).

Severely weathered till comparable to the Thornsgill Till has not been reported in Britain with the possible exception of the Portsoy site in Scotland noted by Peacock (1966). Deep or severe weathering is not uncommon, particularly in association with granitic bedrock, in unglaciated areas; this weathering has, however, usually been ascribed to Tertiary climates or hydrothermal effects, although Linton (1955) does refer to the possibility of it being interglacial. In this context, Eden and Green's (1971) discussion of the origin of the Dartmoor growan is of interest. They refer to the widespread occurrence of sandy material where granite has been weathered under temperate or sub-tropical conditions (e.g. Bakker, 1967). In warmer climates, much greater quantities of clay are produced by hydrolysis of felspar. Thus the sandy character of the Troutbeck Paleosol may be due to the weathering of the dominant andesite and microgranite lithologies, the comparative lack of mudstone as large clasts, and the temperate climates of interglacials during the Middle and Upper Pleistocene.

As has been noted, the survival of the Troutbeck Paleosol is significant in that it implies low rates of glacial erosion during the Late Devensian glaciation. The landscape of the Lake District may be of greater antiquity than hitherto suspected. Recent work in the Pennines of northern England also supports this conclusion (Atkinson *et al.*, 1978; Gascoyne *et al.*, 1983).

Finally, the paleosol is of great interest because it occurs at a number of sites over an area of about 8 km^2. In this area it acts as a valuable stratigraphic marker (Morrison, 1978) and adds considerably to our knowledge of soils and Quaternary landscape evolution prior to the Late Devensian glaciation.

ACKNOWLEDGEMENTS

The author gratefully acknowledges the assistance with pedological and petrological problems provided by Dr P. Bullock, Dr D. A. Jenkins and Dr B. Roberts; practical help was given by Dr P. Wilson (SEM), Mr R. Merriman (XRD) and Mr C. P. Murphy (thin sections). Mr S. Frampton drew the figures. Dr J. A. Catt and Mr J. Rose kindly commented on an earlier version of this paper and the latter is thanked for encouragement and advice over a number of years.

REFERENCES

Atkinson, T. C., Harmon, R. S., Smart, P. L. and Waltham, A. C. (1978). 'Paleoclimatic and geomorphic implications of ^{230}Th/^{234}U dates on speleothems from Britain', *Nature*, **272**, 24–28.

Bakker, J. P. (1967). 'Weathering of granites in different climates, particularly in Europe', in *L'Evolution des Versants* (Ed. P. Macar), pp. 52–68, University of Liege and Academie Royale de Belgique.

Beaumont, P., Turner, J. and Ward, P. F. (1969). 'An Ipswichian peat raft in glacial till at Hutton Henry, Co. Durham', *New Phytol.* **68**, 797–805.

Boardman, J. (1978). 'Grèze litées' near Keswick, Cumbria, *Biul. Peryglac.* **27**, 23–34.

Boardman, J. (1979). 'Pre-Devensian weathered tills near Threlkeld Common, Keswick, Cumbria', *Proc. Cumbs. Geol. Soc.* **4**, 33–44.

Boardman, J. (1980). 'Evidence for pre-Devensian glaciation in the northeastern Lake District', *Nature*, **286**, 599–600.

Boardman, J. (1981a). 'The valleys of Mosedale and Thornsgill, in *Field Guide to Eastern Cumbria* (Ed. J. Boardman), pp. 7–39, Quaternary Research Association.

Boardman, J. (1981b). *Quaternary Geomorphology of the Northeastern Lake District*, Unpublished PhD Thesis, University of London.

Boardman, J. (1983). 'The role of micromorphological analysis in an investigation of the Troutbeck Paleosol, Cumbria, England', in *Soil Micromorphology* (Eds. P. Bullock and C. P. Murphy), pp. 281–288, AB Academic Publishers, Berkhamsted.

Boardman, J. (1984). 'Red soils and glacial erosion in Britain', *Quaternary Newsletter*, **42**, 21–24.

Boardman, J. (1984a). Comparison of soils in midwestern United States and western Europe with the interglacial record, *Quaternary Research* (in press).

Boardman, J., Lowe, J. J., Holyoak, D. and Harkness, D. D. (in preparation). Buried peat in Mosedale, Cumbria.

Boulton, G. S., Jones, A. S., Clayton, K. M. and Kenning, M. J. (1977). 'A British ice-sheet model and patterns of glacial erosion and deposition in Britain', in *British Quaternary Studies: Recent Advances* (Ed. F. W. Shotton), pp. 231–246, Clarendon Press, Oxford.

Brewer, R. (1976). *Fabric and Mineral Analysis of Soils*, Krieger, New York.

British Standards Institution (1975). *Methods of Test for Soils for Civil Engineering Purposes*, BS 1377.

Bullock, P., Carroll, D. M. and Jarvis, R. A. (1973). 'Paleosol features in Northern England', *Nature Phy. Sci.* 242, 53–54.

Bullock, P. and Murphy, C. P. (1976). 'The microscopic examination of the structure of sub-surface horizons in soils', *Outlook on Agriculture*, 8, 348–354.

Carter, P. A., Johnson, G. A. L. and Turner, J. (1978). 'An interglacial deposit at Scandal Beck, N. W. England', *New Phytol.* 81, 785–790.

Connell, E. R., Edwards, K. J. and Hall, A. M. (1982). 'Evidence for two pre-Flandrian paleosols in Buchan, north-east Scotland', *Nature*, 297, 1–3.

Coope, G. R. (1977). 'Quaternary coleoptera as aids in the interpretation of environmental history', in *British Quaternary Studies: Recent Advances* (Ed. F. W. Shotton), pp. 55–68, Clarendon Press, Oxford.

Eden, M. J. and Green, C. P. (1971). 'Some aspects of granite weathering and tor formation on Dartmoor, England', *Geog. Annlr.* 53A, 92–99.

Folk, R. L. (1974). *Petrology of Sedimentary Rocks*, Hemphill, Austin, Texas.

Gascoyne, M., Ford, D. C. and Schwarcz, H. P. (1983). 'Rates of cave and landform development in the Yorkshire Dales from speleothem age data', *Earth Surface Proc. and Landforms*, 8, 557–568.

Gaunt, G. D., Coope, G. R., Osborne, P. J. and Franks, J. W. (1972). 'An interglacial deposit near Austerfield, southern Yorkshire', *Rep. Inst. Geol. Sci.* 72/4.

Harmon, R. S., Ford, D. C. and Schwarcz, H. P. (1977). 'Interglacial chronology of the Rocky and Mackenzie Mountains based upon ^{230}Th-^{234}U dating of calcite speleothems', *Can. J. Earth Sci.* 14, 2543–2552.

Hodgson, J. M. (1974). *Soil Survey Field Handbook*, Soil Survey Techn. Mono. No. 5, Harpenden.

Jackson, D. (1978). 'The Skiddaw Group', in *The Geology of the Lake District* (Ed. F. Moseley), pp. 79–98, Yorkshire Geological Society.

Jarvis, R. A., Allison, J. W., Bendelow, V. C., Bradley, R. I., Carroll, D. M., Furness, R. R., Kilgour, I. N. L., King, S. J. and Matthews, B. (1983). *Soils of Northern England (1:250,000)*, Soil Survey of England and Wales.

King, C. A. M. (1976). *The Geomorphology of the British Isles: Northern England*, Methuen, London.

Krinsley, D. H. and Doornkamp, J. C. (1973). *Atlas of Quartz Sand Surface Textures*, Cambridge University Press, Cambridge.

Linton, D. L. (1955). 'The problem of tors', *Geogrl. J.* 121, 470–487.

Manley, G. (1959). 'The late-glacial climate of north-west England', *Liv. and Man. Geol. J.* 2, 188–215.

Manley, G. (1973). 'Climate', in *The Lake District*, (Eds W. H. Pearsall and W. Pennington), pp. 106–120, Collins, London.

Matthews, B. (1977). *Soils in Cumbria I Sheet NY53 (Penrith)*, Soil Survey Record No. 46.

Millward, D., Moseley, F. and Soper, N. J. (1978). 'The Eycott and Borrowdale Volcanic rocks', in *The Geology of the Lake District* (Ed. F. Moseley), pp. 99–120, Yorkshire Geological Society.

Mitchell, G. H., Penny, L. F., Shotton, F. W. and West, R. G. (1973). *A Correlation of Quaternary Deposits in the British Isles*, Geol. Soc. Lond. Spec. Rep. No. 4.

Morrison, R. B. (1978). 'Quaternary soil stratigraphy — concepts, methods and problems', in *Quaternary Soils* (Ed. W. C. Mahaney), pp. 77–108, Geo Abstracts, Norwich.

Moseley, F. (1978). 'An Introductory Review', in *The Geology of the Lake District* (Ed. F. Moseley), pp. 1–16, Yorkshire Geological Society.

Peacock, J. D. (1966). 'Note on the drift sequence near Portsoy, Banffshire', *Scott. J. Geol.* **2**, 35–37.

Pearsall, W. H. and Pennington, W. (1973). *The Lake District*, Collins, London.

Rose, J. and Boardman, J. (1983). 'River activity in relation to short-term climatic deterioration', *Quat. Studies in Poland*, **4**, 189–198.

Sissons, J. B. (1980). 'The Loch Lomond Advance in the Lake District, northern England'. *Trans. Roy. Soc. Edin. Earth Sci.* **71**, 13–27.

Soil Survey Staff. (1975). *Soil Taxonomy*, U.S. Department of Agriculture, Soil Conservation Service, Agriculture Handbook 436.

Van Vliet Lanoë, B. (1985). 'Frost effects in soils', in *Soils and Quaternary Landscape Evolution* (Ed. J. Boardman), Wiley, Chichester (in press).

Veneman, P. L. M., Vepraskas, M. J. and Bouma, J. (1976). 'The physical significance of soil mottling in a Wisconsin toposequence', *Geoderma*, **15**, 103–118.

Woillard, G. (1979). 'Abrupt end of the last interglacial S.S. in north-east France', *Nature*, **281**, 558–562.

Soils and Quaternary Landscape Evolution
Edited by J. Boardman
© 1985 John Wiley & Sons Ltd.

11

Fabrics of Probable Segregated Ground-Ice Origin in Some Sediment Cores from the North Sea Basin

EDWARD DERBYSHIRE, MICHAEL A. LOVE AND MARTIN J. EDGE

ABSTRACT

Four principal types of macrofabric have been identified in the course of investigations into the origin and properties of sediments of Quaternary age in the North Sea basin. These features are compared with soil fabrics from the European mainland attributed to the development of segregated ground-ice during the Pleistocene.

INTRODUCTION

Widespread evidence of the past and present occurrence of segregated ground-ice and related soil structures in fine-grained sediments of Quaternary age has been reported from many regions of the world. Studies of the mechanisms of segregated ground-ice formation include, for example, the North American work by Black (1954, 1963, 1969), Dionne (1966a and b, 1968, 1969, 1970, 1971a and b), French (1971, 1975), Mackay (1953, 1962, 1966, 1971, 1973), Mackay and Black (1973) and Rampton and Mackay (1971). A similar level of research has developed in the other continents of the world. Despite the early work of Fitzpatrick (1956) and Pissart (1964) in associating 'fragipan' horizons (Smith, 1946; USDA, 1951) with the former existence of permafrost, until recent times very little attention has been directed towards the structural characteristics and physical properties of formerly frozen ground. Exceptions to this include, in North America, the work of De Kimpe and McKeague (1974) in Quebec, Wang *et al.* (1974) in Nova Scotia, Miller *et al.* (1971a and b) in Ohio, and Petersen *et al.* (1970) and Ranney *et al.* (1975) in Pennsylvania.

Recently, widespread evidence of segregated ground-ice fabrics (in fine-grained Pleistocene sediments) on the *mainland* of north-western Europe has been

Soils and Quaternary landscape evolution

considered in some detail (e.g. Pissart, 1970; Haesaerts and Van Vliet, 1973; Van Vliet-Lanoë, 1976; Pissart, 1982; Van Vliet, 1980; Haesaerts and Van Vliet-Lanoë, 1981; Langohr and Van Vliet, 1981; Van Vliet-Lanoë and Flageollet, 1981; Van Vliet and Langohr, 1981; Van Vliet *et al.*, 1981; Van Vliet-Lanoë, 1982; and Van Vliet and Langohr, 1983). However, in relative terms, there are few records of similar features from the British Isles. Exceptions include Fitzpatrick (1956), Stewart (1961), Crampton (1965), Matthews (1976) and Romans (1976), all of which describe indurated horizons (fragipans) whose occurrence is limited to parts of Scotland, Wales, north and south-west England. There appear to be no records in the published literature of segregated ground-ice fabrics in south-east England, nor, as far as we are aware, have any been described from the north-west European continental shelf. Southern England is known to have been subjected to periglacial conditions several times during the Pleistocene (for example, Watt *et al.*, 1966; Ranson, 1968; Paterson, 1971; Sparks *et al.*, 1972; Coope, 1975; Williams, 1975; Shephard-Thorn and Wymer, 1977; Watson, 1977). Terrestrial evidence from southern England and from

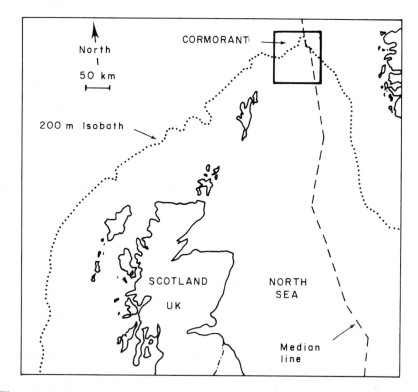

Figure 11.1   Location of sediment cores. Name of area according to British Geological Survey designation

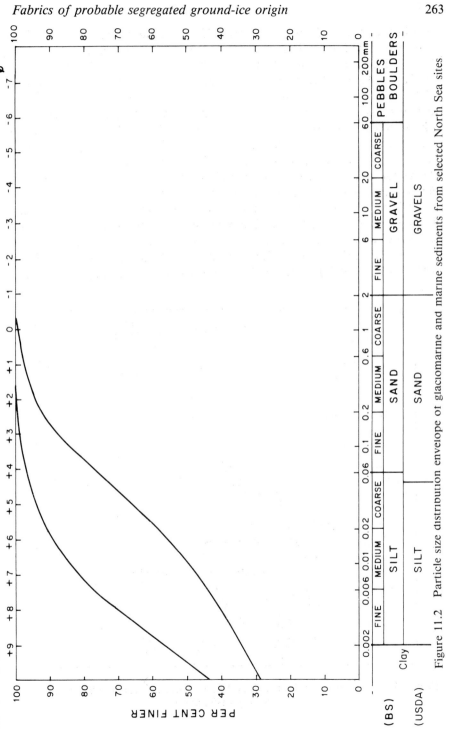

Figure 11.2 Particle size distribution envelope of glaciomarine and marine sediments from selected North Sea sites

mainland north-western Europe suggests the probability that permafrost and seasonal ground-ice growth extended on to the continental shelf both during the emergence of the sea bed as dry land (owing to glacio-eustatic changes in sea-level) and in the form of sub-sea permafrost under shallow water conditions in a manner similar to that currently occurring off the northern coast of Alaska (*cf.*. Lewellen, 1974; Chamberlain *et al.*, 1978; Hollingshead *et al.*, 1978; and Rogers and Morack, 1978).

SOME NORTH SEA SEDIMENTS

Soil fabrics strongly reminiscent of types produced by cyclic ground freezing in a periglacial environment have recently been observed in Quaternary sediment cores of glacial and marine origin at several localities in the North Sea basin (Figure 11.1). This paper records the presence, illustrates the form and disposition of these features for the first time, and outlines some implications for the Pleistocene of the North Sea and adjacent land areas.

The borehole samples were obtained by rotary coring and thin-walled tube push-sampling. They include sediments of marine, glaciomarine, glaciofluvial, glaciolacustrine and glacigenic origin. The glacigenic deposits (tills) are rarely more than a few metres thick. The thickest sequences are marine and glaciomarine deposits which consist predominantly of clay-silts or silty clays, with varying frequencies of sandy or silty partings. They commonly contain shell fragments. A representative particle-size grading envelope, with mean silt and clay fractions of 42.0 and 46.0 per cent respectively, is shown in Figure 11.2. The sediments examined are commonly finely bedded to finely laminated (< 1 mm) and contain occasional coarse sand, gravel and pebble clasts, the latter being interpreted as dropstones derived from floating ice.

The core samples described here were split by inserting a blade longitudinally some 20–30 per cent into the core diameter and pulling the sample halves gently apart. These exposed surfaces were then photographed at up to 1:1 scale. The samples were described with respect to fabric, grain size, and colour, including use of a low-power stereomicroscope. Particle-size characteristics were determined using standard wet-sieving and fixed-pipette methods (British Standards Institution, 1975). Clay mineralogy was determined by means of X-ray diffraction using the finer than $2\,\mu$m fraction. Clay-size minerals were generally of the low activity type, kaolinite, illite/mica, calcite, feldspar and quartz occurring in all samples.

In addition, both macrofabric and microfabric features were examined and photographed using a Cambridge S4 scanning electron microscope. The combination of the high resolution of the SEM, the use of low magnifications, and the compilation of photo-mosaics was designed to ensure maximum representativeness.

Figure 11.3   Subhorizontal planar joints ('macrofissility') produced by compaction and dilation. Sample depth 91 m. Scale in cm

Figure 11.4 Macrofabric transitional from type I to type II. Note low-angle discontinuities cutting across the primary laminations. Sample depth 66 m. Scale in mm

Figure 11.5 Type II macrofabric. Sample depth 51 m

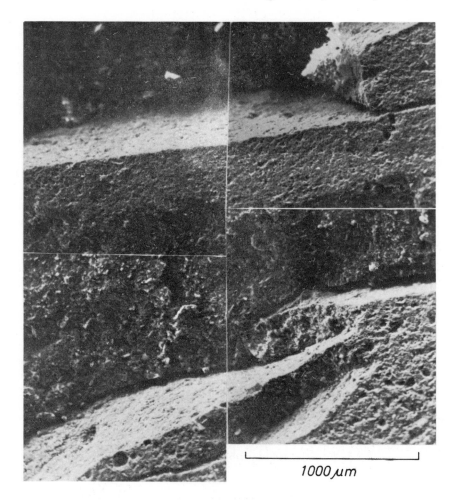

Figure 11.6   SEM photomosaic of a vertical surface showing lenticular peds and their
interfaces. Sample depth 69.5 m

FABRIC

In detail, there is considerable variation in the sediment fabric, but four principal
macrofabric types have been documented over a wide area of the North Sea
in both British and Norwegian sectors: they occur singly and in combination
(Table 11.1).

Table 11.1   Principal site-specific macrofabric types

Type	Characteristics
I	Essentially primary sedimentary deposition fabric (beds, laminae, etc.).
II	Subhorizontal intra-sample joints frequently of lenticular form.
III	Conjugate near-vertical and subhorizontal intra-sample joint sets.
IV	Trans-sample joint sets with high dips (>40°), frequently slickensided and sometimes polished.

└─────────────┘ *2000 μm*

Figure 11.7   Scanning electron micrograph (horizontal surface) of overlapping peds. Type II macrofabric. Sample depth 69.5 m

Fabrics of the primary depositional type occur at all sample depths and are generally unmodified in the upper 10 m of all cores. In addition, some subhorizontal planar partings occur following the lamination planes: these are found only at considerable depths in cores and are attributed to compaction and dilation on unloading (Figure 11.3). Often in deep cores exsolution of gas from the pore water takes place when samples are extruded at atmospheric pressures. Furthermore, rapid changes in effective stress brought about by removal of the existing *in situ* pressures (i.e. bringing a deep sample to the

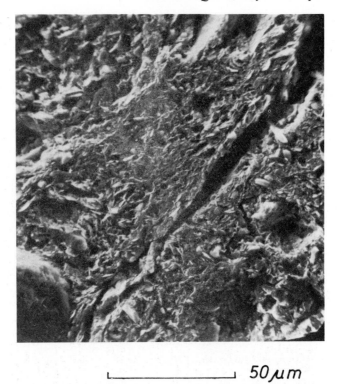

L_____J *50 µm*

Figure 11.8   Scanning electron micrograph (vertical surface) of localized clay fabric
anisotropy adjacent to discontinuity. Sample depth 25.4 m

surface) is often physically evidenced by longitudinal core expansion which may
also lead to an apparent linear core recovery in excess of 100 per cent.

Many of the core samples examined show a series of low angle (10–15°) joints
cutting across the primary laminations and disrupting them to varying degrees,
making them transitional to the type II fabric. Figure 11.4 shows a sample from
66 m below the sea bed in the Cormorant area of the North Sea. Low-angle
joints are quite well developed but the primary laminar fabric is not severely
disturbed. In contrast, the samples shown in Figures 11.5 and 11.6 have a well-
developed type II macrofabric, the lenticular peds masking the primary
laminations. The upper surfaces of the peds often display veneers or cappings
of finer material (as discussed later). The peds here rarely exceed 3 mm in
thickness but they may exceed 1 cm in diameter. The overlapping lenticular units
(Figure 11.7) show 'offsetting' producing somewhat irregular planar voids. The
aggregates readily break down into their constituent lensoid peds which have
relatively high bulk densities (about 1.8 g cc^{-1}).

Macrofabric of the third type consists of a conjugate system of two,

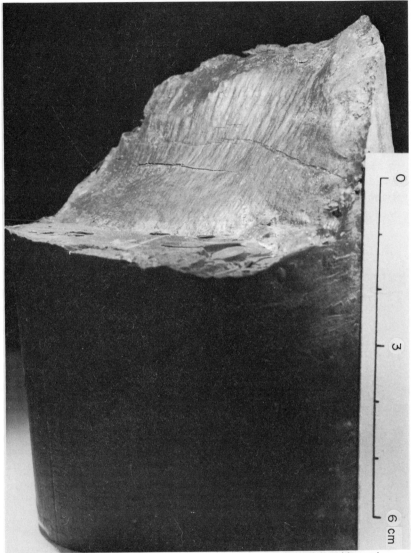

Figure 11.9   Type IV macrofabric showing high-angle polished slickenside cutting across peds of type II macrofabric. Sample depth 96 m. Scale in cm

high-angle (>45°), joint sets intercepting a low-angle set. It is clearly post-depositional as it can be seen to deform laminae adjacent to the discontinuity. The surfaces of the peds show evidence, in the form of marked clay anisotropy, of reorientation by compression and localized shearing. These reoriented zones may be as much as 50 μm wide on either side of the planar voids (Figure 11.8). Sample storage invariably induces a certain amount of moisture loss. This also involves the ingress of air into the sample which is facilitated by the presence of a fissure network. Oxidation readily occurs along these discontinuities and is evidenced by staining of the discontinuity surfaces. Fungal growth may also accompany this process. Fissure surfaces are often coated by uniformly graded material as discussed later.

Trans-sample discontinuities (Type IV) are the largest-scale fabric component identified. The spacing of these features is very difficult to assess for obvious reasons but the potential influence on the outcome of strength test results (laboratory or *in situ*) is an important practical consideration. Despite the paucity of information on their distribution throughout the sedimentary column in the North Sea basin, where present they are steeply inclined (>40°) and usually slickensided or polished (Figure 11.9). In all cases, they transgress and deform to varying degrees the other macrofabric types, clearly indicating an origin during one or more subsequent stages in the post-depositional alteration of the sediment.

DISCUSSION

Several characteristics of the samples with macrofabrics of types II and III have been discussed, notably by Van Vliet in several recent papers (loc cit.), and

Figure 11.10   Veneer or capping of finer material on uppermost ped surface (right)

ascribed to processes operating under periglacial and permafrost conditions. Given the similarity of these fabrics to those found in some of the marine sediments described here (Figure 11.1), it is suggested that similar ground-ice-induced processes operated on parts of the North Sea bed during the Pleistocene.

The migration of finer particles within the matrix during freezing cycles to form veneers or silt caps has long been recognized in terrestrial sediments (Fitzpatrick, 1956; Corte, 1963; Van Vliet-Lanoë, 1976), and similar cappings of finer material on the surfaces of peds are quite common in some horizons within the Pleistocene deposits of the North Sea (Figure 11.10).

The experimental work of Taber (1930), Beskow (1935, 1947), Pissart (1964) and Chamberlain and Gow (1978) has illustrated clearly the development of a fissure network in an otherwise intact sediment as a result of cyclic freezing, the growth of the ice forming these discontinuities being dependent on mineral particle size and shape, void size and distribution, freezing rates, effective stress, orientation of the major principal stress axis, and temperature gradients, as well as the amount of water available. In the soil samples from the North Sea basin described in this paper offsetting occurs between the peds and tends to obscure the polygonal network of discontinuities common in ice-segregated soil fabrics. This type of fabric (termed type II here) has been produced experimentally in the laboratory. Figure 11.11 shows ped formation after a single cycle of freezing and thawing between − 20 °C and + 20 °C (each of 24-hour duration). Both type II (right) and type III macrofabrics can be seen.

Figure 11.11   Experimentally produced freezing fabric formed in a hitherto unstructured clay after one freeze–thaw cycle. Scale in cm

It is widely recognized that the bulk density of the ice-induced peds is increased during this process. Values of $1.65-1.85\,\mathrm{g\,cc^{-1}}$ have been cited for some terrestrial European locations (Van Vliet and Langohr, 1981). The mean value for the peds described in this paper falls well within this range at $1.8\,\mathrm{g\,cc^{-1}}$.

The frequency of subvertical joints within macrofabric types II and III materially influences the permeability and the strength of the soil mass: bulk strength is reduced while compaction of individual peds increases their strength (Chamberlain and Blouin, 1978; Chamberlain and Gow, 1978). Large angular blocky peds (1–5 cm in diameter) cut by pronounced vertical joints have been attributed by Van Vliet and Langohr (1981) to very slow freezing of saturated clay or loamy clay within the upper part of the permafrost zone, the frequency of spacing of the ice veins and of the consequent subvertical joints decreasing with depth into the permafrost. In the cores examined from the North Sea basin, some of these high-angle joints are seen to be slickensided as a result of shear deformation, suggesting the 'fentes septiformes' of Bertouille (1972). Others (type IV) are highly polished and transect macrofabric types I, II and III, suggesting a subsequent and separate process with significantly greater shear movements than are commonly attributed to soil freezing. It is considered probable that processes including differential compaction and glacial loading have produced shear planes on scales from the trans-sample size (Figure 11.9) to those less than 1 mm in length in lightly overconsolidated glaciomarine silty clays (Derbyshire and Love, in press). The sample from the lower Tyne valley in north-east England (Figure 11.12) shows a slickensided and polished set of shear planes developed within glaciolacustrine sediments which are directly overlain by till, raising the possibility of glacially induced shearing. Although this is not the only process known to give rise to shear planes of this kind, the high overconsolidation ratios throughout the North Sea in many sediments with large shear displacements and type IV macrofabric are difficult to explain satisfactorily by other than glacial and periglacial processes. In contrast, macrofabrics of types II and III are believed to be induced by the action of periglacial processes alone.

The fabric evidence presented here suggests that ground freezing occurred both in the active layer (type II) and in the upper part of a permafrost layer (type III), on several occasions represented by samples from depths of between 10 m and about 70 m below sea bed. This represents a depth below *present* sea level of 130–190 m. The relationship of the frost-affected horizons to lower (glacial) sea level stages is unfortunately not yet known in detail in the North Sea.

There remains the clear possibility that these distinctive soil fabrics are widespread in southern England, more so than has been recognized to date. Such fabrics can themselves be used to provide evidence for the existence of cold climates by means of the minimum temperature conditions known to be necessary for their formation. Very recently, a thin diamict interbedded with a loess-like silt and other sediments has been discovered between fossil-bearing

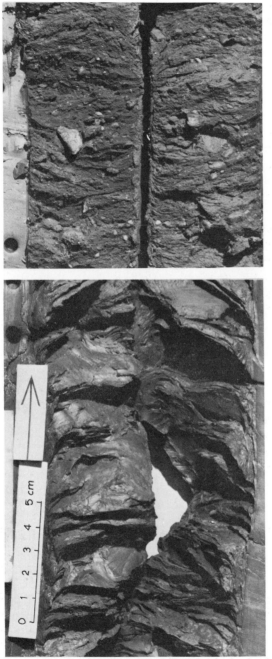

Figure 11.12   Till overlying sheared and polished glaciolacustrine laminates of the lower Tyne valley, north-east England

Figure 11.13   Macrofabric types II and III developed in terrestrial (chalky) diamict from
Marsworth, Hertfordshire. Scale in cm

beds of two different interglacials at Marsworth, near Tring, Hertfordshire
(SP935138). Whether the diamict is a till has yet to be determined, but what
is clear is the well-developed platy fabric (Figure 11.13) found in the upper 20 cm
of this small exposure which, of course, may represent a partial, truncated
profile. The upper surfaces of the smaller peds in this type II and type III
macrofabric bear 'silt caps' and the soil readily breaks up into its constituent
lenticular peds which have a mean bulk density of more than $1.9\,\mathrm{g\,cc^{-1}}$.

The similarity of the Marsworth material to fragipans and to fabric known
to result from segregated ground-ice growth adjacent to and within permafrost
is striking and raises the possibility that ground-ice soil fabrics may be more

widespread in the Pleistocene terrestrial sediments of Britain than is generally recognized. Fabrics strongly reminiscent of those widely regarded as of terrestrial ground-ice origin occur at several levels within the North Sea Pleistocene sequence in both the British and Norwegian sectors. Yet other fabric types may be the product of glacial deformation and consolidation processes. It appears that study of their type, extent and stratigraphy will prove a fruitful source of information on Pleistocene processes and environments and on variations in the engineering performance of fissured marine and glaciomarine sedimentary sequences.

ACKNOWLEDGEMENTS

This paper arises out of a contract research programme partly financed by the Building Research Establishment (Department of the Environment). Some photographs are published by permission of Shell (UK) Exploration Limited. Grateful thanks are due to Dr G. R. Coope for arranging a visit to the site at Marsworth, and to the British Geological Survey for the provision of additional core samples.

REFERENCES

Bertouille, H. (1972). 'Effet du gel sur les sols fins', *Revue de Géomorphologie Dynamique*, **21**, 71–84.

Beskow, G. (1935). 'Tjälbildingen och tjällyftningen', *Sveriges Geol. Undersökning Auh. och uppsater*, ser. C 375 (Arsbok 26 (3)).

Beskow, G. (1947). *Soil Freezing and Frost Heaving with Special Application to Roads and Railroads* (translated by J. O. Osterberg), Northwestern University Technological Institute.

Black, R. F. (1954). 'Permafrost — A Review', *Geol. Soc. Amer. Bull.*, **65**, 839–855.

Black, R. F. (1963). 'Les coins de glace et le gel permanent dans le nord de l'Alaska', *Annales Geog.*, **72**, 391, 257–271.

Black, R. F. (1969). 'Climatically significant fossil periglacial phenomena in north central United States', *Biul. Peryglacjalny*, **20**, 225–238.

British Standards Institution (1975). *Methods of Test for Soils for Civil Engineering Purposes*, BS 1377, London.

Chamberlain, E. J. and Blouin, S. E. (1978). 'Densification by freezing and thawing of fine material dredged from waterways', *Third International Conference on Permafrost*, Edmonton, Alberta, Canada, pp. 623–628.

Chamberlain, E. J. and Gow, A. J. (1978). 'Effects of freezing and thawing on the permeability and structure of soils', *Proceedings of the International Symposium on Ground Freezing*, Ruhr University, Bochum, W. Germany, Vol. 1, pp. 31–44.

Chamberlain, E. W., Sellman, P. V., Blouin, S. E., Hopkins, D. M. and Lewellen, R. I. (1978). 'Engineering properties of subsea permafrost in the Prudhoe Bay region of the Beaufort Sea', *Third International Conference on Permafrost*, Edmonton, Alberta, Canada, pp. 630–635.

Coope, G. R. (1975). 'Climatic fluctuations in north-west Europe since the last interglacial, indicated by fossil assemblages of Coleoptera', in *Ice Ages: Ancient and Modern* (Eds A. E. Wright and F. Moseley), Seel House Press.

Corte, A. E. (1963). 'Particle sorting by repeated freezing and thawing', *Science*, **142**, 499–501.

Crampton, C. B. (1965). 'An indurated horizon in soils of South Wales', *Journal of Soil Science*, **16**, 230–241.

De Kimpe, C. R. and McKeague (1974). 'Micromorphological, physical and chemical properties of a podzolic soil with a fragipan', *Canadian Journal of Soil Science*, **54**, 29–38.

Derbyshire, E. and Love, M. A. *in press*. 'Microshears in diamicts from the North Sea basin', in *Scanning Electron Microscopy in Geology* (Ed. W. B. Whalley), Geo Books, Norwich.

Dionne, J. C. (1966a). 'Formes de cryoturbation fossiles dans le sud-est de Québec', *Cahiers Géog. Québec*, **10**, 89–100.

Dionne, J. C. (1966b). 'Fentes en coin fossiles dans le Québec meridional', *Acad. Sci. (PARIS) Comptes rendus*, **262**, 24–27.

Dionne, J. C. (1968). 'Observations sur les tourbières reticulées du Lac Saint-Jean', *Mimeo*, 16 pp.

Dionne, J. C. (1969). 'Nouvelles observations de fentes de gel fossiles sur la cote sud du Saint-Laurent', *Rev. Géog. Montreal*, **23**, 307–316.

Dionne, J. C. (1970). 'Fentes en coin fossiles dans la region du Quebec', *Rev. Géog. Montreal*, **24**, 313–318.

Dionne, J. C. (1971a). 'Contorted structures in unconsolidated Quaternary deposits, Lake Saint-Jean and Saguenay regions, Québec', *Rev. Géog. Montreal*, **25**, 5–33.

Dionne, J. C. (1971b). 'Vertical packing of flat stones', *Can. Jour. Earth Sci.* **8**, 1585–1595.

Fitzpatrick, E. A. (1956). 'An indurated soil horizon formed by permafrost', *J. Soil Science*, **7**, 248–257.

French, H. M. (1971). 'Ice cored mounds and patterned ground, southern Banks Island, Western Canadian Arctic', *Geografiska Annale*, **53A**, 32–38.

French, H. M. (1975). 'Pingo investigations and terrain disturbance studies, Banks Island, N.W.T., Canada', *Can. Jour. Earth Sci.* **12**, 132–144.

Haesaerts, P. and Van Vliet, B. (1973). 'Evolution d'un permafrost fossile dans les limons du dernier glaciaire, à Harmignies (Belgique)', *Bull. de l'Association Francaise pour l'étude du Quaternaire*, **3**, 151–164.

Haesaerts, P. and Van Vliet-Lanoë, B. (1981). 'Phénomènes périglaciaires et sols fossiles observés à Maisieres Canal, à Harmignies et à Roucourt', *Biul. Peryglacjalny*, **28**, 291–325.

Hollingshead, G. W., Skjolingstad, L. and Rundquist, L. A. (1978). 'Permafrost beneath channels in the Mackenzie Delta, N.W.T., Canada', *Third International Conference on Permafrost*, Edmonton, Alberta, Canada, pp. 407–412.

Langohr, R. and Van Vliet, B. (1981). 'Properties and distribution of Vistulan permafrost traces in today's surface soils of Belgium with special reference to the data provided by the soil survey', *Biul. Peryglacjalny*, **28**, 137–148.

Lewellen, R. I. (1974). 'Offshore permafrost of Beaufort Sea, Alaska', in *Proceedings of a Symposium on Beaufort Sea Coast and Shelf Research* (Eds J. C. Reed and J. E. Slater), Vol. 1, pp. 417–437, Arctic Institute of North America.

Mackay, J. R. (1953). 'Fissures and mud circles on Cornwallis Island, N.W.T.', *Canadian Geographer*, **3**, 31–37.

Mackay, J. R. (1962). 'Pingos of the Pleistocene Mackenzie River Delta', *Geographical Bulletin*, **18**, 21–63.

Mackay, R. R. (1966). 'Segregated epigenetic ice and slumps in permafrost, Mackenzie Delta area, N.W.T.', *Geographical Bulletin*, **8**, 59–80.

Mackay, J. R. (1971). 'Ground ice in the active layer and the top portion of permafrost, 26–30', in *Proceedings of a seminar on the permafrost active layer, National Research Council Canada, Technical Memo*, **103**, (Ed. R. J. E. Brown), pp. 26–30.

Mackay, J. R. (1973). 'The growth of pingos, Western Arctic Coast, Canada', *Can. Jour. Earth Sci.* **10**, 979–1004.

Mackay, J. R. and Black, R. F. (1973). 'Origin, composition, and structure of perennially frozen ground and ground ice: a review', *Permafrost: North American contribution, Second International Permafrost Conference, Yakutsk, USSR. Nat. Acad. Sci. Publ.* **2115**, 185–192.

Matthews, B. (1976). 'Soils with discontinuous induration in the Penrith area of Cumbria', *Proceedings of the North of England Soils Discussion Group* (1974) **11**, 11–19.

Miller, F. P., Wilding, L. P., and Holowaychuk, N. (1971a). 'Cranfield silt loam, a Fragiudalf. I. Macromorphological, physical, and chemical properties', *Soil Science Society of America Proceedings*, **35**, 319–324.

Miller, F. P., Wilding, L. P. and Holowaychuk, N. (1971b). 'Cranfield silt loam, a Fragiudalf. II. Micromorphology, physical and chemical properties', *Soil Science Society of America Proceedings*, **35**, 324–331.

Paterson, K. (1971). 'Weichselian deposits and fossil periglacial structures in North Berkshire', *Proceedings Geologists' Association, London,* **82**, 455–468.

Petersen, G. W., Ranney, R. W., Cunningham, R. L., and Matelski, R. P. (1970). 'Fragipans in Pennsylvania soils: a statistical study of laboratory data', *Soil Science Society of America Proceedings* **34**, 719–722.

Pissart, A. (1964). 'Contribution experimentale à la genèse des sols polygonaux', *Soc. Géol. Belgique Annales*, **87**, 214–223.

Pissart, A. (1970). 'Les phénomènes physiques essentielles liés au gel, les structures périglaciaires qui en resultant et leurs signification climatique', *Soc. Géol. Belgique Annales*, **93**, 7–49.

Pissart, A. (1982). 'Déformation de cylindres de limon entoures de gravieres sous l'action d'alternance de gel-dégel. Expériences sur l'origine des cryoturbations', *Biul. Peryglacjalny*, **29**, 275–285.

Rampton, V. N. and Mackay, J. R. (1971). 'Massive ice and icy sediments throughout the Tuktoyaktuk Peninsula, Richards Island, and nearby areas, District of Mackenzie', *Geol. Surv. Can. Paper*, 71–21.

Ranney, R. W., Ciolkosz, E. J., Cunningham, R. L., Petersen, G. W. and Matelski, R. P. (1975). 'Fragipans in Pennsylvania soils: properties of bleached prism face materials', *Soil Science of America Proceedings*, **39**, 695–698.

Ranson, C. E. (1968). 'Ice-wedge casts in the Corton Beds', *Geological Magazine*, **105**, 74–75.

Rogers, J. C., and Morack, J. L. (1978). 'Geophysical investigation of offshore permafrost, Prudhoe Bay, Alaska', *Third International Conference on Permafrost*, Edmonton, Alberta, Canada, 561–566.

Romans, J. C. C. (1976). 'Indurated layers', *Proceedings of the North of England Soils Discussion Group* (1974), **11**, 20–30.

Shephard-Thorn, E. R. & Wymer, J. J. (Eds) (1977). 'South East England and the Thames Valley', *Guidebook for Excursion A5, International Union for Quaternary Research, X. Congress Birmingham*, Geo. Abstracts.

Smith, G. D. (1946). 'Fragipan horizons in soils: A bibliographic study and review of some of the hard layers in loess and other materials', cited in J. J. Smalley and J. E. Davin (1982), *N. Z. Soil Bureau. Bib. Rept.* **30**, 122 pp.

Sparks, B. W., Williams, R. B. G. and Bell, F. G. (1972). 'Presumed ground ice depressions in East Anglia', *Proc. Roy. Soc. Lond.* **A 327**, 329–343.

Stewart, V. L. (1961). 'A permafrost horizon in the soils of Cardiganshire', *Welsh Soils Discussion Group Report*, **2**, 19–22.

Taber, S. (1930). 'The mechanics of frost heaving', *Journal of Geology*, **38**, 303–317.

United States Department of Agriculture. (1951). *Soil Survey Manual*, U.S. Dept. Agric. Handbk 18, Soil Survey Staff, Bureau Plant Industry, Soil and Agriculture.

Van Vliet B. and Langohr R. (1981). 'Correlation between fragipans and permafrost — with special reference to Weichsel silty deposits in Belgium and Northern France', *Cantena*, **8**, 137–154.

Van Vliet B. and Langohr, R. (1983). 'Evidence of disturbance by frost of pore ferri-argillans in silty soils of Belgium and Northern France', in *Soil Micromorphology* (P. Bullock and C. P. Murphy), pp. 511–518, AB Academic Publishers, Berkhamsted.

Van Vliet-Lanoë, B., Coque-Delhuille, B. and Valadas, B. (1981). 'Les structures dérivées de la formation de glace de ségrégation dans les arénes deplacées. Analyse et application à la Margeride Occidentale', *Physio-Géo*. **2**, 17–38.

Van Vliet-Lanoë, B. (1976). 'Traces de ségrégation de glace en lentilles associées aux sols et phénomènes périglaciaires fossils', *Biul. Peryglacjalny*, **26**, 41–45.

Van Vliet-Lanoë, B. (1980). 'Corrélations entre fragipan et permagel. Application aux sols lessivés loessiques', *C. R. Groupe de Travail 'Régionalisation du Periglaciare'*, Strasbourg, **5**, 9–22.

Van Vliet-Lanoë, B. (1982). 'Structures et microstructures associées à la formation de glace de ségrégation: leurs conséquences', *Proceedings Fourth Canadian Permafrost Conference*, Calgary, Alberta, 1981, 116–122.

Van Vliet-Lanoë, B. and Flageollet, J. C. (1981). 'Traces d'activité périglaciaire dans les Vosges Moyennes', *Biul. Peryglacjalny*, **28**, 209–219.

Wang, C., Nowland, J. L. and Kodama, H. (1974). 'Properties of two fragipan soils in Nova Scotia including scanning electron micrographs', *Canadian Journal of Soil Science*, **54**, 159–170.

Watson, E. (1977). 'The periglacial environment', *Phil. Trans. Roy. Soc. Lond.* **B280**, 183–198.

Watt, A. S., Perrin, R. M. S. and West, R. G. (1966). 'Patterned ground in Breckland: structure and composition', *Jour. Ecol.* **54**, 239–258.

Williams, R. G. B. (1975). 'The British climate during the last glaciation; an interpretation based on periglacial phenomena', in *Ice Ages: Ancient and Modern*, 95–120, (Eds A. E. Wright and F. Moseley), pp. 95–120, Seel House Press.

Soils and Quaternary Landscape Evolution
Edited by J. Boardman
© 1985 John Wiley and Sons Ltd.

12

Soil Development in Flandrian Floodplains: River Severn Case-Study

MARK HAYWARD

ABSTRACT

The Severn alluvium in the Shropshire lowlands is an example of a lowland temperate floodplain in a region of long-established agriculture. A number of cutbank exposures are used as stratigraphic sections and soil profiles in order to examine properties to greater depths than is customary in soil pits. Data are presented on soil structure, free iron and organic carbon. In profiles showing moderate to strong pedality the amounts of pyrophosphate-extractable iron and organic carbon decline from the surface in association with a diminution in subangular blocky structure. Although these trends suggest that recognizable sola have developed in the soils of intermediate (loamy) and finer texture, the texture of the alluvium partly controls structural development, and grain size also plays an important role in determining chemical attributes. The possible influences on soil-profile development of sediment additions and burial are also considered.

INTRODUCTION

Recently there has been a considerable increase in interest in various aspects of floodplains, for example in relation to sedimentology, sediment budgets and stratigraphy. The soils of floodplains are often mentioned in passing, but there remains much scope for original research, particularly into what might be called synsedimentary soil formation, that is, pedogenesis concurrent with floodplain formation. Alluvial soils should not be viewed simply as the result of weathering on stable land surfaces, because soil formation overlaps with sedimentation, and is influenced by the varying rates and variable chemistry of the latter.

The River Severn floodplain in the Shropshire lowlands is an example of a floodplain in a lowland temperate region of long-established agriculture (Figure 12.1). The mean January and July temperatures at Shrewsbury are, respectively, 3.5 °C and 16.4 °C, and mean annual precipitation is 636 mm. Although the

281

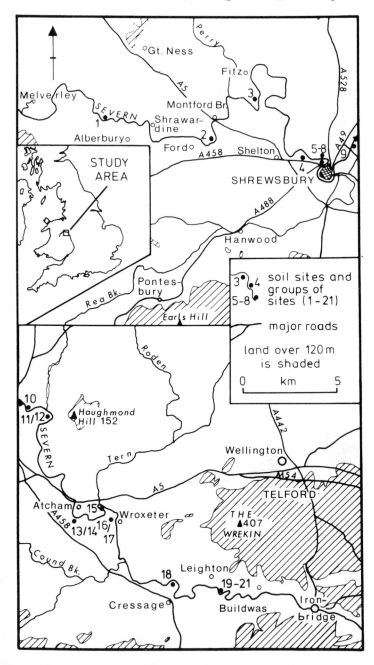

Figure 12.1   The study area and soil sample sites. Reproduced by permission of John Wiley & Sons

Figure 12.2   Severn floodplain from Wroxeter bluff. Brompton section (site 16) in background with figures for scale. Site 15 is on the crown of the bend where the river bifurcates round the ait. Floodplain extends to field boundary in middle distance

Severn is among the larger British rivers, with a channel width in the study area from 30 m to about 70 m, and a mean annual discharge near Montford Bridge (Figure 12.1) of 40.2 m^3 s^{-1}, its floodplain is usually narrow. Except for upstream of Shrawardine the alluvial tract varies in width from 100 m to 1 km, with 200–300 m a common dimension (Figure 12.2).

Alluvial morphology suggests widespread slow channel shift during the deposition of the Flandrian fill. That channel migration has been slow during historical times is confirmed by the survival of Medieval ploughmarks on the floodplain and fishweir sites in the channel.

A number of cutbanks expose alluvium ranging in texture from silty clay to loamy sand (Figure 12.3). An alluvial sequence has been proposed which comprises, from older to younger, silty, loamy and sandy alluvium (Hayward, 1982). The stratigraphy and slow channel shift imply that much of the floodplain has built up by vertical accretion; Lewin (1983) suggests that this may be typical of slowly migrating channels.

At present there is very little dating available for the Severn in Shropshire. However, D. J. Pannett (pers. comm.) records the Roman road opposite Wroxeter (Figure 12.2) resting on silty-clay alluvium with ferrimanganiferous nodules, and overlain by about 1 m of brown loamy alluvium with soil structure.

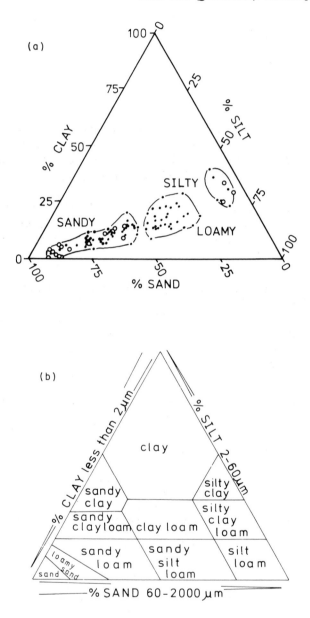

Figure 12.3 (a) Texture triangle summarizing available data. Dots: floodplain soils; circles: recent channel and flood sediment. Labels refer to site groups — small number in 'silty' category is compensated by thick units at Brompton and Fitz, whereas many 'sandy' units and horizons are thin. (b) Texture classes (Soil Survey of England and Wales)

This division corresponds to that at the cutbank site here referred to as Brompton (site 16, Figure 12.1).

SEVERN FLOODPLAIN SOILS

Methods

As part of a study relating floodplain geomorphology, stratigraphy and soil formation (Hayward, 1982) a number of cutbank sites along the River Severn were examined. Up to 4 m of alluvium is exposed, providing sections to rather greater depths than conventional soil pits. The sites are located on Figure 12.1 and listed in Table 12.1.

Soil structure was seen best in uncleaned faces, but samples were taken from as far back from the surface as possible to minimize the effects of weathering of the section. Laboratory methods are cited at the appropriate points below.

Table 12.1   List of sample sites in the study area

Number	Name		Site characteristics
1	Cae Howel	L	Raised channel belt of wide upstream floodplain
2	Montford Bridge	L	Proximal floodplain ridge
3	Fitz	L	Long-ploughed floodplain; cutbank at riffle
4	The Mount	L	Close to upstream end of bluff in Devensian deposits
5	Coton Hill	L	Proximal ridge in floodplain at mouth of cutoff loop
6	Frankwell A	R	Close to downstream end of same bluff; possible plaggen component in soil (bone, brick fragments)
7	Frankwell B	R	Fill in former barge gutter
8	Frankwell C	R	Proximal floodplain ridge (very sandy)
9	Underdale	R	Straight reach
10	Pimley	L	Proximal floodplain ridge
11	Uffington bar	R	Sandy channel marginal sediment (vegetated)
12	Uffington	R	Sandy, narrow floodplain
13	Atcham B	R	Alluvium/bluff colluvium on edge of low terrace
14	Atcham A	R	Meandering reach (featureless: long-ploughed?)
15	Near Tern Outfall	R	Proximal floodplain ridge on meander
16	Brompton	R	Long-ploughed floodplain
17	Brompton 'new'	R	Recently accreted sandy floodplain segment
18	Cressage	R	Proximal floodplain ridge
19	Leighton C	L	Example of older floodplain ridge
20	Leighton B	L	Example of younger floodplain ridge
21	Leighton A	L	Youngest, proximal floodplain ridge

R, Right bank site; L, left bank site
Vegetation is permanent pasture or long ley throughout

Soil structure

At the most elementary level two main categories of soil profile can be recognized among the studied soils. First, profiles showing pedality to a depth of at least 0.5 m; and second, profiles with shallow or weak soil structure. These two groups correspond closely with the 'oldest' and 'intermediate', and 'youngest' stratigraphic groups of Hayward (1982).

Fine subangular blocky structure to a depth of 1 m is found in the medium textured (loamy) soils associated with short meandering reaches near Atcham (represented by site 14) and Leighton (site 19). The deep development of structure includes numerous earthworm burrows, some of which extend to greater depths than ped formation. Below a depth of about 2 m in this alluvium sedimentary laminae are more frequently preserved.

In the silty alluvium at Brompton (site 16) and Fitz (site 3) stronger fine subangular blocky structure was recorded to slightly shallower depths. Surface pedality at Brompton is the strongest recorded (Figure 12.4). Although the subangular blocky structure at both sites is not apparent below 0.7 m, the alluvium below has irregular blocky aggregates 20–50 mm across. Bioturbation appears to have obliterated sedimentary structures throughout this alluvium.

The most striking structures were recorded at The Mount (site 4). From 0.5 to 1.2 m depth irregular coarse prismatic peds occur in alluvium containing

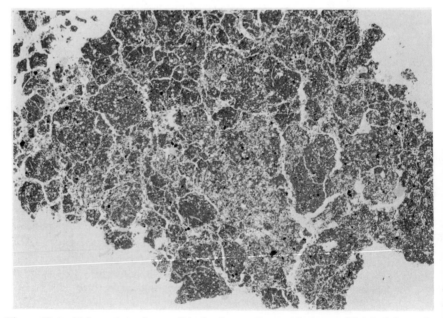

Figure 12.4   Thin section of part of Ap horizon at Brompton (natural light). Subangular blocky structure. Frame length: 50 mm; sample depth: 15 cm

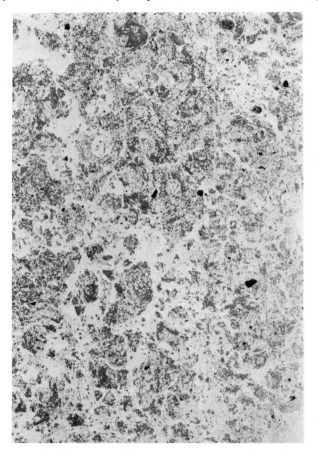

Figure 12.5   Thin section of part of Ah horizon at The Mount (natural light). Crumb structure. Frame length: 50 mm; sample depth: 12 cm

about 40 per cent clay. Overlying this horizon the profile has an Ah horizon with a fine crumb structure in contrast to the common subangular blocky peds (Figure 12.5).

Relatively strong and deep pedality was also recorded at Atcham B (site 13), Cae Howel (site 1), Coton Hill (site 5), Frankwell A (site 6) and Underdale (site 9). At Underdale the section exposes alluvium in which the usual fine subangular blocky structure is replaced at a depth of about 0.3 m by a coarse blocky structure. The ped faces at 0.9 m were recorded in the field as having possible clay coats and there is evidence from thin sections taken at 0.6 m and 1.1 m that illuvial clay is present.

At Coton Hill the aggregates are friable and irregular, and may be caused by intensive bioturbation and worm-casting (Figure 12.6). This type of structure

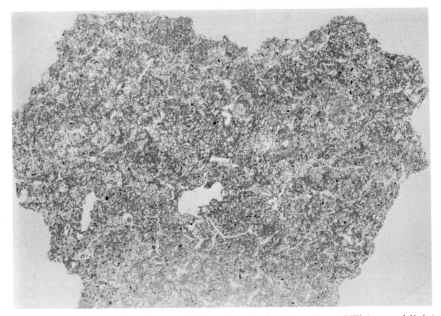

Figure 12.6   Thin section of part of bioturbated subsoil at Coton Hill (natural light).
Frame length: 50 mm; sample depth: 45 cm

was not recorded in other relatively recent ('proximal') floodplain ridges but
clay percentages at Coton Hill are about twice those at comparable locations
(14–20 per cent against 6–12 per cent).

Most of the other profiles examined occur on floodplain ridges adjoining the
channel, and are probably of more recent age. Few structures, sedimentary or
pedogenic, were recorded at Montford Bridge (site 2). Bioturbation may have
destroyed sedimentary structures and few aggregates have formed in the sandy
alluvium. The alluvium of most proximal ridges does have sedimentary
structures, however, usually small-scale cross-stratification and horizontal
lamination. Among the sites on sandy proximal ridges a further distinction can
be drawn between those with a prominent cap of finer sediment (49 per cent
and 67 per cent silt and clay in the surface 40 cm at Leighton A and B—
sites 20 and 21 — respectively), and those where such a feature does not occur
(Cressage (site 18), 'Near Tern Outfall' (site 15) and Pimley (site 10), in addition
to Coton Hill and Montford Bridge). Pedogenic structures are weakly developed
in the surface alluvium at Leighton A and B, but at greater depth the sedimentary
structures are quite well preserved, except where thin siltier layers show evidence
of bioturbation. At Cressage and Pimley distinct units 15–25 cm thick of siltier
alluvium occur and these are more bioturbated than the intervening sandier
alluvium.

A few sites remain to be considered. At Frankwell B (site 7) the soil profile in the fill of an old barge gutter may be less than 100 years old. (A barge gutter is the navigation channel around a former fishweir site, being either an artificial cut or part of a naturally divided channel). The neighbouring site, Frankwell C (site 8), is on a sandy floodplain ridge with no evidence of peds in a profile yielding over 80 per cent sand. Opposite Uffington, site 11 is on a small, well-vegetated, marginal bar emergent at low stages. Its stratification is well preserved. The floodplain site (site 12) has a very weakly developed profile in sandy loam alluvium which is similar to other right bank profiles between Underdale and Uffington.

Marked differences in soil structure are found among the sites. Well-drained, bioturbated loams exhibit fine subangular blocky peds to about 1 m, but the sandiest soils on relatively young floodplain ridges show weak or no pedogenic structure. Apart from their relatively recent formation, this must be due to the lack of cohesion in the sandiest soils, and the limited scope for the formation in them of soil crumbs in the manner envisaged by Emerson (1959). Further interpretation is deferred until the Discussion below.

Free iron in the profiles

European researchers have emphasized the role of iron oxides as visible signs of, and as active participants in, soil development. For example, Duchaufour and Souchier (1978) present a synthesis, based on studies of soils on the Vosges granite, which illustrates the role of free iron in directing soil development towards either brunification or podzolization. Iron and clay, they propose, determine the type of humification, and this in turn controls the mobility of complexes containing iron.

Pyrophosphate- and residual dithionite-extractable iron were determined according to the method of Bascomb (1968), except that a Lovibond colour comparator was used to measure the amount of iron in the extract. Iron was calculated as the percentage weight of Fe_2O_3 in the sample. This is a conventional expression of the data as the Fe_p (pyrophosphate) extract will contain iron from organo-metal complexes, and possibly from hydroxides and in ferrous form. The pyrophosphate extract is taken as a measure of the organically bound iron in these soils, following the recent confirmation by Farmer *et al.* (1983) that Fe_p is a fairly reliable measure of organically bound iron. It is possible that iron from amorphous hydroxides was extracted, or that finely divided amorphous material was peptized (Bascomb, 1968; Jeanroy and Guillet, 1981) in addition. In general terms the Fe_p extract can be regarded as the chemically active soil iron. If the better developed Severn floodplain soils are analogous to the Brown soils of stable sites, then Fe_p values should decline from the surface.

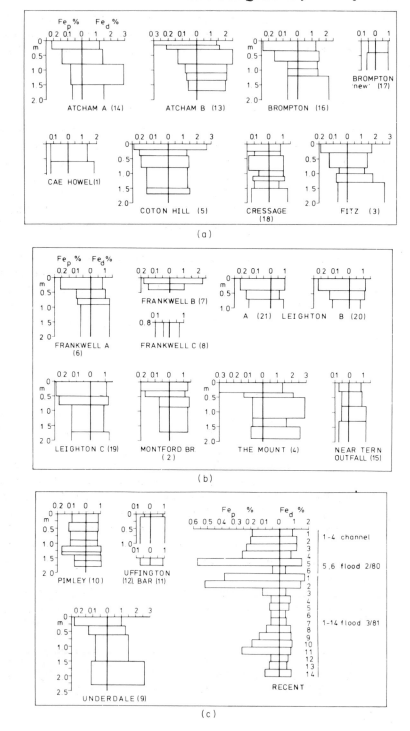

The residual dithionite extract is taken as an estimate of the amount of iron present as, for example, goethite and lepidocrocite, and which may be involved in gleying. These hydroxides of iron may be inherited from earlier episodes of weathering.

Field observations of profile mottling

The river stage usually falls quickly after a high flow to leave 3–4 m of alluvium above water level. In consequence, many of the soil profiles are well drained. Gleying is seldom in evidence above 60 cm at the cutbank sites, and augering indicates that floodplain ridge soils are unmottled above 60 cm. However, swales between ridges are usually sites of alluvial gley soils and alluvial soils with gleyic features. Cases were recorded of mottling declining from the surface possibly where standing water is retained by fine-textured surface alluvium in depressions.

The best-developed segregation features were recorded at Brompton and Fitz. The lowest unit of alluvium at these sites contains numerous ferrimanganiferous nodules up to pea size. Segregation in this silty clay appears to have reached a stage in which large areas have been depleted of iron. The sediment contains numerous old rootlet channels, but they tend not to be zones of iron accumulation: the nodules are matrix-centred.

Profile distribution of free iron

Figure 12.7a–c presents the extractable iron data. It is clear that many sites, including some proximal floodplain ridges, display a down-profile decline in pyrophosphate-extractable iron. This decline probably reflects the diminishing amounts of organic matter as depth increases, and possibly the change of amorphous hydroxides of iron to a microcrystalline form. Exceptions to this trend occur mainly on sandy floodplain ridges, in particular Cressage and Pimley. At Leighton A and B the highest Fe_p value occurs in the surface horizon which probably represents a capping of finer suspended sediment laid down on a point bar. At other sandy ridge sites, however, there appears to be no textural influence. It is impossible to explain empirically all of the variation in soil properties.

Alluvial soil Fe_p values vary within the textural range of recent sediments, the finest textured of which have the highest, and the sandiest the lowest, Fe_p values.

Figure 12.7 (a–c) Profile distributions of free iron. Pyrophosphate-extractable (Fe_p and residual dithionite-extractable (Fe_d) iron expressed as per cent weight Fe_2O_3 of soil. Sampling units are soil horizons identified in sections

In twelve profiles there is a tendency for Fe_d values to increase with depth. The soils with most clay have an Fe_d peak in the finest horizon or sedimentary unit and the sandiest soils yield the lowest values. Texture is therefore an important influence on the amount of residual dithionite-extractable iron. Textural controls are examined further in the following section.

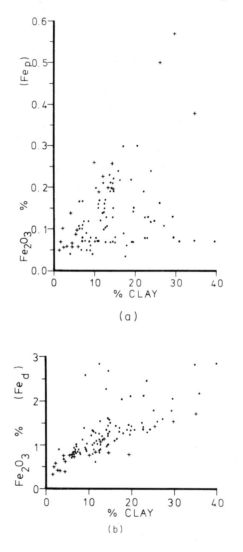

Figure 12.8   (a) Scattergram of per cent weight of Fe_p against per cent weight of clay for all samples. Dots: floodplain soils; crosses: recent channel and flood sediment. (b) Scattergram of per cent weight of Fe_d against per cent weight of clay. Symbols as for Figure 12.8a

Systematic aspects of free iron in the soils

Figure 12.8a shows the relationship of Fe_p and percentage clay. The direct association breaks down for alluvium at depth and this could be due to the tendency for the pyrophosphate extractable Fe to diminish with time after burial (Dormaar and Lutwick, 1983). If the Fe_p extract is partly represented by finely divided ferruginous particles (Jeanroy and Guillet, 1981), then some form of association of Fe_p and percentage weight of clay may be expected. This relationship is strongest for recent channel and flood sediments, sandy floodplain ridges, and A horizons with subangular blocky structure.

There is a fairly close association between Fe_d per cent and percentage clay, but four points lie well outside the main scatter (Figure 12.8b).

In recent sediments there is a weak direct association between the amounts of Fe_p and Fe_d extracted (Figure 12.9). In the older alluvium, surface horizons have relatively high values of Fe_p, with Fe_d values from 1.0 to 1.5 per cent; there is a trend towards low Fe_p and high Fe_d at depth. This may be due to the growth of goethite as a weathering product, or to originally different iron geochemistry.

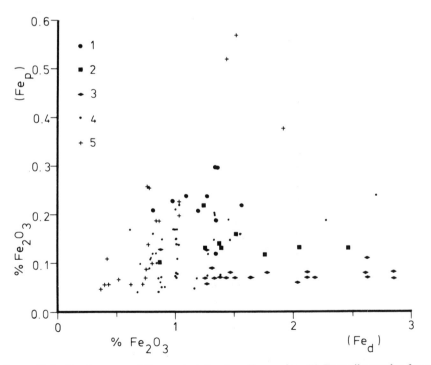

Figure 12.9   Scattergram of Fe_p against Fe_d for all samples. (1) Topsoil samples from soils with strong pedality. (2) Subsoil samples from the same soils. (3) Deeper alluvium from the same. (4) Sandy alluvium. (5) Recent channel and flood sediment

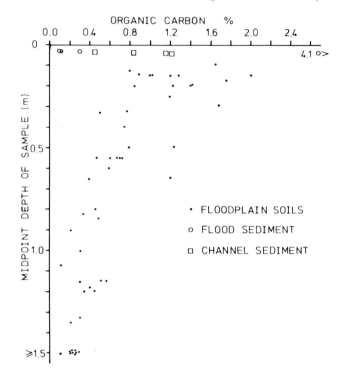

Figure 12.10   Depth plot of organic carbon. Depths shown are midpoints of sampled horizons

Organic carbon declines down the profiles (Figure 12.10); this property was determined by the method of Walkley and Black (1934). The trend is parallel to that of Fe_p, and the direct association between these two properties is shown in Figure 12.11.

Finally, the importance of texture in controlling other soil properties is shown in Figure 12.12. Exchangeable calcium was determined in an ammonium acetate extract, and the CEC was estimated according to the method of Bower *et al.* (1952). Both, especially the CEC, show a positive relationship with percentage clay, and in both plots the recent channel and flood sediments tend to yield higher values of CEC and exchangeable calcium than floodplain soils with the same amount of clay.

DISCUSSION

The better developed soils of the Severn floodplain show horizon development comparable to that of other alluvial soils in Britain. Soil structure, organic carbon and pyrophosphate-extractable iron decline with depth and the profiles

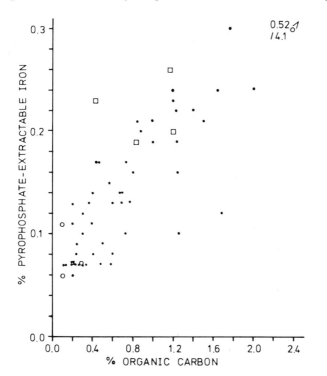

Figure 12.11   Scattergram of Fe_p against organic carbon. Symbols as in Figure 12.10

appear to be simple pedogenic sequences passing down from more-weathered to less-weathered material. They are Brown alluvial soils according to the Avery (1980) classification, and their base status renders them Fluventic Dystrochrepts according to Soil Survey Staff (1975). Soils on the sandy proximal ridges are Ranker-like alluvial soils (Typic Udifluvents).

The tendency for subangular blocky structure to weaken with depth is commonly observed in soils which are not affected by additions of material to the surface. However, in floodplain soils, where floods may deposit fresh sediment, the weakening of structure at depth may be due to reversion of the material to an unstructured state after burial. Possibly, this can be explained by the breakdown of iron-organic complexes, or the dehydration or segregation of ferric hydroxides at the same time as the oxidation of organic matter. Duchaufour and Souchier (1978) have stressed the importance of iron and clay in structural stability. These post-burial changes require more investigation especially with regard to palaeosols (Fenwick, 1985).

Evidence has been presented for some textural control of soil properties but two other points arising from the scattergrams require attention. First, recent channel and flood sediments appear to differ from floodplain soils of the same

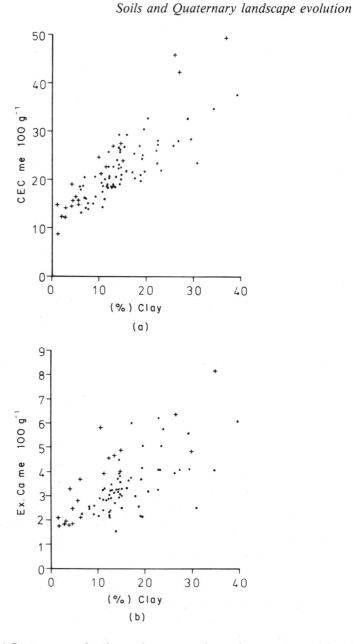

Figure 12.12   (a) Scattergram of cation-exchange capacity against per cent weight of clay. CEC by method of Bower *et al.* (1952) expressed as milliequivalents per 100 g of oven dry soil. Dots: floodplain soil; crosses: recent channel and flood sediment. (b) Scattergram of calcium in ammonium acetate extract against per cent weight of clay. Units and symbols as in Figure 12.12a

texture in respect of some chemical attributes. It was shown that the former tend to have higher values of Fe_p, exchangeable calcium and CEC than floodplain soils with the same amount of clay. The differences may be due to soil formation (for example, changes in the nature of extractable iron, and leaching of calcium), but with the present evidence the contrast could equally be related to the selective enrichment and depletion phenomena noted by Walling (1983) when comparing river sediments and their source areas.

The second qualification to the influence of textural control, and the more important one, is the ability of the organic carbon — pyrophosphate-extractable iron relationship to override the influence of grain size. The better-developed soils show an important association between these properties and soil structure.

The assertion that such alluvial materials contain simple soil profiles cannot be sustained without analyses of actual weathering phenomena. In non-calcareous soils such as those of the Severn floodplain future work should attempt to demonstrate transformations within horizons and transfers between them. Also, alluvial soils should be studied with the nature of floodplain sedimentation in mind. Aside from heterogeneity within and between sedimentary units, floodplain surfaces receive additions of sediment which interfere with soil development. Other, non-pedogenic processes, related to floodplain sedimentation might therefore form soils with similar properties to those recorded on the Severn floodplain. Three assumptions or hypotheses can be made *a priori*, necessitated by the rising alluvial profile, but which may be revised in the light of further work.

The first assumption is that floodplain soil materials go through a succession which might be summarized as 'sediment-topsoil-subsoil-deposit'. In the 'topsoil' stage the soil will inherit some properties of the original sediment and these may include relatively high values of organic matter (pending an accumulation by floodplain surface vegetation), pyrophosphate-extractable iron and cation-exchange capacity. As burial proceeds the material is isolated from the source of organic matter ('subsoil' stage), and related properties should decline, but weathering continues. Finally, the 'deposit' stage is marked by low values of organic carbon (around 0.2 per cent) and other chemical properties may be determined by the clay minerals. Weathering continues although it might properly be called early diagenesis.

The second assumption takes account of the rising alluvial profile with respect to water table. Although if the channel rises during a long period of aggradation the water table will rise as well, for individual floodplain ridges and portions of floodplain, later additions of sediment will be higher in relation to the mean water table than earlier sediment. In consequence, the processes of gleying may give way to, for example, oxidative and hydrative weathering, decalcification (where appropriate), and clay translocation. The nature and intensity of processes in the stages outlined above will change over time.

The third assumption relates sedimentation rates to soil profile development

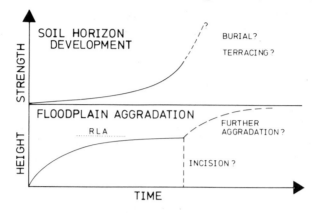

Figure 12.13   Hypothetical relationship between intensity of soil-forming processes (or horizon development) and sedimentation rate (expressed as elevation of floodplain surface). RLA is a regime level of aggradation imposed by hydrological and sedimentological controls

(Figure 12.13). In many floodplains the rising alluvial surface, or the elevation of successive floodplain ridges, will follow a curve similar to the lower one in Figure 12.13, which is based on Figure 24 of Nanson (1980). It is suggested that the intensity of soil-forming processes will vary, in consequence, according to the upper curve, which is derived in part from Figure 8.17 in Birkeland (1974), with a suggested alternative trend based on recent investigations by Birkeland (1985). Thus, as the floodplain surface approaches a hypothetical regime level of aggradation (imposed by hydrological and sedimentological controls), the intensity of soil-forming processes should increase markedly. Again, there are implications for the sequence described earlier in this section. The result of the development of floodplain soils in the way shown in Figure 12.13 would be greater horizon development at the top of the alluvial sequence.

CONCLUSIONS

Twenty-one cutbank sections along the River Severn in the Shropshire lowlands were described and sampled and data are presented here on soil structure and chemistry. Alluvium of variable age and texture is exposed. Texture (specifically percentage clay) appears to be an important control on the amount of residual dithionite-extractable iron, cation-exchange capacity, possibly exchangeable calcium, and, in surface horizons, pyrophosphate-extractable iron. Soil structure also is probably partly controlled by grain size, but in soils of loamy and silty texture the fine subangular blocky structure becomes weaker at depth as values of Fe_p and organic carbon decline. The parallel decline of strength of structural development, Fe_p and organic carbon is empirical evidence that the soils are now true sola, rather than stacks of heterogeneous sediment. Nevertheless, three mechanisms are suggested which might reproduce the observed trends in a rising alluvial profile.

ACKNOWLEDGEMENTS

Fieldwork in Shropshire was carried out during the tenure of a University of Reading Postgraduate Scholarship. Accommodation in the field area was provided by Preston Montford Field Centre. I am grateful to Jill Verran for supplementary analyses and Sara Wheeler for photographic reproduction. Bob Evans made useful comments on the manuscript.

REFERENCES

Avery, B. W. (1980). *Soil Classification for England and Wales (higher categories)*, Soil Survey Technical Monograph No. 14.
Bascomb, C. L. (1968). 'Distribution of pyrophosphate-extractable iron and carbon in soils of various groups', *Journal of Soil Science*, **19**, 251–268.
Birkeland, P. W. (1974). *Pedology, Weathering and Geomorphological Research*, Oxford University Press, New York.
Birkeland, P. W. (1985). 'Quaternary soils of the western United States', in *Soils and Quaternary Landscape Evolution* (Ed. J. Boardman), Wiley, Chichester (in press).
Bower, C. A., Reitemeier, R. F. and Fireman, M. (1952). 'Exchangeable cation analysis of saline and alkali soils', *Soil Science*, **73**, 251–261.
Dormaar, J. F. and Lutwick, L. E. (1983). 'Extractable Fe and Al as an indicator for buried soil horizons', *Catena*, **10**, 167–173.
Duchaufour, P. and Souchier, B. (1978). 'Roles of iron and clay in genesis of acid soils under a humid, temperate climate', *Geoderma*, **20**, 15–26.
Emerson, W. W. (1959). 'The structure of soil crumbs', *Journal of Soil Science*, **10**, 235–244.
Farmer, V. C., Russell, J. D. and Smith, B. F. L. (1983). 'Extraction of inorganic forms of translocated Al, Fe and Si from a podzol Bs horizon', *Journal of Soil Science*, **34**, 571–576.
Fenwick, I. (1985). 'Paleosols: problems of recognition and interpretation', in *Soils and Quaternary Landscape Evolution* (Ed. J. Boardman), Wiley, Chichester (in press).
Hayward, M. (1982). *Floodplain Landforms, Sediments and Soil Formation: the River Severn, Shropshire*, Unpublished PhD thesis, University of Reading.
Jeanroy, E. and Guillet, B. (1981). 'The occurrence of suspended ferruginous particles in pyrophosphate extracts of some soil horizons', *Geoderma*, **26**, 95–105.
Lewin, J. (1983). 'Changes of channel patterns and floodplains', in *Background to Palaeohydrology, a Perspective* (Ed. K. J. Gregory), pp. 305–319, Wiley, Chichester.
Nanson, G. C. (1980). 'Point bar and floodplain formation of the meandering Beatton River, northeast British Columbia, Canada', *Sedimentology*, **27**, 3–29.
Soil Survey Staff (1975). *Soil Taxonomy*, US Department of Agriculture, Soil Conservation Service, Handbook No. 436.
Walkley, A. and Black, I. A. (1934). 'An examination of the Degtjareff method for determining soil organic matter and a proposed modification of the chromic acid titration method', *Soil Science*, **37**, 29–38.
Walling, D. E. (1983). 'The sediment delivery problem', in *Scale Problems in Hydrology* (Guest Eds I. Rodriguez-Iturbe and V. K. Gupta), *Journal of Hydrology*, **65**, 209–237.

APPLICATIONS: OTHER AREAS

Soils and Quaternary Landscape Evolution
Edited by J. Boardman
© 1985 John Wiley & Sons Ltd.

13

Quaternary Soils of the Western United States

PETER W. BIRKELAND

ABSTRACT

A variety of deposits and soils of various ages exist in different
environmental settings across western USA. In places, soils have
an estimated age of more than 1 my. The rate of formation of various
soil features varies with environment. Bt horizons form most rapidly in
aridic areas, but they can form in about 10 000 yr in some alpine
areas. Soil formation also seems to be more rapid in alpine areas than in
the warmer, drier areas downvalley in the same ranges. Six morphological
stages of carbonate accumulation are recognized in semi-arid to arid
regions and the rate at which each stage is attained varies with region.
Pedogenic carbonate and at least some clay seems to be of atmospheric
origin. Soils can be used to date deposits and faults, but soil dates are not
always very accurate. Soil features not at depths predicted by soil-water
movement models can be used to suggest former soil-moisture regimes, and
possibly climatic change.

INTRODUCTION

Soils have played a role in the dating of Quaternary deposits in western USA.
One of the reasons for this is that commonly there are no materials for numerical
dating methods in the correct stratigraphic position. Hence, some workers resort
to relative dating methods to estimate age (Colman and Pierce, 1977; Birkeland
et al., 1979), and soils are one such method. This review will discuss some of
this work using the estimated ages and a chronosequence format. The soil
chronosequences occur in varied environments of western USA such as alpine
areas with glacial and periglacial deposits, desert basins with fluvial deposits,
and marine-terrace deposits in a mediterranean climate. The results of these
studies also have been used to suggest past climatic conditions and to help date
faults. Examples are chosen to reflect the breadth of the kinds of studies being

made and my familiarity with the areas. Further details of the included works can be found in the cited references and in Birkeland (1984).

Horizon nomenclature

The horizon nomenclature used is that of the Soil Conservation Service (Guthrie and Witty, 1982) with some modification (Birkeland, 1984). Three B horizons are common. Bw horizons have markedly redder hue or higher chroma than the assumed parent material. Bt horizons have an accumulation of pedogenic clay (pedogenic clay and carbonate show evidence for translocation and accumulation by pedological processes) that is generally greater in content than the assumed parent material, but not necessarily greater in content than that of the A horizon. Many of these Bt horizons meet the requirements of the argillic horizon (Soil Survey Staff, 1975). Bk horizons are characterized by an accumulation of pedogenic carbonate, and some of these meet the requirements of the calcic horizon (Soil Survey Staff, 1975). Accumulations of carbonate great enough to coat most of the grains, and thus impart a white colour to the horizon, are K horizons (Gile *et al.*, 1965). Gile *et al.* (1966) recognized four morphological stages of carbonate accumulation, later expanded to six stages by Machette (in press) (Figure 13.1). Stages I and II are Bk horizons, and the morphology for each is influenced by the parent material. Gravelly parent materials have thin discontinuous carbonate coats on clast bottoms at stage I, and thicker and continuous coats at stage II as well as carbonate present in the matrix. In contrast, non-gravelly parent materials have carbonate filaments at stage I, and nodules or filaments in greater concentrations at stage II (Holliday, 1982). Stage III morphology is the K horizon and is somewhat similar for both kinds of parent materials. Stage IV carbonate is cemented and water movement is inhibited; repeated solution and reprecipitation of carbonate produces an upper laminar subhorizon. The next stages (V and VI) are recognized by degree of repeated brecciation and recementation. Finally, the C horizon is subdivided into a slightly oxidized horizon (Cox) and a visually unaltered horizon (Cu).

Harden index of profile development

It is useful in studies using soils to estimate ages of deposits or geomorphological events, to be able to condense descriptive field data to a numerical scheme that depicts the overall development of the soil profile. One can appreciate the problem more when it is remembered that each soil horizon consists of at least ten properties, and that each profile may have three or more horizons. It is a difficult task to sort through this complex data set for each soil of interest and attempt to use the information for age estimation or correlation.

Harden (1982b; Harden and Taylor, 1983) has developed a useful profile development index using field morphological data. Ten soil properties of each

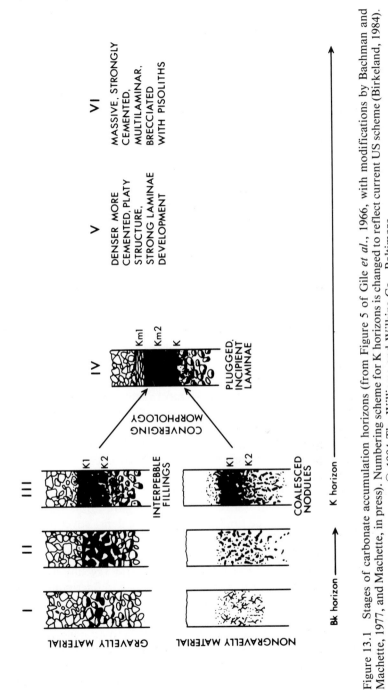
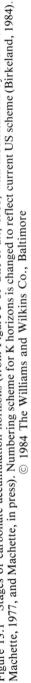

Figure 13.1 Stages of carbonate accumulation horizons (from Figure 5 of Gile *et al.*, 1966, with modifications by Bachman and Machette, 1977, and Machette, in press). Numbering scheme for K horizons is changed to reflect current US scheme (Birkeland, 1984). © 1984 The Williams and Wilkins Co., Baltimore

horizon are used to calculate the index. The properties are rubification, colour paling, total texture, clay films, structure, dry and moist consistence, melanization (value decrease), and colour lightening (value increase). Individual properties of each horizon are compared to those of the parent material, the latter being approximated by the Cu horizon or by material being laid down in the contemporary depositional system. Deviations from the parent material are assigned points, with more points for greater pedological differences from the parent material. Data are normalized and summed to obtain a horizon value, and a series of other calculations results in a weighted mean value for each profile. Ideally, higher values correspond with greater profile development, and commonly also with greater age of surface or deposit. Not all ten properties are necessary to calculate an index value.

A problem with the Harden index in many areas is that younger eolian materials make up a distinct surface layer. The properties of these materials at the time of deposition are unknown and would be difficult to estimate, as they are presently masked by pedological properties. These introduce an error in the index, and the error is usually on the side of greater index values.

Factors of soil formation

In order to better understand soil genesis, it is important to recognize the soil-forming factors of each area, and the role of each factor in soil development (Jenny, 1941, 1980; Birkeland, 1984). The main factors are climate, organisms, topographic setting, parent material and time. One problem in the factorial approach is that it is difficult to keep most factors constant over the duration of soil formation and in the ideal case one could include a change in any factor in explaining soil development. An example would be past change in soil-moisture regime. Several workers are doing this (for example, McFadden, 1982; Reheis, 1984), and their work will be mentioned later.

The areas to be discussed comprise a wide variety of climates and vegetation. Mean annual precipitation ranges from 10 to 100 cm, and mean annual temperature ranges from $-4\,°C$ to $+24\,°C$ (Figure 13.2). Vegetation follows the climatic trends. Alpine tundra is present as a dense cover on the older soils above treeline in Colorado. The only areas presently under forest, mainly various species of pine, are the mountains of Idaho, and parts of Beaver, Utah. The other areas are characterized by semi-arid to arid vegetation associations.

In most places igneous and metamorphic rocks and finer-grained components of dominantly granitic composition make up a substantial amount of the parent materials. The main exception to this is the volcanic rocks of San Clemente Island. Much of the clay and carbonate, at least in some areas, is of atmospheric origin (Gile et al., 1981; Muhs, 1983b; Machette, in press). Perhaps this accounts for at least some of the pedogenic clay and carbonate over the entire region (for example Colman, 1982).

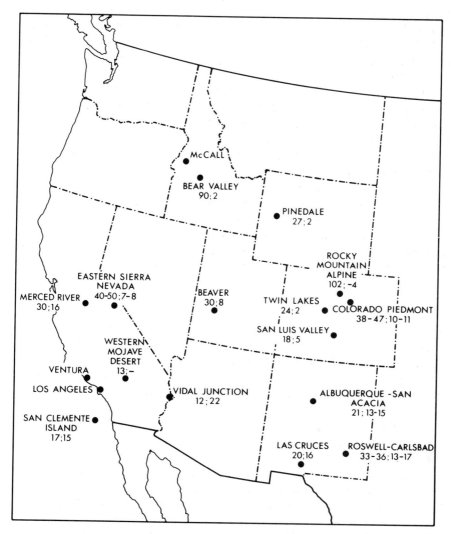

Figure 13.2   Localities mentioned in text and climatic data. First number is mean annual precipitation (cm), and second is mean annual temperature (°C). Data for Southern California transect west of Vidal Junction are given in text

All soils are studied at topographically stable sites where there is little erosion or deposition (other than from the atmosphere). The one exception to this is the till catena study, initiated to determine the degree of erosion on seemingly stable moraine crests.

Absolute dating of the deposits is poor in most of the areas. Indeed, soils and other relative-dating methods (Birkeland *et al.*, 1979) have been used by some workers to suggest ages that are related to the oxygen-isotope stages of Shackleton and Opdyke (1974). Holocene tills in cirques are dated by radiocarbon and relative-dating methods. For tills of the major alpine glaciations, Colman and Pierce (1981) used weathering rinds on clasts to suggest times of deposition in some places. The marine-terrace sequence is dated by correlation with the oxygen-isotope stages, and assuming a uniform uplift rate (Muhs, 1983a). Volcanic ashes of known age are used to suggest ages for some fluvial deposits in the southwest (Machette, in press), and they and river incision rates are used to date deposits in Montana (Reheis, 1984).

SOIL CHRONOSEQUENCES FORMED FROM TILL

Tills of various ages are common in the large valleys of the high mountain ranges across western USA (Burke and Birkeland, 1983a; Porter *et al.*, 1983). Chronosequences have been studied in all environmental settings from the alpine conditions where Holocene cirque moraines are present to the semi-arid conditions at the foot of the ranges where Pleistocene moraines are present.

Holocene alpine soil chronosequence, Colorado Rocky Mountains

Several studies have been done on alpine chronosequences formed on tills and periglacial deposits in the Rocky Mountains (Shroba and Birkeland, 1983). In the Rocky Mountain alpine (Figure 13.2), soils up to several hundred years old have no greater development than thin A/Cu profiles, with Cu horizon colour hues of 5 Y to 10 YR. Soils approximately 1000–2000 years old have weakly developed A/Cox/Cu profiles, usually 10 YR colour hues in the Cox horizon, and little pedogenic accumulation of silt or clay. In contrast, soils approximately 3000–5000 years old have A/Cox/Cu or A/Bw/Cox/Cu profiles, a maximum colour hue of 10 YR in the Cox or Bw horizon, and an increase in fines (silt and/or clay) in the A horizon that is ascribed to an eolian origin. A small portion of the clay might have been translocated to the B horizon. Finally, soils about 10 000 years old have A/Bw/Cox/Cu or A/Bt/Cox/Cu profiles, a maximum colour hue of 7.5 YR in the B horizon and an accumulation of 3–10 per cent more clay and 1–10 per cent more silt relative to the underlying horizons. Much of the A horizons and parts of some B horizons consist of a considerable silt-and-clay aeolian component (see also Burns and Tonkin, 1982), and the pedogenic translocation of these is indicated by deposits of fines on the tops of clasts (Burns, 1980). The contribution of clay from *in situ* weathering has not been evaluated.

Clay minerals and extractive chemistry suggest considerable alteration in alpine soils during the Holocene. The main clay-mineral trends are the weathering of mica and the progressive increase in mixed-layer (10–18 Å) clays with age for those soils that probably always have been above treeline. In contrast, soils presently in the alpine but within about 100 m of treeline show considerably greater mica weathering and an increase in mixed-layer (10–18 Å) clays, vermiculite and hydrobiotite. These more extensive clay-mineral changes could constitute pedological evidence for local treeline having been at least 100 m above its present position some time during the Holocene. For the Wind River Mountains of Wyoming (east of Pinedale, Figure 13.2), an area similar to the Colorado Rocky Mountains, there is considerable pedogenic accumulation of oxalate-extractable iron and aluminium in Bw horizons, as well as depletion of acid-extractable phosphorus (Birkeland, 1984, Figure 8.11).

Late Pleistocene soil chronosequence, Rocky Mountains

Soils developed in bouldery sandy tills of the major moraines formed during the more extensive glaciations of the Rocky Mountains from Wyoming to New Mexico display some trends with time (Shroba, 1977; McCalpin, 1982; Shroba and Birkeland, 1983; Birkeland, 1984, Figure 8.10). Although considerable debate exists as to the age of the deposits, for the purposes of discussion they are here assigned two approximate ages: the younger is assigned a 20 000 years age, and the older a 140 000 years age. Possible age ranges will be discussed later. Soils discussed are presently under both conifer forest and sagebush vegetation. Although there is considerable overlap in the soil data for each age group (Meierding, in press), trends mentioned here are for the better-developed soils of each group.

Soils formed in 20 000-years-old tills commonly are weakly developed. Subsurface soil properties relate best to age; these vary from Cox to Bw to Bt horizons. Soils in the semi-arid areas under sagebrush vegetation typically have Bt horizons that lack clay films, and average 5 per cent clay increase over parent material values. This latter relationship is best where there is a surface accumulation of aeolian fines. Colours are reddest (7.5 YR hue) under sagebush vegetation, and where pedogenic carbonate is present it usually has stage I morphology. In contrast, under conifer forest vegetation Bw and Cox horizons are most common beneath the A horizons, colour hues are usually 10 YR, and pedogenic carbonate is absent.

Soils formed from 140 000-years-old tills are moderately developed. The best developed soils have Bt horizons with clay films, and a 5–13 per cent clay increase relative to the parent material. Colour hues are 10 YR or 7.5 YR and, if pedogenic carbonate is present, it usually has stage II morphology. Other than the presence of carbonate in some semi-arid soils under sagebrush, these soils show less variation with vegetation than do the 20 000-years-old soils.

Soils formed from still older till, of unknown age, display greater development (Shroba, 1977). One at Twin Lakes, Colorado, has a 92-cm-thick Bt horizon, a maximum of 28 per cent clay, of which at least 17 per cent is pedogenic, and a 7.5 YR hue.

Late Pleistocene soil chronosequence, Sierra Nevada

Soils developed from late Pleistocene tills of the Sierra Nevada display pedogenic trends somewhat similar to those in the Rocky Mountains (Burke and Birkeland, 1979; Shroba and Birkeland, 1983; Birkeland, 1984, Figure 8.10). Soils formed from 20 000-years-old till have a subsurface horizon that is either a Cox or Bw horizon and 10 YR hue. Granitic clasts are usually unweathered. Soils formed from 140 000-years-old till commonly are similar to the above younger soils, but included granitic clasts are weathered to grus. Burke and Birkeland (1979) report one locality of the older till with a moderately developed Bt horizon, and we have subsequently found other localities on moraine crests with Bt horizons. Hence, except for the slightly redder B horizons in the Rocky Mountains, the better developed soils in both areas are quite similar for the

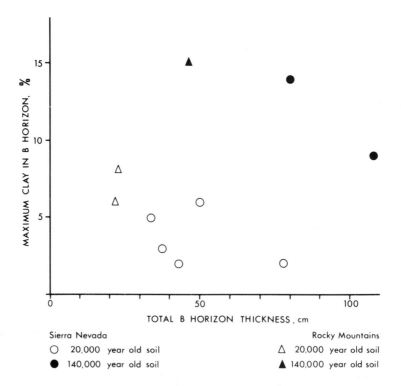

Figure 13.3   Thickness and maximum clay content of B horizons, Twin Lakes area, Colorado (Shroba, 1977), and Mammoth Lakes area, eastern Sierra Nevada, California (Burke and Birkeland, 1979)

above ages (Figure 13.3). Still older soils exist, and although they are difficult to date, the best developed of these has a 145-cm-thick Bt horizon with 40 per cent clay and 5 YR hue (Birkeland *et al.*, 1979). The deposit from which this latter soil formed could be about 0.75 million years.

Moraine catena chronosequence, Rocky Mountains and Sierra Nevada

Work is in progress studying the development of soils at several positions on the slopes of moraines of different ages. The primary reason for the study was that colleagues were questioning the use of soils formed on moraine crests for age estimation because if erosion takes place on the crests the soils may exhibit minimal development for the age of the moraine. Many studies have shown that moraines flatten with time (Meierding, in press); hence if the crests of the older ones have been lowered, soils must have been eroded, and materials eroded from crests should accumulate in downslope positions.

Soil catenas on moraines of about 20 000 and 140 000 years old in diverse environments have similar general morphologies. Areas studied include Pinedale, Wyoming (Swanson, 1984), Bear Valley, Idaho (Berry, 1984) and the eastern Sierra Nevada (Burke and Birkeland, 1983b) (Figure 13.2). Soils are studied at summit, shoulder, backslope, footslope and toeslope positions. Commonly, soils on the younger deposits are A/Bw/Cox profiles at all positions, the soils are formed in bouldery till at the summit, shoulder and backslope positions, but they are formed in sorted sandy colluvium in the footslope position and some have cumulic A horizons. In contrast, soils at equivalent positions on the older moraines are better developed than those of the younger catena. Bt horizons are usually present and are best expressed and thickest in the footslope or toeslope positions. All data reflect these age and downslope trends, including the Harden profile development index, accumulation of pedogenic clay, dithionite-extractable Fe and Al, clay mineralogy, and the weathering of feldspar in the silt fraction. The trends result from a combination of both hillslope geomorphological and pedogenic processes. Finally, the results are interpreted to indicate general regional hillslope stability on the older moraines prior to and during the glaciation that produced the younger moraines because the older soils are so well preserved and only overlain in places by a thin layer of young (last glacial?) colluvium. Although there is good evidence for periglacial activity at relatively low altitude sites in parts of the Rocky Mountains (Mears, 1981), evidence for such activity was not expressed in any of these catena soils.

SOIL CHRONOSEQUENCES FORMED FROM FLUVIAL DEPOSITS

River terraces underlain by sand and gravel deposits are common throughout western USA. Dating of these always has been difficult, and in places has to

Soils and Quaternary landscape evolution

depend on the predicted response of streams to late Quaternary environmental change (Scott, 1965; Schumm, 1965, 1977; Baker, 1983). Here I will review studies in California using the Harden profile index to correlate soils both within the state and elsewhere, discuss a pedogenic gradient along a climatic gradient in the southern part of the state, and show how pedogenic carbonate can be used to estimate ages of fluvial deposits across the south-western states.

Harden profile index: relationship with time and use in correlation of profile development

One of the better soil chronosequences in California is that of the Merced River deposits, San Joaquin Valley of California, for the factors of soil formation seem fairly constant and in places the older soils are approximately 3 million years old (Harden, 1982a, b). This is the area in which Harden (1982b) developed her index of profile development. Soils formed from deposits 10 000 years old or younger are characterized by A/Cox profiles, whereas those 40 000 years old and older have Bt horizons. With increasing age the hues of the Bt horizons are 10 YR at 40 000 years, 7.5 YR at 130 000 years, 5 YR at 330 000 years, 2.5 YR at 600 000 years and 10 R at 3 million years. All other field properties, for example B-horizon thickness, clay films and structure increase in state of development with age of deposit. The Harden profile index for these soils shows a good trend with age (Figure 13.4a).

Harden index data can be used to suggest short- and long-range correlation of Quaternary deposits. As regards the former, Ponti (in press) has studied a sequence of similar soils in the south-western Mojave Desert, and uses the Harden profile index to assign approximate ages to the deposits. Correlations with the Merced River deposits, discussed above, based on the Harden index values are quite good, at least for deposits less than about 250 000 years old.

Harden and Taylor (1983) have shown that the Harden profile index can also be used to suggest long-range correlation of deposits. Soils formed in different soil-moisture regimes were compared (Figure 13.2): xeric-inland (San Joaquin Valley data discussed above); xeric-coastal in California (Ventura); aridic in New Mexico (Las Cruces); and udic in Pennsylvania. The Pennsylvania soils are formed from till, but the rest are formed from fluvial deposits. Although the Harden profile index is made up of ten soil properties, it can be calculated using any number of properties less than ten. For the four chronosequences listed above, an index value was calculated using the eight properties that best correlate with time. Three of the chronosequences give a similar broad trend with time; only data for Las Cruces are separable and form a parallel trend at lower index values. However, when only the four best properties (not always the same four) are used to calculate the index for each chronosequence, the trends for the four areas are overlapping and similar (Figure 13.4b). More thought has to be given to the pedological explanation of the similar trends,

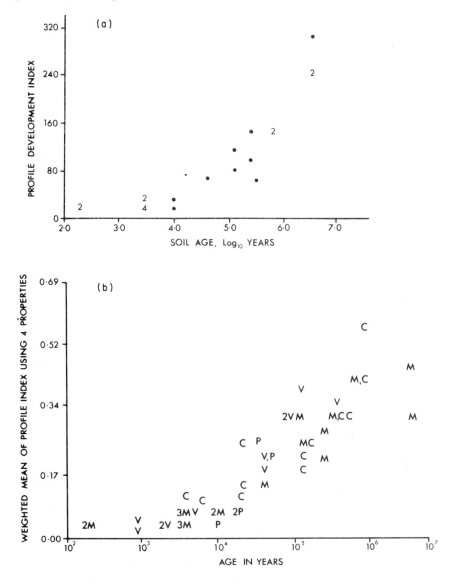

Figure 13.4(a)   Harden profile development index versus age, Merced River soils, San Joaquin Valley, California (Figure 6B of Harden, 1982b). Numbers indicate the number of profiles at that position. Reproduced by permission of Elsevier Science Publishers, Amsterdam. (b) Harden profile development index versus age for soils in four climatic regimes. Numbers are same as above. Letters indicate locality: V, Ventura, California; M, Merced River, San Joaquin Valley, California; C, Las Cruces, New Mexico; P, Pennsylvania. Figure 5C of Harden and Taylor, (1983); reproduced by permission of University of Washington

but of interest to geological studies is the development of a pedological technique useful in assigning approximate ages and correlating deposits in areas of contrasting environment.

Pedogenic gradient in Southern California

There is a pronounced east–west bioclimatic gradient in Southern California, and soils change along the gradient. McFadden (1982) has studied soil chronosequences along the bioclimatic gradient; the change in soil processes and properties along the bioclimatic gradient constitute a pedogenic gradient (Tedrow, 1977). Climate varies from arid (<12 cm mean annual precipitation; $>22\,°C$ mean annual temperature) near Vidal Junction in the eastern Mojave Desert, to semi-arid (12–25 cm; 15–22 °C) and xeric (25–75 cm; 12–22° C) in the ranges surrounding the Los Angeles Basin (Figure 13.2). Parent materials are river deposits and ages of some could approach 1 million years. On a mass basis (gm cm^{-2} profile^{-1}) both pedogenic clay (Birkeland, 1984, Figure 11.10) and dithionite-extractable iron form log- or power-function curves with time; that is, the rate of accumulation decreases with time. Bioclimate is the factor controlling the amounts of both constituents over some specified time, with low amounts in the arid soils, slightly greater amounts in the semi-arid soils, and much greater amounts in the xeric soils. Carbonate has accumulated only in soils in the arid and semi-arid climates.

Atmospheric additions and weathering during soil formation help explain the pedogenic clay, iron and carbonate along the above Southern California transect (McFadden, 1982). Several lines of evidence suggest that aeolian additions are important to the clay content and mineralogy of the arid soils. One is that atmospheric additions to soils are common east of this transect, in areas discussed later. Another is that the relatively high clay contents only in the A horizons of Holocene soils and the presence of translocated clay in Pleistocene soils suggests that time is important in moving aeolian clays within the soils. Finally, the presence of montmorillonite in the Holocene A horizons and in the Bt horizons of Pleistocene soils adds support that the clays are aeolian additions. Some of the montmorillonite in some of the older aridic soils seems to have been converted to other clay minerals; in particular, in the drier areas part could have been converted to palygorskite following the scheme proposed by Bachman and Machette (1977). In contrast, the main clay minerals in the older soils in the semi-arid and xeric climates are vermiculite and kaolinite and, although some of these clay minerals could result from the transformation of either illuviated aeolian clay or parent material clay, part are due to the *in situ* weathering of parent material minerals, notably biotite and feldspar. The long-term mean for carbonate accumulation in the aridic soils approximates $0.10\,g\,cm^{-2}/100$ years with profile amounts increasing in a linear fashion with time. Calculations suggest that carbonate derivation from weathering is insignificant and that about

20 per cent may be derived from Ca^{2+} in rainwater; hence, the bulk must be atmospheric in origin. In contrast to the above soil constituents, iron is released during the weathering of relatively less-stable mafic minerals and subsequently accumulates in the soils. In all environments the ratio of oxalate-extractable iron: dithionite-extractable iron initially increases with time and then declines. Before declining, the ratio reaches the highest values in the more moist soil environments and the lowest values in the drier soil environments.

Carbonate soils near Las Cruces, New Mexico, and correlation with carbonate soils elsewhere

The study of Gile and colleagues, summarized in Gile *et al.* (1981), was a landmark study of pedogenesis under aridic conditions. Many workers call upon processes elucidated in this work to explain many soil features in the arid and semi-arid environments. Indeed, the work serves as an example of soil-geomorphic research that could be done in any environment.

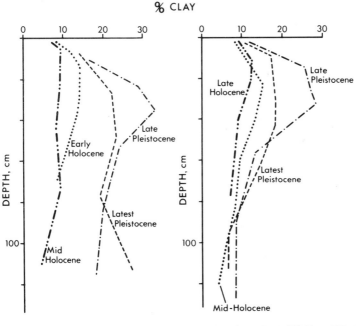

Figure 13.5   Variation in clay content with depth, age and gravel content, Las Cruces, New Mexico (from Figure 42 of Gile and Grossman, 1979, and Table 28 of Gile *et al.*, 1981)

Of greatest interest to workers using soils to date landscapes is that the area is that in which the stages of carbonate morphology were defined, and the amounts of both pedogenic clay and carbonate increase progressively with time. For example, clay accumulation is discernible in Holocene soils (Figure 13.5), and 10 000-years-old soils have clay profiles not unlike some formed from till in western USA mountains that are possibly ten times older (Birkeland, 1984, Figure 8.10). Such rapid accumulation rates might not have been predicted for aridic environments. For the attainment of carbonate stages and K horizons in gravelly materials, stage III K horizons form in < 100 000 years, and stage IV K horizons in about 150 000 years (Figure 13.6). Gravel percentage helps determine the rapidity at which clay accumulates and particular carbonate stages are reached, for both are more rapid in gravelly materials. Because gravel makes up a large volume of the mass, less absolute clay is required to produce a notable clay increase (percentage by weight of < 2 mm fraction) over a fixed thickness in a gravelly soil than in a non-gravelly one. Finally, to form a K horizon, most grains have to be coated with carbonate. This can be attained more rapidly in gravelly materials because they have a lower surface area per unit volume than do non-gravelly ones.

Aeolian influx was suspected and shown to be important in explaining the clay and carbonate accumulations (Gile *et al.*, 1981). First, derivation from *in situ* weathering had to be investigated. Calculations by Gardner (1972) have shown that unreasonable thickness of materials have to be weathered to provide the Ca present in many K horizons. Hence, dust traps were set out in the field near Las Cruces to determine the importance of aeolian additions (Gile *et al.*, 1981). The conclusion is that carbonate dust is an important source for pedogenic carbonate, but potentially greater amounts are available as Ca^{+2} in rainfall. Substantial clay was also deposited in the traps, enough probably to account for much of the clay in the soils. This atmospheric origin for most pedogenic clay and carbonate in aridic soils is well accepted by most workers, but more work is desirable on quantifying regional trends.

Machette (in press) has reported on regional trends in the attainment of carbonate stages for south-western USA (Figure 13.6; see Figure 13.2 for localities mentioned). Although reliability of dating varies, it is best where volcanic ashes of known ages are associated with deposits and soils. The results indicate that areas in southern New Mexico display the most rapid rates of carbonate morphology stage development, the aridic area (Vidal Junction) in Southern California (McFadden, 1982) has a relatively low rate of development, and the high mountain valley at Twin Lakes, Colorado (Shroba and Birkeland, 1983) and the Colorado Piedmont, have the lowest rates for those areas that were compared. The San Luis Valley, Colorado, also is an area with a relatively low rate of carbonate accumulation (McCalpin, 1982), with a rate similar to that at Vidal Junction. The above ranking of the areas on the rate of development of carbonate morphology is matched by the rate of accumulation

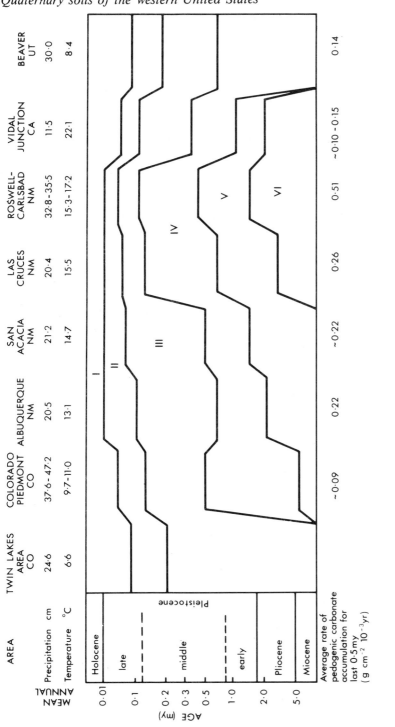

Figure 13.6 Approximate time required to attain various morphological stages of carbonate accumulation and average rate of carbonate accumulation for localities in the western United States (from Machette, in press and written communication, 1983; some data for Vidal Junction from McFadden, 1982; also Figure 8.14 of Birkeland, 1984)

of carbonate (g cm^{-2} profile^{-1}) in each area (Figure 13.6), and this probably reflects both nearby sources of carbonate dust and the amount of Ca^{2+} in rainwaters. Gypsum dust also can supply some Ca^{2+} toward the total amount in pedogenic carbonate (Lattman and Lauffenberger, 1974).

SOIL CHRONOSEQUENCE,
MARINE TERRACES OF SOUTHERN CALIFORNIA

Soil chronosequences formed from marine-terrace deposits have been difficult to locate along the California coast probably because many deposits are overlain by younger fan deposits derived from abandoned sea cliffs. One exception to this is the study by Muhs (1982) on San Clemente Island, which lies off the coast of Southern California (Figure 13.2). Twelve marine terraces are recognized (Muhs, 1983a), and the oldest could be about 1 million years old. The climate is arid mediterranean. Many soil properties increase in a linear fashion with age; these include solum thickness, the g cm^{-2} profile^{-1} for clay, soluble salts, profile averages for the smectite/mica ratio in the clay fraction, and quartz/plagioclase ratio in the silt fraction. In contrast, data on dithionite-extractable Al best fit a power function. In terms of morphology, soils less than about 200 000 years old are Mollisols and Alfisols with natric horizons (Bn), whereas older soils are Vertisols or soils with vertic properties. Although many chronosequences elsewhere display greater horizon differentiation with time (reviewed in Birkeland, 1984), here is a chronosequence that, because of the great amounts of clay in the older soils, eventually loses some of the distinct horizonation with time because of shrinking and swelling pedoturbation processes.

Many properties of soils on San Clemente Island seem to have an atmospheric origin (Muhs, 1982, 1983b). The presence of a silty A horizon for most soils, a clay content probably higher than can be accounted for by the weathering of the parent materials, non-clay minerals in soils that are not present in the parent materials, and satellite imagery that shows dust plumes extending from the mainland to the island, all combine to strongly indicate that much of the silt and clay is of atmospheric origin mainly from the mainland. The pedogenic soluble salts, however, probably originate as marine aerosols.

APPLICATIONS

There are many applications of soils information in western USA. These include dating deposits to provide information on fault movements, and interpretation of past climates.

Using soils to date deposits

Many of the chronosequence studies referred to above have been using soils information to provide an estimation of the age of the parent deposit; hence, these will not be repeated. Success in using soils in this manner has varied depending upon the regional environment and kind of soils. Experienced workers from different areas have been in general agreement on age estimations at various field conferences. In one recent study, soils were used to suggest that a till correlated with an oxygen-isotope stage of about 140 000 years, and uranium-trend data (see Fenwick, 1985) confirmed an age of about 130 000 years (Shroba *et al.*, 1983).

The study of Colman and Pierce (1983) illustrates the problem of accurately dating moraines using soils. They independently dated moraines near McCall, Idaho (Figure 13.2) using rock-weathering rinds (Colman and Pierce, 1981), and examined the soils. Moraines are approximately 14 000, 20 000, 60 000 and 140 000 years old. Using a variety of soil properties, including some of the index data of Harden (1982b), they find that soils formed on the younger two moraines are different, whereas those formed on the older two moraines are quite similar. These kinds of results should encourage workers to seek dated sites to further test the use of soils in dating deposits, but also caution workers as to the possible age plus-minus for particular spans of time when using soils to date deposits. For example, this study points out that soils that we sometimes suggest are 140 000 years old, could be only about half as old.

It would be helpful to provide a plus-minus figure for age estimates based on soils, but this is hard to do. Rather, estimates can be made on the number of deposits one might be able to differentiate over various time spans on the basis of soil properties. For the Holocene, at least four glacial and periglacial deposits can be recognized in alpine areas (Miller and Birkeland, 1974; Shroba and Birkeland, 1983), and I doubt if a substantially greater number can be recognized in other environments. In contrast, for the age span from 10 000 to 150 000 years it seems that two and at the most three deposits can be recognized (Burke and Birkeland, 1979; Colman and Pierce, 1983; Shroba and Birkeland, 1983). Soils greater than 140 000 years are still better developed, but I doubt if more than one deposit per 100 000-year duration could be recognized solely on soils data. Hence, a plot of the number of recognized deposits per unit time versus time would follow a log or power function.

Using soils to date faults

Western USA is an area of widespread active faulting; it is important to obtain an estimate on the ages of many faults because of their proximity to urban areas or to proposed structures such as nuclear reactor power plant sites. Soils are important in approximately dating these faults because materials for numerical dating methods often are absent.

The most common pedological method is to use soils to estimate the ages of deposits associated with faulting, and thus be able to put approximate age brackets on the time of faulting. This has been done with varying success in a wide variety of environments (Borchardt *et al.*, 1980; Douglas, 1980; Swan *et al.*, 1980; McCalpin, 1982).

Recurrent movement on a single fault also can be estimated on soil data. Such a study was made by Machette (1978; see also Birkeland, 1984, Figures 12.8 and 12.9) near Albuquerque, New Mexico, an area of carbonate soils. Several buried soils are present on the down-dropped side of the fault and these vary both in carbonate morphological stages and in total carbonate (g cm^{-2} profile^{-1}). The assumption was made that following fault movement, material was rapidly eroded from the fault scarp, burying the soil on the down-dropped side, and that most of geologic time was one of soil development. Thus, repeated movement on the fault with varying lengths of time between fault events resulted in buried soils, each with a different morphology and carbonate content. Estimations of the regional rate of accumulation of pedogenic carbonate (Figure 13.6) allowed for an estimation of the time between movements on the fault, and four such movements have occurred in the past 0.5 million years.

Pedological evidence for past climatic change

Ample evidence exists for past climatic change in the geological record of western USA. Criteria for such change based on soils has been slow in coming, but recent workers are beginning to interpret anomalous soil features in terms of different past climates.

An example of climatic change from soil data is the soil-forming intervals of Morrison (1978). Morrison and Frye (1965) advocated climatic change as necessary to form some soils in the desert basins but recent work suggests another interpretation for the same phenomena. The geological and pedological record associated with Pleistocene lake deposits suggested to them that well-developed soils formed in a remarkably short time. These short durations during which relatively well-developed soils formed were termed soil-forming intervals. Climatic change was used to explain such rapid soil development. Alternative interpretations have been put forward to explain the same soils or phenomena: one was to call upon more time to form them (Birkeland and Shroba, 1974). Recent work by Scott *et al.* (1983) indicates that much more time is available to form the soils in question. Hence, soil evidence other than a postulated short time span for formation will have to be used to propose climatic change in the region.

Much current work is focusing on features within soil profiles that seem to require a greater depth of water movement than is available under the present climate. It is common to compare present water movement, as deduced from water-balance calculations (Arkley, 1963), with depths of various soil features. For example, in Southern California the lower parts of Bt horizons are considered to extend to depths no longer reached by the contemporary soil-moisture regime (Torrent *et al.*, 1980a and b). In Montana, complex relationships between the depths of pedogenic clay and carbonate in a soil chronosequence are difficult to explain in all cases, but some of the features may relate to past conditions of greater soil moisture (Reheis, 1984). Using an array of geological evidence for past climate, Reheis (1984) calculated the glacial soil-moisture regime for this last example and found that the leaching index (precipitation (P)-potential evapotranspiration (ET) for those months in which $P > ET$: Arkley, 1963) could have been 23 cm compared to the present-day index of 10 cm. Finally, in the aridic area example in Southern California, McFadden (1982) notes that carbonate has accumulated at two depths in Pleistocene soils. The deeper accumulation is considered to reflect greater soil moisture during glacial times, and the shallower accumulation the post-glacial soil-moisture regime. The contrast in leaching index is 3.7 cm for the glacial climate and 0.1 cm for the post-glacial. Computer models using these leaching index values duplicate the field carbonate-depth distribution fairly well.

In some places soil data can be used to suggest that soil-moisture conditions have not changed drastically over relatively long intervals of time. For example, in some aridic environments pedogenic palygorskite seems to require a long time to form (Bachman and Machette, 1977) and this suggests that the long-term soil-moisture regime has probably remained aridic (McFadden, 1982). Likewise, large accumulations of pedogenic gypsum are best explained in the same manner (Reheis, 1984). Finally, the lack of pedogenic iron trends with time in some Sierra Nevada soils could be taken as evidence that past climates, including those related to at least the last major glaciation, have not been such that soil-moisture conditions were appreciably wetter than the present (Shroba and Birkeland, 1983). In all of these climatic change considerations, it might be best to restrict discussion to changes in soil-moisture regime rather than in climate, because the connection between the latter is quite complex.

ACKNOWLEDGEMENTS

I thank S. L. Forman for criticizing the entire manuscript, and L. D. McFadden and D. R. Muhs for reading those parts pertaining to their research.

REFERENCES

Arkley, R. J. (1963). 'Calculation of carbonate and water movement in soil from climatic data', *Soil Sci.* **96**, 239–248.
Bachman, G. O. and Machette, M. N. (1977). *Calcic Soils and Calcretes in the Southwestern United States*, US Geol. Surv. Open-File Rept. 77–794.

322					*Soils and Quaternary landscape evolution*

Baker, V. R. (1983). 'Late-Pleistocene fluvial systems', in *Late Quaternary Environments of the United States, the Late Pleistocene*, (Ed. S. C. Porter), pp. 115–129, University of Minnesota Press, Minneapolis.

Berry, M. E. (1984). *Morphological and Chemical Characteristics of Soil Catenas on Pinedale and Bull Lake Moraine Slopes, Bear Valley (Salmon River Mountains), Idaho*, Unpublished MS Thesis, University of Colorado, Boulder.

Birkeland, P. W. (1984). *Soils and Geomorphology*, Oxford University Press, New York.

Birkeland, P. W. and Shroba, R. R. (1974). 'The status of the concept of Quaternary soil-forming intervals in the western United States', in *Quaternary Environments* (Ed. W. C. Mahaney), pp. 241–276, York University Geog. Mono. No. 5, Toronto, Canada.

Birkeland, P. W., Colman, S. M., Burke, R. M., Shroba, R. R. and Meierding, T. C. (1979). 'Nomenclature of alpine glacial deposits, or, what's in a name?', *Geology*, 7, 532–536.

Borchardt, G., Rice, S. and Taylor, G. (1980). *Paleosols Overlying the Foothills Fault System Near Auburn, California*, California Div. Mines and Geology Spec. Rept. 149.

Burke, R. M. and Birkeland, P. W. (1979). 'Reevaluation of multiparameter relative dating techniques and their application to the glacial sequence along the eastern escarpment of the Sierra Nevada, California', *Quaternary Res.* 11, 21–51.

Burke, R. M. and Birkeland, P. W. (1983a). Holocene glaciation in the mountain ranges of the western United States', in *Late Quaternary Environments of the United States, the Holocene* (Ed. H. E. Wright, Jr.), pp. 3–11, University of Minnesota Press, Minneapolis.

Burke, R. M. and Birkeland, P. W. (1983b). 'Toposequences and soil development as a relative dating tool on a two-fold chronosequence of eastern Sierra Nevada morainal slopes, California', *Geol. Soc. Amer. Abst. with Prog.*, 15, 327.

Burns, S. F. (1980). *Alpine Soil Distribution and Development, Indian Peaks, Colorado Front Range*, Unpublished PhD Thesis, University of Colorado, Boulder.

Burns, S. F. and Tonkin, P. J. (1982). 'Soil-geomorphic models and the spatial distribution and development of alpine soils', in *Space and Time in Geomorphology* (Ed. C. E. Thorn), pp. 25–43, Allen and Unwin, London.

Colman, S. M. (1982). 'Clay mineralogy of weathering rinds and possible implications concerning the sources of clay minerals in soils', *Geology*, 10, 370–375.

Colman, S. M. and Pierce, K. L. (1977). *Summary Table of Quaternary Dating Methods*, US Geol. Surv. Misc. Field Studies Map MF–904.

Colman, S. M. and Pierce, K. L. (1981). *Weathering Rinds on Basaltic and Andesitic Stones as a Quaternary Age Indicator, Western United States*, US Geol. Surv. Prof. Pap. 1210.

Colman, S. M. and Pierce, K. L. (1983). *The Glacial Sequence Near McCall, Idaho: Weathering Rinds, Soil Development, Morphology and Other Relative-Age Criteria*, US Geol. Surv. Open-File Rept. 83–727.

Douglas, L. A. (1980). 'The use of soils in estimating the time of last movement of faults', *Soil Sci.* 129, 345–352.

Fenwick, I. (1985). 'Paleosols: problems of recognition and interpretation', in *Soils and Quaternary Landscape Evolution* (Ed. J. Boardman), Wiley, Chichester (in press).

Gardner, L. H. (1972). 'Origin of the Mormon Mesa Caliche, Clark County, Nevada', *Geol. Soc. Ann. Bull.* 83, 143–155.

Gile, L. H., Peterson, F. F. and Grossman, R. B. (1965). 'The K horizon—a master soil horizon of carbonate accumulation', *Soil Sci.* 99, 74–82.

Gile, L. H., Peterson, F. F. and Grossman, R. B. (1966). 'Morphological and genetic sequences of carbonate accumulation in desert soils', *Soil Sci.* 101, 347–360.

Gile, L. H. and Grossman, R. B. (1979). *The Desert Project Soil Monograph*, US Dept. Agri.

Gile, L. H., Hawley, J. W. and Grossman, R. B. (1981). *Soils and Geomorphology in the Basin and Range Area of Southern New Mexico — Guidebook to the Desert Project*, New Mexico Bureau of Mines and Mineral Resources, Mem. 39.

Guthrie, R. L. and Witty, J. E. (1982). 'New designations for soil horizons and layers and the new Soil Survey Manual', *Soil Sci. Soc. Amer. J.* **46**, 443–444.

Harden, J. W. (1982a). *A Study of Soil Development Using the Geochronology of Merced River Deposits,* Unpublished PhD Thesis, University of California, Berkeley.

Harden, J. W. (1982b). 'A quantitative index of soil development from field descriptions: examples from a chronosequence in central California', *Geoderma*, **28**, 1–28.

Harden, J. W. and Taylor, E. M. (1983). 'A quantitative comparison of soil development in four different climates', *Quaternary Res.* **20**, 342–359.

Holliday, V. T. (1982). *Morphological and Chemical Trends in Holocene Soils at the Lubbock Lake Archeological Site, Texas*, Unpublished PhD Thesis, University of Colorado, Boulder.

Lattman, L. H. and Lauffenburger, S. K. (1974). 'Proposed role of gypsum in the formation of caliche', *Zeit. fur Geomorph. N.F., Suppl. Bd.* **20**, 140–144.

Jenny, H. (1941). *Factors of Soil Formation*, McGraw-Hill, New York.

Jenny, H. (1980). *The Soil Resource*, Springer-Verlag, New York.

Machette, M. N. (1978). 'Dating Quaternary faults in the southwestern Unites States by using buried calcic paleosols', *US Geol. Surv. J. Res.* **6**, 369–381.

Machette, M. N. (in press). 'Calcic soils of the American Southwest', *Geol. Soc. Amer. Spec. Pap.*

Mears, B., Jr. (1981). 'Periglacial wedges and the late Pleistocene environment of Wyoming's intermontane basins', *Quaternary Res.*, **15**, 171–198.

McCalpin, J. P. (1982). Quaternary geology and neotectonics of the West Flank of the Northern Sangre de Cristo Mountains, South-Central Colorado, *Colorado School of Mines Quarterly*, **77**, 3.

McFadden, L. D. (1982). *The Impacts of Temporal and Spatial Climatic Changes on Alluvial Soils Genesis in Southern California*', Unpublished PhD Thesis, University of Arizona, Tucson.

Meierding, T. C. (in press). 'Correlation of Rocky Mountain Pleistocene deposits by relative dating methods: A perspective', in *Quaternary Chronologies* (Ed. W. C. Mahaney), Geobooks, Norwich.

Miller, C. D. and Birkeland, P. W. (1974). 'Probable pre-Neoglacial age of the type Temple Lake moraine, Wyoming: Discussion and additional relative-age data', *Arc. Alp. Res.* **6**, 301–306.

Morrison, R. B. (1978). 'Quaternary soil stratigraphy — concepts, methods, and problems', in *Quaternary Soils* (Ed. W. C. Mahaney), pp. 77–108, Geo Abstracts, Norwich.

Morrison, R. B. and Frye, J. C. (1965). *Correlation of the Middle and Late Quaternary successions of the Lake Lahontan, Lake Bonneville, Rocky Mountain (Wasatch Range), Southern Great Plains, and Eastern Midwest Areas*, Nevada Bur. Mines Rept. 9.

Muhs, D. R. (1982). 'A soil chronosequence on Quaternary marine terraces, San Clemente Island, California', *Geoderma*, **28**, 257–283.

Muhs, D. R. (1983a). 'Quaternary sea-level events on northern San Clemente Island, California, *Quaternary Res.* **20**, 322–341.

Muhs, D. R. (1983b). 'Airborne dust fall on the California Channel Islands, U.S.A.', *J. Arid Environ.* **6**, 223–238.

Ponti, D. J. (in press). 'The Quaternary alluvial sequence of the Antelope Valley, California', *Geol. Soc. Amer. Spec. Pap.*

Porter, S. C., Pierce, K. L. and Hamilton, T. D. (1983). 'Late Wisconsin mountain glaciation in the western Unites States', in *Late Quaternary Environments of the United States, the Lake Pleistocene*, (Ed. S. C. Porter), pp. 71–111, University of Minnesota Press, Minneapolis.

Reheis, M. C. (1984). *Chronologic and Climatic Control on Soil Development, Northern Bighorn Basin, Wyoming and Montana*, Unpublished PhD Thesis, University of Colorado, Boulder.

Schumm, S. A. (1965). 'Quaternary paleohydrology', in *The Quaternary of the United States* (Eds. H. E. Wright, Jr. and D. G. Frey), pp. 783–794, Princeton University Press, Princeton, NJ.

Schumm, S. A. (1977). *The Fluvial System*, Wiley, New York.

Scott, G. R. (1965). 'Nonglacial Quaternary geology of the Southern and Middle Rocky Mountains', in *The Quaternary of the United States* (Eds H. E. Wright, Jr. and D. G. Frey), pp. 234–254, Princeton University Press, Princeton, NJ.

Scott, W. E., McCoy, W. D., Shroba, R. R. and Rubin, M. (1983). 'Reinterpretation of the exposed record of the last two cycles of Lake Bonneville, western United States', *Quaternary Res.* **20**, 261–285.

Shackleton, N. J. and Opdyke, N. D. (1974). 'Oxygen isotope and paleomagnetic stratigraphy of equatorial Pacific core V28–238: Oxygen isotope temperatures and ice volume on a 10^5 and 10^6 year scale', *Quaternary Res.* **3**, 39–55.

Shroba, R. R. (1977). *Soil Development in Quaternary Tills, Rock-glacier Deposits, and Taluses, Southern and Central Rocky Mountains*, Unpublished PhD Thesis, University of Colorado, Boulder.

Shroba, R. R. and Birkeland, P. W. (1983). 'Trends in late-Quaternary soil development in the Rocky Mountains and Sierra Nevada of the western United States', in *Late Quaternary Environments of the United States, the Late Pleistocene* (Ed. S. C. Porter), pp. 145–156, University of Minnesota Press, Minneapolis.

Shroba, R. R., Rosholt, J. N. and Madole, R. F. (1983). 'Uranium-trend dating and soil B horizon properties of till of Bull Lake age, North St. Vrain drainage basin, Front Range, Colorado', *Geol. Soc. Amer. Absts. with Prog.* **15**, 431.

Soil Survey Staff (1975). *Soil Taxonomy*, US Department of Agriculture, Soil Conservation Service, Agriculture Handbook 436.

Swan, F. H., III, Schwartz, D. P. and Cluff, L. S. (1980). 'Recurrence of moderate to large magnitude earthquakes produced by surface faulting on the Wasatch fault zone, Utah', *Bull. Seis. Soc. Amer.* **70**, 1431–1462.

Swanson, D. K. (1984). *Soil Catenas on Pinedale and Bull Lake Moraines, Willow Lake, Wind River Mountains, Wyoming*, Unpublished MS Thesis, University of Colorado, Boulder.

Tedrow, J. C. F. (1977). *Soils of the Polar Landscapes*, Rutgers University Press, New Brunswick, NJ.

Torrent, J., Nettleton, W. D. and Borst, G. (1980a). 'Clay illuviation and lamella formation in a Psammentic Haploxeralf in Southern California', *Soil Sci. Soc. Amer. J.* **44**, 363–369.

Torrent, J., Nettleton, W. D. and Borst, G. (1980b). 'Genesis of a Typic Durixeralf of Southern California', *Soil Sci. Soc. Amer. J.* **44**, 575–582.

Soils and Quaternary Landscape Evolution
Edited by J. Boardman

14

Holocene Soil-geomorphological Relations in a Semi-arid Environment: The southern High Plains of Texas

VANCE T. HOLLIDAY

ABSTRACT

Studies of Holocene pedogenesis and geomorphology on the semi-arid southern High Plains of north-west Texas and eastern New Mexico provide sufficient data for an initial assessment of Holocene soil-geomorphological relations in the region. The primary settings for deposition in the Holocene are in draws (ephemeral drainages), playas (ephemeral lake basins), dune fields, and dunes on the lee-sides of playas. In the draws soils formed during the period from 8500 to 5500 radiocarbon years (ry) BP developed in lacustrine sediments and marshy environments, these factors determining the characteristics of the soils. The soils have mollic epipedons, often with gleyed subhorizons but seldom with diagnostic subsurface horizons. Aeolian sediments fill the draws between 5500 and 4500 ry BP. The draws were stable landscapes for the rest of the Holocene except for minor aeolian and slope wash sedimentation within the past 1000 years. The soils formed in this setting include Ochrepts, Ustalfs, and Ustolls. Soils in the dune fields are generally Psamments, although soils in the early Holocene sediments are often Ustalfs. Soils in the most recent lee-side dune sediments are Entisols. The older soils are either Ochrepts or Ustalfs. Time appears to be the dominant variable producing differences in the soils, although contrasts in parent material also exist: permeable sand in the dune fields, highly calcareous silt in the lee-side dunes. The playa fill of Holocene age is a thick clay that appears to have aggraded slowly in a lacustrine environment over the past 10 000 years with some aeolian additions in the mid-Holocene. The soils in these playas are deep, dark grey Vertisols. The clayey parent material and lacustrine environment of the playas are the dominant influences on soil formation.

Wind appears to be the dominant soil-geomorphological agent of the Holocene on the southern High Plains. Aeolian sedimentation occurred in discrete but significant intervals: prior to about 7000 ry BP, at about 5000 ry BP, and several times since about 2000 ry BP. Aerosolic clay and carbonate are then mechanically infiltrated into the aeolian sediments, promoting rapid formation of argillic and calcic horizons. Lacustrine sedimentation is an important depositional process and lacustrine deposits and micro-environments are significant soil-forming factors along draws in the early Holocene and in playas throughout the Holocene.

INTRODUCTION

During the past 15 years a number of investigations have been conducted into the genesis of soils of Holocene age on the southern High Plains of north-west Texas. These studies have been carried out on soils formed in several landscape settings and in a variety of clayey to sandy parent materials. However, there has been relatively little research into the genesis of Holocene soils in semi-arid environments and even that done in neighbouring areas (e.g. Gile (1977), in southern New Mexico and Machette (1975 a and b), in east-central Colorado) has concentrated primarily on soils formed in gravelly alluvial deposits. Although there has been a longer history of and considerably more research into the geomorphology of the region (e.g. Melton, 1940; Reeves, 1965, 1966, 1970, 1972; Reeves and Parry, 1969; Hawley et al., 1976; Woodruff et al., 1979), there have been few attempts to integrate the pedogenic and geomorphic investigations. The following is a summary of available information concerning the morphology and genesis of Holocene soils and associated landscapes on the southern High Plains.

The theoretical approach to be taken in this paper follows that of Jenny (1941, 1980) and Birkeland (1974) whereby soil formation is considered to be a function of the influence of various factors (climate, organisms, relief, parent material, time, and unspecified local phenomena such as dust). Because the paper focuses on soil-geomorphological relationships the soils will be reviewed in the context of the different landscape settings in which they formed (e.g. valley versus dune). Within a given landscape setting an attempt will be made to determine which, if any, of the factors is the dominant variable. Differences in the soils in the given setting may then be ascribed to that variable or variables.

In the following review, soils are referred to either by informal name or formal series name as defined by county soil surveys. All soils are classified according to *Soil Taxonomy* (Soil Survey Staff, 1975).

LOCATION AND SETTING

The southern High Plains of north-west Texas and eastern New Mexico, also known as the Llano Estacado ('Staked Plains'), is an extensive plateau covering approximately $130\,000\,km^2$. It is the southernmost portion of the High Plains section of the Great Plains physiographic province (Figure 14.1). The plateau is defined by escarpments along the west, north, and east sides.

The climate of the southern High Plains is continental and semi-arid; classified as dry, mid-latitude, and semi-desert (Strahler and Strahler, 1983, Plate C.2).

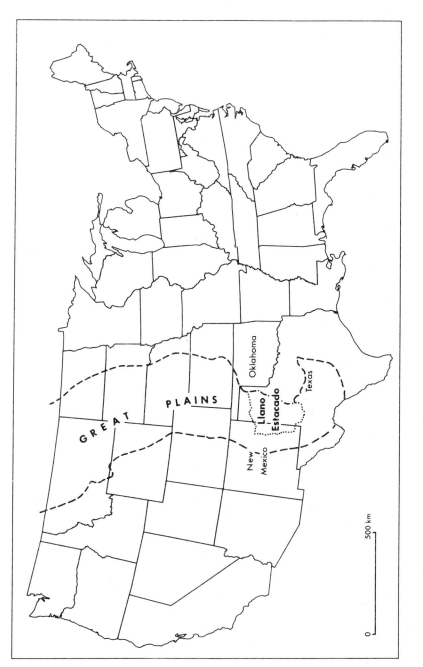

Figure 14.1 Map of the USA with the location of the Great Plains with the Llano Estacado (based on Hunt, 1974)

The mean annual precipitation of the region ranges from less than 35 cm in the south-west to over 50 cm in the north-east (Carr, 1967). However, as is typical for a semi-arid, continental region, the annual precipitation varies considerably from year to year. Temperature extremes are also common with summer temperatures being quite high. The MAT for July and August generally ranges from 25 to 30 °C. Wind is an important climatic feature, blowing almost constantly across the region throughout the year with average annual wind speeds ranging from 16 to 24 kph and speeds well in excess of 80 kph not uncommon (Lotspeich and Everhart, 1962; NOAA, 1982).

The natural vegetation of the Llano Estacado is a mixed-prairie grassland (Wendorf, 1961; Blair, 1950; Lotspeich and Everhart, 1962). The dominant plant community is short-grass which includes types of grama (*Bouteloua* spp.) and buffalo grass (*Buchloe dactyloides*). Trees are absent except along the escarpments and re-entrant canyons. The floristic composition varies from north to south due to changes in climate and soil texture. The native plant communities of the region occur in very few areas today because most of the southern High Plains is under cultivation.

The environment of the Llano Estacado imposes several constraints on soil formation. The high temperatures and winds of the region combined with relatively high evaporation together reduce the effectiveness of the low precipitation (Lotspeich and Everhart, 1962). Leaching, therefore, is minimal and salts and especially carbonates tend to persist and accumulate in the sola, and weathering of primary minerals is probably subdued. The dominantly grassland vegetation has deep roots which extract a significant amount of moisture from the soil, further reducing the effectiveness of the limited rainfall (Lotspeich and Everhart, 1962). The wind is also important in scouring the surface, depositing material in dunes and basins, and creating considerable amounts of dust.

The Llano Estacado has a virtually featureless, constructional surface; one of the largest of its kind in the United States (NOAA, 1982). It has a gentle regional slope from north-west to south-east, with elevations ranging from about 750 m to 1500 m. This surface was formed by deposition of thick, widespread, aeolian sediments, derived primarily from the west and south-west, during the Pleistocene. These 'coversands' or Blackwater Draw Formation, unconformably overlie the Ogallala Group, a thick, extensive, Miocene-Pliocene deposit of alluvial and aeolian sediments derived from the Rocky Mountains and Sacramento Mountains of New Mexico and, locally, the Blanco Formation, a Pliocene lacustrine unit inset against the Ogallala (Figure 14.2). The uppermost portions of the Ogallala and Blanco Formations commonly exhibit strongly indurated, pedogenic calcretes. The surface of the Llano Estacado has been modified locally by formation of several dune fields, thousands of small, ephemeral lake basins or playas formed by deflation and often associated with dunes on the lee side of the basin, and several north-west–south-east-trending

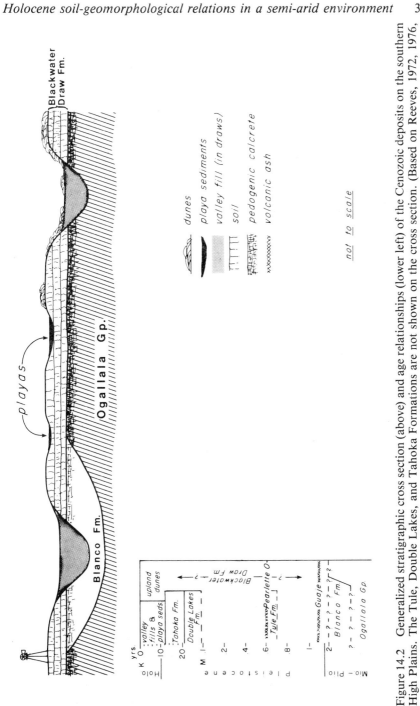

Figure 14.2   Generalized stratigraphic cross section (above) and age relationships (lower left) of the Cenozoic deposits on the southern High Plains. The Tule, Double Lakes, and Tahoka Formations are not shown on the cross section. (Based on Reeves, 1972, 1976, and Hawley *et al.*, 1976)

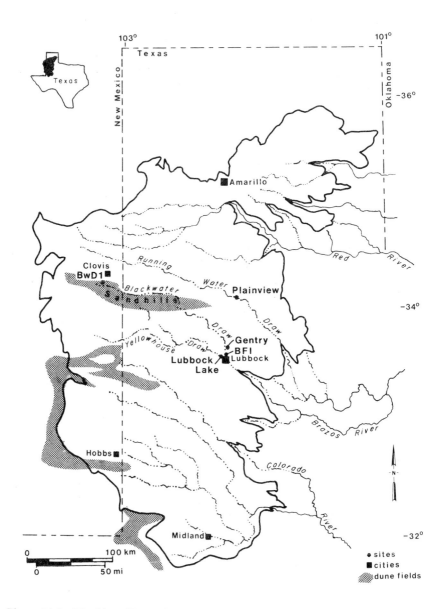

Figure 14.3   The Llano Estacado with the locations of the principal cities of the region
and the draws, stratigraphic sites (BwD 1 is Blackwater Draw Locality 1, the Clovis site),
and dune fields mentioned in the text

ephemeral drainages of 'draws', which are tributaries of the Red, Brazos, and Colorado Rivers, and none of which have flowing water today (Hawley *et al.*, 1976; Reeves, 1972, 1976) (Figures 14.2 and 14.3).

Almost all of the Holocene deposition on the southern High Plains has been in the draws, dune fields, dunes on the lee sides of the playas, and in the playas. It is in these deposits that the nature and rates of Holocene pedogenesis on the southern High Plains, as well as the history of Holocene landscape evolution, can be investigated. The following review of Holocene soil formation, therefore, is organized under the headings of draws, dunes, and playas.

DRAWS

The age of the draws incised into the surface of the Llano Estacado is uncertain. They probably formed in the late Pleistocene but some may be superimposed on earlier Pleistocene or pre-Pleistocene drainages (Reeves, 1970; Clanton and Reeves, 1975). The valleys attained a maximum relief ranging from 10 to 20 m (Stafford, 1977). The draws began to aggrade prior to 11 000 ry BP and have been filling intermittently throughout the Holocene. They generally contain from 3 to 5 m of well-stratified alluvial, lacustrine and aeolian fill.

Stratigraphic, pedological and archaeological research has been conducted at a number of localities along Yellowhouse, Blackwater, and Running Water Draws (Figure 14.3). The archaeological investigations have been of particular significance in providing time diagnostic artefacts and radiocarbon dates that can be used to establish good age controls for the late Quaternary sediments.

Yellowhouse Draw contains one of the most widely known and intensively studied archaeological sites in the region—Lubbock Lake. This site also contains the best-documented Holocene soils in a draw situation (Holliday, 1982) and will, therefore, be discussed in some detail. Comparison will then be made with available data from other draw sites.

Yellowhouse Draw

The Lubbock Lake site is located in an entrenched meander of Yellowhouse Draw in the city of Lubbock, Lubbock County, Texas (Figure 14.4). The draw is a tributary of the Brazos River and in the Lubbock area has cut through the Blackwater Draw Formation and into the Blanco Formation.

Lubbock Lake was discovered in 1936 following excavations for a large, U-shaped reservoir along the inside of the meander (Figure 14.4). The excavation cut completely through the late Quaternary fill in the Draw, thus exposing the stratigraphic sequence. Archaeological and limited geological investigations took place at the site intermittently from the late 1930s until the middle 1960s (Black, 1974; Holliday, in press a). The current research programme has operated continuously since 1973 and includes the disciplines of archaeology, geology,

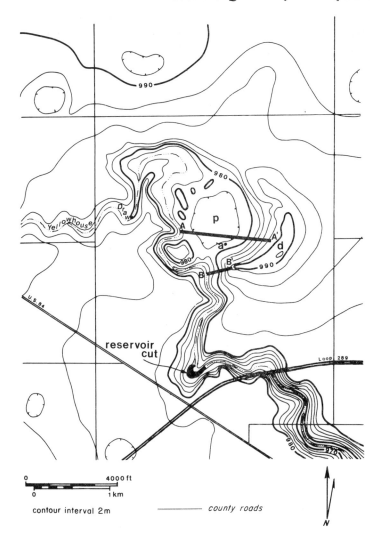

Figure 14.4   Topographic map of Yellowhouse Draw in the area of the Lubbock Lake site showing the locations of the reservoir cut at the site, Cone playa (p), which is intersected by the draw and contains the Arch soil profile (a), and the large lee-side dune (d, the Dune site) and associated trenches (A–A', B–B'). Note the other small closed basins in the area

pedology, paleontology, and paleobotany. Geological investigations were begun by Johnson (1974) and carried on by Stafford (1977, 1978, 1981). The author has continued the geological research, initiated an investigation of the pedology of the site, and expanded this research to include other localities on the Llano Estacado. The data presented below is summarized from Holliday (1982, 1983, in press a,b,c) and Holliday *et al.* (1983, in prep.). The geological and pedological research at Lubbock Lake has been carried out through the excavation of over 150 backhoe trenches along Yellowhouse Draw at and in the vicinity of Lubbock Lake. The interdisciplinary research thus far carried out at Lubbock Lake demonstrates that the site contains a cultural and biological record in a well-stratified context spanning the past 11 000 years.

The chronology at Lubbock Lake is well-established by over 100 radio-carbon ages (Holliday *et al.*, 1983). These ages have been determined by two laboratories on a variety of materials (charcoal, wood, bone, shell, and the NaOH-soluble and -insoluble fractions of organic-rich sediment and buried A horizons). Given the possibilities of inter-lab variations and the vagaries associated with dating of different materials, the ages are generally quite consistent. All principal geologic units and subunits and facies thereof, periods of soil formation and cultural features are well-dated. All ages are in radiocarbon years (ry) before present (BP). Age ranges for geologic units are the times during which the soil developed. In a few situations it is convenient to refer to a soil as being 'X years old'. This means that the soil had X-years to form.

Five principal stratigraphic units (numbered oldest to youngest) and soils formed therein (which have been named) have been identified at Lubbock Lake (Stafford, 1977, 1978, 1981; Holliday, 1982, 1983) (Figure 14.5). Stratum 1 (12 000–11 000 ry BP) is a basal alluvial deposit of sand and gravel up to 1 m thick, representing a meandering stream. Stratum 2 (11 000–8 500 ry BP) is a clayey, organic, often diatomaceous, lacustrine deposit, with some sandy, aeolian additions along the valley margin, averaging about 1 m thick. The Firstview Soil (8 500–6 500 ry BP) is developed in these clayey lacustrine and sandy aeolian sediments. Stratum 3 (6 500 to about 6 000 ry BP), averaging 1 m thick, is composed of calcareous, lacustrine, valley-axis and sandy, aeolian, valley-margin facies. The Yellowhouse Soil (about 6 000–5 500 ry BP) is developed in both facies. Stratum 4 (5 500 to greater than 4 500 ry BP) consists of some sandy alluvial, and clayey lacustrine sediments along the valley axis (less than 1 m thick) and sandy aeolian sediments across the draw (up to 2 m thick). The Lubbock Lake Soil has formed in stratum 4, beginning about 4 500 ry BP. Where stratum 4 and the Lubbock Lake Soil are buried by stratum 5 the soil had about 3 500 years to develop. Where the soil has not been buried it continues to form, hence it has had about 4 500 years to form. The youngest deposit at Lubbock Lake is stratum 5 (up to 1 m thick). Which consists of discontinuous, episodic deposits of slopewash and aeolian sediments and is subdivided into 5A and 5B. The Apache Soil has developed in 5A and the Singer Soil is forming in 5B. Deposition of 5A lasted from about 800 to 450 ry BP followed by development of the Apache Soil. Where 5A and the Apache Soil are buried by 5B the Apache Soil had 200 years to form. Otherwise it continues to form and is about 450 years old. Substratum 5B was deposited beginning about 250 ry BP. Deposition ceased and the Singer Soil began to form within the past 100 years. Substratum 5A

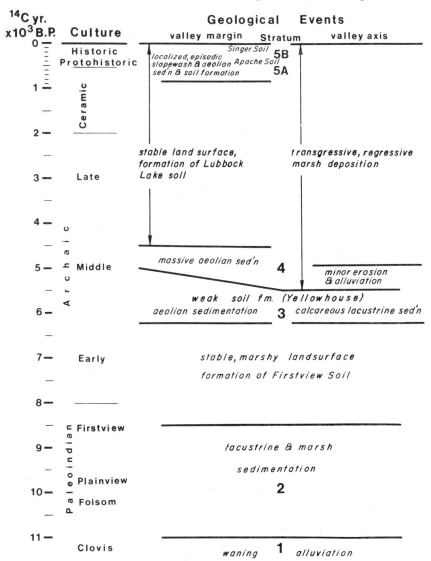

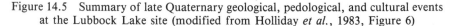

Figure 14.5   Summary of late Quaternary geological, pedological, and cultural events
at the Lubbock Lake site (modified from Holliday *et al.*, 1983, Figure 6)

buries only a portion of stratum 4 and the Lubbock Lake Soil and substratum
5B buries only part of 5A and the Apache Soil.

Most of the parent material was derived from the Blackwater Draw
Formation. The mineralogy of the sand and silt fractions is dominantly quartz
with minor amounts of feldspar. The clay mineralogy is dominated by illite and

mixed-layer illite-smectite, with lesser amounts of smectite and kaolinite. The alluvial deposits also contain carbonates apparently derived from the calcretes in the Ogallala and Blanco Formations.

The Firstview Soil, the oldest buried soil at Lubbock Lake, exhibits O/A/C and A/C profiles and consists of two facies. The marsh facies, found along the valley axis, has a relatively high organic matter content, contains abundant silicified root remains, and, commonly, exhibits a gley horizon immediately below the A horizon. These characteristics indicate that the soil formed in a marsh with the water table at or just below the surface. Toward the valley margin, the soil becomes coarser-grained, better drained, and weakly calcareous, reflecting the facies change in the parent material. All Firstview Soils are classified as Entisols. The field morphology and geologic setting of stratum 2 and the Firstview Soil suggest, however, that sometime during formation of the soil it was a Histosol. It no longer qualifies as such due to post-burial alteration.

The Yellowhouse Soil has A/C (valley-axis and valley-margin facies) and sometimes A/B/C (valley-margin only) profiles. The highly calcareous valley-axis facies exhibits relatively high organic carbon content and minimal leaching of carbonate in the C horizon. This suggests that the water table along the axis of the draw remained high during deposition of the parent material and throughout pedogenesis. The valley-margin facies of the Yellowhouse Soil includes both Haplustolls and Ustorthents. The valley-margin facies of the soil is an Ustorthent.

The Lubbock Lake Soil is the best developed Holocene soil observed on the southern High Plains. The soil has an unburied facies that is 4 500 ry old and a buried facies that is about 3 500 ry old. The soil exhibits mollic (unburied) or ochric (buried) epipedons, usually an argillic and, in some positions, a partially gleyed B horizon, and Stage I or II, often multiple, calcic horizons (following the classification system for calcic horizons of Gile *et al.*, 1966). Classification of the Lubbock Lake Soil varies, depending on presence or absence of mollic, argillic, and calcic horizons. It is classified as either a Calciustoll, Haplustoll, Haplustalf, or Ustocrept.

The Apache Soil has an unburied facies 450 ry old and a buried facies 200 ry old. The soil has mollic (unburied) or ochric (buried) epipedons, and a cambic or possibly argillic B horizon with Stage I calcic horizon. The Apache Soil, like the Lubbock Lake Soil, varies in classification. It occurs as either an Ustochrept, Haplustoll, or minimally developed Haplustalf or Calciustoll.

The Singer Soil is the youngest soil at Lubbock Lake, both in terms of the beginning (100 years ago) and length (100 years) of pedogenesis. The soil has an ochric epipedon and either a C or a cambic B subhorizon. Weak Stage I horizons of carbonate accumulation are common but never qualify as calcic. The Singer Soil is classified as either an Ustorthent or Ustochrept.

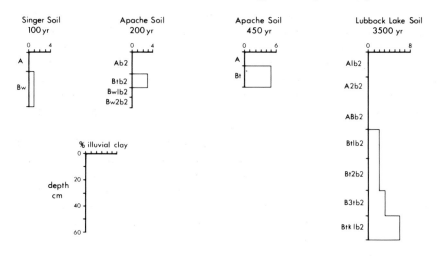

Figure 14.6   Maximum amounts of illuvial clay (calculated from particle-size distribution) in the Singer, buried (200 yr) and unburied (450 yr) Apache, and buried Lubbock Lake soils (modified from Holliday, 1982)

Several observations can be made concerning soil development in the Holocene along Yellowhouse Draw in the area of Lubbock Lake. The nature of the factors of soil formation vary during the Holocene. At the beginning of the Holocene the climate was somewhat cooler than today with increased effective precipitation. By the mid-Holocene the climate was drier and probably warmer than today. However, the water table in the draw remained high and the morphology of the soils formed along the draw reflect the lacustrine microenvironment although the lacustrine parent materials vary (organic-rich clay prior to about 6 500 ry BP and calcareous silt from 6 500 to 5 500 yr BP).

From about 4 500 ry BP to the present the climate of the area has generally been that of today and the soils formed during this period have done so in similar (loamy) parent material. The primary variable among the soil forming factors has, therefore, been time. A chronosequence based on classification at the Great Group level (Soil Survey Staff, 1975) can be established for the soils. They begin as Entisols but very rapidly (within 100 years), Inceptisols and Mollisols form. By 450 years all soils are Inceptisols, Mollisols, or Alfisols.

Mollic epipedons form within 100 years. Equilibrium is reached within a few thousand or even a few hundred years. Evidence suggest that argillic horizons may develop within 200 years, based on minimum amounts of illuvial clay (Figure 14.6) and thin sections. Calcium carbonate accumulation occurs rapidly with horizons meeting the qualification for calcic horizons forming within 200 years.

Through time there has been considerable influence on pedogenesis from the local factors of airfall from dust and burial. The rapid development of argillic and calcic horizons is considered to be the result of mechanical infiltration of aerosolic clay and carbonate. Some evidence for this is suggested by the presence of clay films and a slight increase in clay content in the A horizons of unburied soils. The clay film development in the unburied A horizons also increases with time. The clay and carbonate is considered to be derived from dust, a common meteorological phenomenon on the southern High Plains, of which clay and calcium carbonate are important constituents (Warn and Cox, 1951; Laprade, 1957; Holliday, 1982). This material appears to be injected into the soil as a slurry. This occurs when rain falls through a dust storm (the 'mud rain' of Warn, 1952) or when the water hits the ground and mixes with dust already there.

Burial of the soils is also an important local factor of soil development, affecting organic carbon content and classification along the valley margins. Upon burial the organic carbon content of the buried A horizon decreases. The result is a buried epipedon that meets the colour and thickness requirement for mollic but not the requirement for organic carbon content. Before burial the soils were Mollisols, but within 1000 years after burial the epipedon no longer contains sufficient organic carbon to qualify as mollic.

Blackwater Draw

On lower Blackwater Draw recent excavations at the BFI landfill have revealed excellent exposures of late Quaternary valley fill (Holliday, 1983). The site is in northern Lubbock County about 5 km northeast of Lubbock Lake (Figure 14.3) and exhibits a stratigraphic sequence generally similar to that described for Lubbock Lake (Figure 14.7). The basal valley fill is an alluvial deposit of latest Pleistocene age. The stratum 2 equivalent is a localized, calcareous deposit composed of interbedded sand and clay. The stratum 3 equivalent is composed of two facies: a calcareous lacustrine silt (overlying the fluvial unit) with a soil formed in it identical to stratum 3 and the Yellowhouse Soil at Lubbock Lake and a calcareous aeolian loam (overlying the bedded stratum 2 equivalent) which has a soil formed in it with an A-Bk profile (Ustorthent). Burying the stratum 3 equivalent is a deposit with associated soil identical to and considered equivalent of stratum 4 and the Lubbock Lake Soil at the Lubbock Lake site.

In the upper reaches of Blackwater Draw considerable stratigraphic and limited pedological data are available from the well-known Clovis archaeological site, in Roosevelt County, New Mexico, also known as Blackwater Draw Locality 1 (Figure 14.3). The site is in a basin that drains into Blackwater Draw itself and the area is within the Sandhills dune field (discussed below).

Haynes and Agogino (1966) and Haynes (1975) provided the basic stratigraphic data, and Holliday (in press d) added some geochronological information and pedological data (Figure 14.7). Spring sediments (Unit C) were

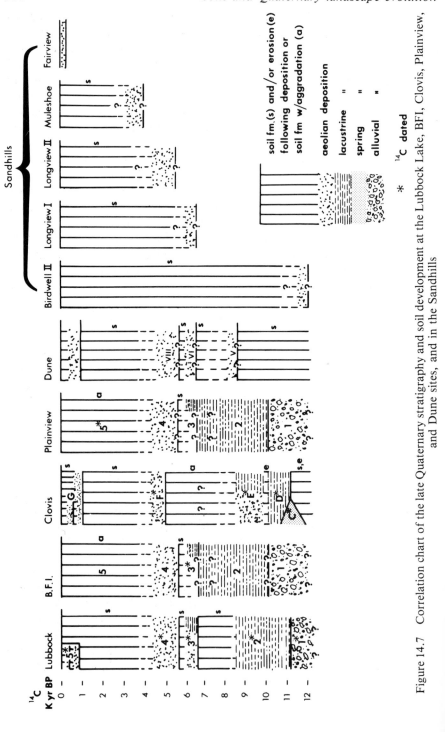

Figure 14.7   Correlation chart of the late Quaternary stratigraphy and soil development at the Lubbock Lake, BFI, Clovis, Plainview, and Dune sites, and in the Sandhills

deposited in the basin between 11 500 and 11 000 ry BP, at the same time as erosion occurred along the draw. From 11 000 to 10 000 ry BP diatomaceous earth and carbonaceous silt (Unit D) was deposited in a lacustrine environment followed by deposition of calcareous aeolian sand along the basin margin and more carbonaceous silt in the basin (Unit E) beginning about 10 000 ry BP and continuing until after 8 500 ry BP. Beginning sometime after 8 500 ry BP soil formation began in Unit E. In some areas of the basin slopewash and aeolian sedimentation during pedogenesis resulted in formation of cumulic, A/C profiles. In other parts of the basin the soil exhibits an A-Bk profile.

Unit E is buried by Unit F, which is also a non-calcareous aeolian unit. This deposit is very sandy and appears to be related to the Sandhills dune field. Unit F, which was probably emplaced a little less than 5 000 ry BP, has a relatively well-developed soil formed in it. This Soil has an ochric epipedon and commonly an argillic B horizon and is classified as a Haplustalf.

The youngest deposit at the site is Unit G, another non-calcareous, very sandy unit traceable into the dunes of the Sandhills. In some exposures it is composed of several deposits. Unit G is probably no more than 500 years old. It contains the modern surface soil which is weakly developed with an A-C profile classified as an Ustipsamment.

There are several basic similarities between the stratigraphy at the Clovis site and Lubbock Lake (Figure 14.7). Both contain diatomaceous, early Holocene lacustrine deposits (Unit D at Clovis, stratum 2 at Lubbock). There are also two early and middle Holocene aeolian deposits (Units E and F at Clovis; strata 3 and 4 at Lubbock) and the younger unit at each site contains a moderately well-developed soil profile. Both sites also have evidence for relatively minor aeolian sedimentation since 1000 to 500 ry BP.

Running Water Draw

Stratigraphic and pedological data are available from middle Running Water Draw from several large quarries excavated along the floor of the draw in the vicinity of the city of Plainview, Hale County, Texas (Figure 14.3) including the Plainview archaeological site. The stratigraphy and pedology along middle Running Water Draw is very similar to that seen in lower Blackwater Draw at the BFI pits.

The Holocene stratigraphy at all localities is quite similar (Holliday, 1983, in press d) and therefore, the depositional and pedologic history of middle Running Water Draw can be summarized as follows (Figure 14.7). The basal valley fill is a deposit of alluvial sand and gravel (Unit 1) up to 1 m thick. Deposition of this unit terminated by about 10 000 ry BP. Unit 1 is overlain by a localized, organic, clayey, lacustrine deposit (Unit 2), usually less than 1 m thick. It is probably time-equivalent to upper stratum 2 at Lubbock Lake.

Overlying Unit 2 is Unit 3, which is composed of two facies. Along the valley axis the deposit is composed of a massive, highly calcareous, apparently lacustrine deposit, not more than 1 m thick. Along the valley margin Unit 3 is a sandy, much less calcareous deposit that is probably of aeolian origin. In a few exposures Unit 3 exhibits a weakly developed soil (A-C profile). Unit 3 in Running Water Draw is very similar to and is considered an equivalent of stratum 3 at Lubbock Lake.

Above Unit 3 is a massive, sandy, deposit (Unit 4), about 1 m thick, that is also probably aeolian. This unit was deposited over 3500 ry ago and is correlated with stratum 4 at Lubbock Lake. Unit 4 exhibits a moderately developed soil which has a mollic epipedon, cambic or argillic subhorizon and a Stage I calcic horizon. The exact nature of the soil varies depending on the nature of the overlying sediments (Unit 5).

Unit 5 is composed of two facies. Along the valley axis Unit 5 is a clayey marsh deposit that represents the surficial sediments on the floor of the draw today. Along the valley margin Unit 5 is composed of several sandy, aeolian deposits separated by a weakly developed soil. Unit 5 was deposited within the past several thousand years.

The rapidity with which Unit 5 sediments were deposited apparently influenced pedogenesis in Unit 4. The clayey, marsh facies of Unit 5 probably aggraded slowly, allowing an argillic horizon to form in Unit 4. As Unit 5 thickened it became a cumulic, mollic A horizon. Eventually the lower portion of the cumulic A horizon was converted into the upper B horizon, forming a composite soil profile. The sandy, aeolian facies of Unit 5 was probably deposited quickly, sealing off Unit 4 and terminating soil formation perhaps 3000 ry BP. In these situations the soil exhibits a mollic epipedon but only a cambic or weak argillic B horizon.

Summary of sediments and soils in draws

There are several significant similarities and differences in the latest Pleistocene and Holocene deposits and soils in Yellowhouse, Blackwater, and Running Water Draws (Figure 14.7). Prior to 11 000 ry BP there was fluvial sedimentation in most draws except in upper Blackwater Draw where either spring sedimentation or erosion occurred. Between 11 000 and 10 000 ry BP fluvial deposition continued along reaches of some draws (lower Blackwater and middle Running Water) while lacustrine sedimentation took place in other parts of the draws (lower Yellowhouse, upper Blackwater). Lakes or marshes occupied the floors of the draws in the earliest Holocene with marsh soils forming in at least a few localities (lower Yellowhouse, upper Blackwater draws). By about 6500 ry BP calcareous lake sediments were deposited along the valley floors with significant aeolian additions along the valley margins. Weakly developed soils then formed in these sediments. Massive aeolian sediments filled the draws between 5500 and 4500 ry BP. A moderately well-developed soil began forming in these aeolian sediments from about 4500 ry BP. This soil commonly has a mollic epipedon, an argillic B horizon, and an incipient or weak calcic horizon (Stage I) with a Stage II horizon in exposures of the soil observed in lower Yellowhouse Draw. Localized, episodic aeolian sedimentation occurred in the late Holocene along with minor slope-wash deposition.

The data available suggest that the genesis and morphology of the soils formed in the Holocene sediments that fill the draws was controlled by the same influences affecting the soils at Lubbock Lake. In the early Holocene and up to about 5000 ry BP the soils are significantly influenced by lacustrine parent materials and microenvironments. In the late Holocene the soils form in well-drained aeolian sediments and gain significant amounts of mechanically infiltrated, aerosolic clay and carbonate. Time is the only significant variable among the factors of soil formation in the late Holocene.

DUNES

Dunes on the southern High Plains occur either as relatively extensive dune fields or as localized deposits on the lee sides of many of the playa basins. These two types of dunes have different origins, mineralogy, and lithology and therefore exhibit different soil morphologies.

Dune fields

There are three areas of the southern High Plains with dune fields (Hawley *et al.*, 1976) generally on its western side (Figure 14.3). On the north-west Llano Estacado is a linear, west-to-east-trending belt of dunes locally known as the 'Sandhills'. On the west-central Llano is a set of west-to-east trending, anastomosing, linear dune fields which represent an eastward extension of the Mescalero dune field of the Pecos River Valley. On the south-western Llano Estacado are several small dune fields that represent eastward extensions of the Monahans dune field and south-eastern-most Mescalero Dunes.

The dune fields appear to be entirely of late Quaternary age, and are composed of non-calcareous, quartzose sand. The Sandhills were apparently derived from the Portales Valley, an ancient (early to middle Pleistocene) channel of the Brazos River. The other dune fields were ultimately derived from the Pecos River Valley and occur in the lee of gaps in the High Plains escarpment, which apparently allowed the aeolian material to move from the valley to the Plains. The only stratigraphic and pedological data available are for the Sandhills.

A series of Holocene sediments, soils, and surfaces have been identified in a portion of the central Sandhills (Gile, 1979, 1981, 1983). Each sedimentary unit and associated soil and geomorphic surface has been named. Ages have been determined by correlation with nearby archaeological sites. The Holocene

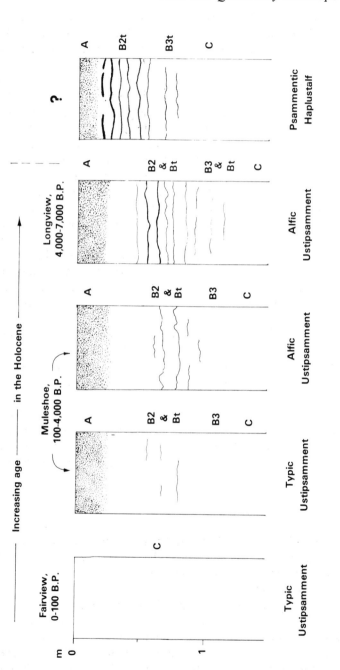

Figure 14.8 Generalized sequence of development of Holocene soils in the Sandhills (from Gile, 1979, Figure 3; reproduced from *Soil Science Society of America Journal* (1979), **43**, 994–1003 by permission of the Soil Science Society of America). The querried profile is the Birdwell II (11 500–7000 ry BP; Gile, 1983)

soil-geomorphological units include: Birdwell II (11 500–7000 ry BP); Longview (7000–4000 ry BP), subdivided into Longview I and Longview II; Muleshoe (4000 ry BP to AD 1800); and Fairview (AD 1880 to present) (Figures 14.7 and 14.8).

Gile (1979) has used the chronosequence in the Sandhills to investigate the nature and rates of formation of the soils including specific pedological features. One of the most noticeable characteristics of the soils are clay-rich bands, defined as 'thin, roughly horizontal, and dominantly to wholly pedogenic accumulations of silicate clay' (Gile, 1979). The clayey bands are no more than a few millimeters thick and are observed to crosscut bedding in the dunes. The clay is considered to have been deposited from clay in suspension on slowing of downward movement of the wetting front. The clay bands are observed to increase in number, thickness, and extent through time and are the most characteristic feature of pedogenic development (Figure 14.8).

It is also noted that the Muleshoe and Longview soils often have somewhat higher silt and clay content in their surface horizons in comparison with immediately underlying horizons. Gile (1979) postulates that this is the result of mechanical infiltration of fines derived from dust and that this is the source of the clay in the clay bands.

In the Sandhills the differences in soil morphology are due primarily to differences in the duration of soil formation. This is most apparent in the development of the clay bands. Through time, however, the parent material and aerosolic additions have exerted considerable influence. The highly permeable nature of the sediments appears to allow infiltration of dust-derived clay along the wetting front, resulting in formation of clay bands. The climate has varied during the Holocene but this appears to manifest itself in the deflation and migration of the dunes. The periods of landscape stability and soil formation seem to be coincident with those periods when the climate was essentially that of today.

Lee-side dunes

Many of the playa lake basins of the southern High Plains have dunes on their down-wind side (Reeves, 1965, 1966). The dunes are generally silty and calcareous and are associated with those playas with calcareous lacustrine sediments on their floors.

The stratigraphy, geochronology, and pedology of the lee-side dunes is known from only one site (Holliday, 1982, 1983). This locality, known as the Dune site, is about 2 km north of the Lubbock Lake site. It is a large dune whose associated playa (Cone playa) is intersected by Yellowhouse Draw (Figure 14.4). Data are available from several trenches (Figures 14.4 and 14.9; Table 14.1) and adjacent exposures.

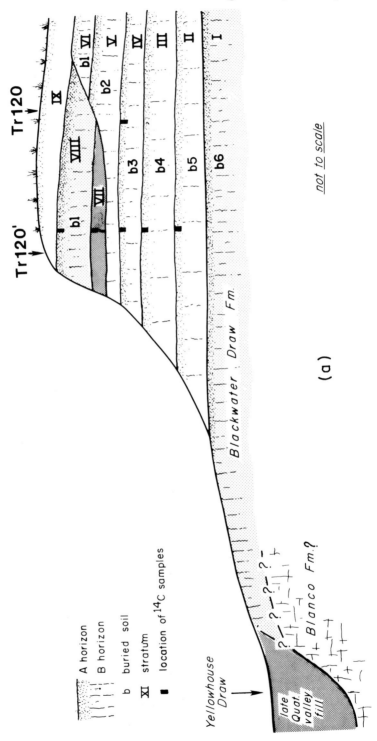

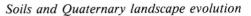

(a)

not to scale

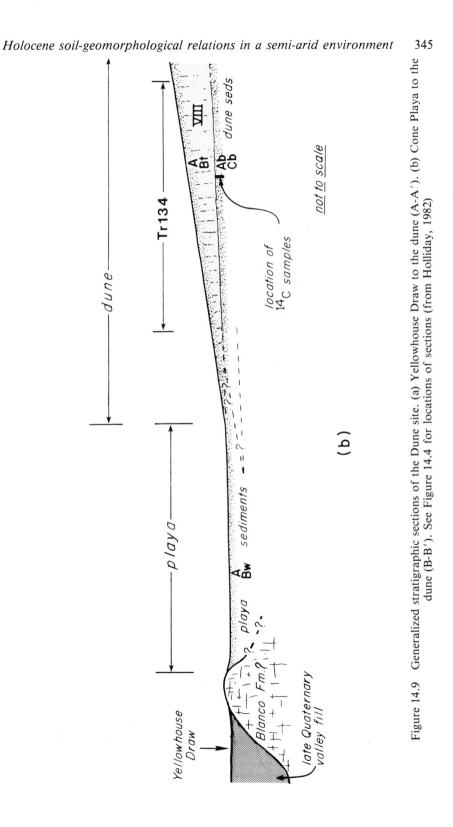

Figure 14.9 Generalized stratigraphic sections of the Dune site. (a) Yellowhouse Draw to the dune (A-A'). (b) Cone Playa to the dune (B-B'). See Figure 14.4 for locations of sections (from Holliday, 1982)

Table 14.1

Trench	Stratum	Horizon	Age, years BP	
134	VII	Ak1b1	6,695 ± 80	(SI-4586)
	VII	Ak2b1	8,320 ± 90	(SI-4587)
120, 120′	VIII	Ab1	1,335 ± 75	(SI-4939)
	VII	Akb2 (upper)	6,115 ± 190	(SI-4169)
	VII	Akb2 (lower)	6,980 ± 215	(SI-5168)
	IV	Ab3	8,855 ± 100	(SI-4977)
	IV	Ab3	12,080 ± 200	(SI-4941)
	III	Ab4	29,080 ± 1030	(SI-4978)
	II	Ab5	33,750 ± 3600	(SI-4979)

Aeolian sediments of two different origins have been identified at the Dune site. Most of the dune is composed of grey, silty, highly calcareous material, apparently derived directly from the floor of the neighbouring playa. The sand and silt mineralogy is mostly quartz and calcite. The dominant clay minerals are sepiolite and illite with minor quantities of smectite, mixed-layer illite-smectite clays and kaolinite. Within this material seven stratigraphic units have been identified (strata II-VII, IX) totalling over 7 m thick (Figure 14.9a).

Because of the similarity of most of the dune deposits the stratigraphic units were differentiated primarily on the basis of buried soils.

The other type of sediment (stratum VIII) is found on the windward side of the dune, inset against the above described deposits (Figure 14.9). This single deposit is a sandy clay loam up to 2 m thick. Limited field data suggest that it was deposited in an arcuate pattern on the windward side of the dune. The deposit is distinctly redder than the typical dune sediments but this is probably in part inherited from a well-developed soil formed in it. The sand and silt mineralogy is dominantly quartz with some calcite. The clay mineralogy is essentially the same as the finer-grained, calcareous material.

A number of radiocarbon ages, determined on the humin (NaOH-insoluble) fractions of the buried A horizons, are available from the dune site. The samples were usually taken from the top of the A horizon so the resulting age is a minimum one for soil formation and a maximum one for the overlying sediments. Two radiocarbon ages are available from a buried A horizon below stratum VII in Trench 134 (Figure 14.9b) and seven radiocarbon ages are available from the buried A horizons formed in dune materials in Trenches 120 and 120′ (Figure 14.9a; Table 14.1).

Stratigraphic and pedological correlations between the trenches, and the deposits and soils in the fill in Yellowhouse Draw have not been confirmed but the following correlations are suggested by the available data (Table 14.2).

Table 14.2

Dune Strata	Soil in top of deposit	Lubbock Lake Strata
IX	Singer	5
VIII	Lubbock Lake	4
VII	Yellowhouse	3
V	Firstview	2

The data suggest that stratum VIII is material derived primarily from the High Plains surface (i.e. it is of the same origin as stratum 4 in Yellowhouse Draw) and deposited against the material derived from the playa. The sepiolite in stratum VIII probably represents material deflated from the playa and mixed in with 4.

The Firstview Soil equivalent in the dune is moderately well developed. The A horizon is often missing. An Ab2 was described (Tr 120′, Table 14.3) but it appears to be primarily an organic, interdune deposit (Figure 14.9a). In thin sections from the B horizon argillans are commonly developed on skeleton grains, but not sufficiently to qualify the B horizon as argillic. Secondary $CaCO_3$ is pervasive in the subhorizons and all B horizons qualify as calcic. It is difficult to classify this soil in the absence of an epipedon but based on other soils observed in the dune it was probably not mollic and the soil, therefore, is an Ustochrept.

Data are not yet available for the Yellowhouse Soil equivalent. Cursory field examination, however, indicates that it is morphologically similar to the Firstview Soil equivalent.

The Lubbock Lake Soil equivalent in the dune is the best developed soil observed in any dune exposure. The epipedon is ochric. Thin sections indicate that the B horizon qualifies as argillic. There is considerable secondary $CaCO_3$ in the B horizons, almost 40 per cent by weight and they qualify as calcic. The Lubbock Lake Soil equivalent is classified as a Haplustalf.

The Apache Soil equivalent in Trench 120 is a very weakly developed soil. It exhibits an ochric epipedon and a calcic subhorizon. The soil is classified as an Ustochrept.

Development of the soils in the dune is difficult to compare with equivalent soils in the draw. The Lubbock Lake Soil equivalent in the dune is quite similar to its draw counterpart but the Firstview, Yellowhouse, and Singer equivalents are considerably different from their counterparts. This is probably because the factors of parent material and relief (specifically the landscape position) vary considerably between these settings. Within the dune soils, however, the principal variable in the Holocene has been time. The climate varied in the early Holocene but it is not yet possible to determine the influence this variable may have had.

Table 14.3(a) The Dune site, trenches 120 and 120'. Field descriptions

Equivalent soil	Substrat	Horizon	Depth (cm)	Munsell Colour Dry	Munsell Colour Moist	Texture	Structure	Dry Consistence	CaCO$_3$	Bndy	Comments
Trench 120 Singer	IX	A	0–8	10 Yr 6/3	10 Yr 4/4	SL	2fp1	sh	ev	cw	
		Bk	8–130	10 Yr 8/2	10 Yr 7/4	SCL	1csbk	sh	ev	cs	
	VIII	Abl	130–160	7.5 Yr 6/3	7.5 Yr 5/3	SCL	1cpr& 2csbk	sh	ev	cs	
		Btkbl	160–195	10 Yr 6/3	10 Yr 4.5/4	SCL	2cpr& 2csbk	h	ev	cw	
Lubbock Lake		Bklbl	195–243	10 Yr 8/2	10 Yr 6/3	CL	1mpr& 2csbk	h	ev	gs	
	VI	Bk2bl	243–415	10 Yr 8/1	10 Yr 7/3	SiC	2cpr& 2csbk	h	ev	aw	
		Btkb2	415–435	10 Yr 7/3	10 Yr 5/4	SiC	3mpr& 2csbk	h	ev	gw	
Firstview	V	Bkb2	435–532	10 Yr 8/2	10 Yr 7/3	SiC	m	vh	ev	cs	Surface soil removed by erosion at this location
Trench 120' Lubbock Lake	VIII	Abl	0–11	7.5 Yr 6/3	7.5 Yr 5/3	SCL	1cpr& 2csbk	sh	ev	cw	
		Btbl	11–107	7.5 Yr 5/3	7.5 Yr 6/4	SCL	2mpr& 2csbk	vh	ev	as	
?		Ab2	107–120	10 Yr 5/3	10 Yr 3/3	SCL	1mpr& 2csbk	vh	ev	cs	
	VII	Akb2	120–152	10 Yr 6/2	10 Yr 4/3	CL	3mab	eh	ev	aw	
Firstview	V	Bklb2	152–189	10 Yr 7/3	10 Yr 5/4	CL	3mpr& 2csbk	h	ev	gw	
		Bk2b2	189–260	10 Yr 8/2	10 Yr 7/3	SiC	m	vh	ev	cs	

Abbreviations for descriptions follow Soil Survey Staff (1951).

Table 14.3 (b) The Dune site, trenches 120 and 120'. Laboratory data*

Horizon	Particle-size distribution			OC (%)	CaCO$_3$ (%)
	Sand (%)	Silt (%)	Clay (%)		
Trench 120					
Ak	68.5	22.0	9.5	0.8	11.1
Bk	67.8	21.6	10.6	0.4	18.4
Abl	73.4	10.6	16.0	0.6	14.5
Btkbl	55.3	18.0	26.7	0.4	7.9
Bklbl	44.2	16.5	39.3	1.2	36.3
Bk2bl	54.1	22.8	23.1	0.6	39.5
Btkb2	50.9	22.3	26.8	1.3	14.2
Trench 120'					
Btbl	60.9	18.4	20.7	0.3	8.0
Ab2	63.3	14.2	22.5	0.2	4.3
Akb2	49.0	19.9	31.1	0.2	2.6

*Particle-size distribution follows Day (1965); sand by sieving, silt and clay by pipette on organic carbon and CaCO$_3$-free basis. Organic carbon (OC) by Walkley-Black method after Allison (1965). CaCO$_3$ by Chittick apparatus, after Dreimanis (1962) and Bachman and Machette (1977)

The stratigraphic and pedological sequence preserved in the lee-side dunes may be important climatic indicators. Playas commonly deflate during dry seasons and droughts. The sediments in the dunes may be indicators of lower effective moisture, resulting in dry lake beds. The soils would then represent wetter periods when the lakes held water, thus preventing deflation. This, combined with increased vegetation cover, would result in dune stability. Such an approach to the lee-side dunes must be done with caution, however, because fluctuations in the water table, which may be only indirectly linked to climate, may affect the presence or absence of water in the lake basins.

PLAYAS

The ephemeral-lake basins or playas of the southern High Plains include small (generally less than 4 km^2), fresh-water depressions that number in the thousands (e.g. Figure 14.4) and large (tens of km^2), saline lake basins of which there are about twenty-five (Reeves, 1965, 1966, 1972; Hawley *et al.*, 1976). Pedological studies have been conducted only in the small, fresh-water basins (e.g. Allen *et al.*, 1972; Harris *et al.*, 1972).

The origin of the small lake basins has been considered to be the result of wind deflation (Reeves, 1966). These basins and lacustrine sediments therein are generally of late Quaternary age.

Two distinct types of surface sediments are found in the small deflation basins. The most common sediment is a clayey, dark grey, usually non-calcareous (at least in the upper portion of the unit) deposit. The soil series mapped on this unit is the Randall clay (e.g. Blackstock, 1979) and therefore the sediment is locally known as the Randall clay. The other type of deposit is a silty to loamy, light grey, highly calcareous unit. Soils formed in this unit are mapped as the Arch loam (e.g. Blackstock, 1979) and the sediments are so referred to, informally.

Virtually no research has been done comparing the distribution, origins, and ages of the two types of playa fill. A few general comments can be made, based on field observations and examination of county soil surveys and topographic maps. The basins with the Arch loam are generally larger than those with the Randall clay and seem to be more numerous on the western and north-western portions of the Llano Estacado. Overall, however, the Randall clay is the more common surficial unit in the playas. At least some of the Arch sediments are of late Pleistocene age. The basal sediments at the Dune site, which were apparently deflated from Arch deposits in the adjacent playa (Figure 14.4), date to about 33 000 ry BP (discussed earlier). The playa sediments must, therefore, be somewhat older. Several radiocarbon ages have been secured from the Randall clay in a spectacular exposure of playa sediments at the Gentry pit in north central Lubbock County about 10 km north-east of Lubbock Lake (Figure 14.3). The playa sediments are about 4 m thick. The organic clay at the base of

Table 14.4 Arch loam, Cone playa, field description*

Horizon	Depth (cm)	Munsell colour Moist	Texture	Structure	Consistence	CaCO₃	Bndy	Comments
Ap	0–18	10 Yr 4/3	SiCL	1msbk	fr	em	cs	
AB	18–30	10 Yr 4/3	SiCL	1msbk	fr	es	gw	
Bw1	30–48	10 Yr 5/2	C	1msbk	fr	es	gw	
Bw2	48–88	10 Yr 6/2	C	1mpr& 1csbk	vh	es	cs	at 82–88 cm fine granules w/many fine and very fine carbon concretions
2Bg1	88–104	2.5 Yr 7/2	SiCL	1msbk	vh	e	cw	
2Bg2	104–122	5 Yr 6/2 (5 Yr 6/3)	CL	2fabk part.	vh	non-em	cw	common, medium, faint mottles () the k bifurcates to the south in
2Bk	122–145	10 Yr 8/1	CL	1–2f bk	vh	em	cw	middle of east wall
2Bg	145+	5 Yr 7/2 (5 Yr 7/4) common (2.5 Yr 6/4) few	CL	1fabk	vh	e		common, faint mottles (); few, very fine, dendritic Mn stains

*Below 88 cm parent material is pre-Arch (early Wisconsin?) lacustrine sediment.

the section has been dated to about 10 000 radiocarbon years BP and an age of about 1000 years BP was secured on organic material about 1 m below the surface (Warren Wood, pers. comm.). The Randall clay in this playa, then, is of Holocene age. Very limited evidence from other sources indicates that the Randall clay in most playas is of Holocene age (C. C. Reeves, pers. comm.).

It is also commonly observed that reddish, aeolian sands are found as a thin wedge within the Randall clay on the upwind side of the playas. This is seen at the Gentry pit. The position of the wedge in about the middle of the dated Randall clay suggests that it is possibly related to the mid-Holocene aeolian sediments that fill the draws, and accumulate in dunes.

One other age relationship between the Arch and Randall sediments is often observed. The Arch sediments commonly occur as benches around the edges of playa basins with the Randall clay in the central part of the basin inset against the Arch material (e.g. Blackstock, 1979). This further indicates that in at least some situations the Arch sediments predate the Randall and were severely deflated prior to Randall deposition; the deflation forming the basin for the clay.

The Arch soils generally are weakly developed (Table 14.4). They have ochric epipedons, a calcic horizon and sometimes cambic B horizons. The subhorizons are highly calcareous, reflecting the parent material. The Arch is classified as a Calciorthid (Blackstock, 1979). The principal influence affecting formation of the Arch soil appears to be its parent material. Soil formation is also probably affected by deflation and periodic accumulation of water on the surface of the playa.

The Randall Clay is a deep, dark, cumulic soil (Allen *et al.*, 1972; Blackstock, 1979). It has a mollic epipedon and no diagnostic subsurface horizons. The horizons are quite thick with gradual transitions from A to AC to C. Gleyed horizons are common in the lower profile. The soils are classified as Pellusterts. This is the result of the high clay content of the parent material which is dominated by smectite and mixed-layer illite-smectite (Allen *et al.*, 1972; Harris *et al.*, 1972). The most significant influence on the genesis of the Randall soil is the slowly aggrading, fine-grained parent material in the lacustrine microenvironment of the lake basin.

In comparing the Arch and Randall soils certainly one of the most significant variables in the factors of their formation is parent material. Very limited data suggests that time may also be an important variable. It appears that the parent material for the Arch was in place by the beginning of the Holocene and the soil formed on a relatively stable landscape, although with some deflation. In contrast, the parent material for the Randall slowly accumulated throughout the Holocene with concomitant pedogenesis. The Arch exhibits somewhat stronger soil horizonation but whether this is due to the length of time available for development or variations in parent material or both is not yet understood.

SUMMARY AND CONCLUSIONS

The surface of the southern High Plains is a Pleistocene feature that has been modified by wind deflation and water erosion in the late Quaternary. These geomorphological agents have produced the principal settings for Holocene deposition. These include draws (ephemeral drainage ways), dunes, including large dune fields and dunes on the lee sides of lake basins, and playas (ephemeral lake basins), formed by deflation and often with lee-side dunes.

The types and environments of deposition vary depending on setting and age. Along the draws early Holocene deposition is primarily lacustrine. Between about 6000 and 4500 ry BP the draws are filled with considerable thicknesses of aeolian sediments. Most of the rest of the Holocene is marked by landscape stability with only minor aeolian deposition along with some slope wash within the past 1000 years. The dunes are characterized by several discrete intervals of activity separated by relatively long periods of stability and soil formation. There were several cycles of aeolian activity between about 6000 and 4500 ry BP with some minor activity between about 3000 and 2000 ry BP and within the past few hundred years. The playas are characterized by gradual, fairly continuous deposition of lacustrine sediments throughout the Holocene with deposition of some aeolian sediments in the mid Holocene.

The factors influencing the formation of soils in the Holocene also vary depending on the setting and period of soil formation. Pedogenesis along the draws in the early Holocene, and up to about 5000 BP, was influenced by marsh microenvironments. Specifically, this would include lacustrine parent material (organic clay or calcareous silt) and a high water table. Variations in the late Holocene soils is primarily a function of variation in the length of time available for pedogenesis. Through time these soils are subjected to considerable influence by additions of aerosolic clay and carbonate. Within the dune fields and lee-side dunes variations in soil morphology throughout the Holocene is also a function of variation in length of time for development. The soils in the dune fields, and apparently to a lesser degree those in the lee-side dunes, are influenced by aerosolic clay and carbonate. In comparing the soils in these two types of dunes the principal variable in the factors affecting their formation is parent material. The dune-field sediments are sandy, permeable and non-calcareous and the lee-side dunes are composed of silty, highly calcareous deposits.

The data suggest that the Holocene geomorphological history of the southern High Plains was dominated by relatively long periods of soil formation, interrupted only by discrete, though intense, periods of aeolian activity. This took the form of deflation in dune fields and playas and sedimentation in dunes and draws, primarily in the mid Holocene between about 6000 and 4500 ry BP.

The dominant variables in the factors of soil formation in the Holocene in the region is time. Through time there is considerable influence from aerosolic additions of clay and carbonate. The principal exception is the variation in lacustrine parent materials and microenvironments in playas throughout the Holocene and along draws in the early Holocene.

The above demonstrates that aeolian activity has been a very important process on the southern High Plains in the late Quaternary. Wind has scoured the playa basins, forming the most ubiquitous setting for Holocene deposition. The wind then built the dune fields and lee-side dunes and partially filled the draws, adding to the topography of the area and providing the parent material for the Holocene soils. And, finally, airborne clay and carbonate has been injected into the Holocene sediments, resulting in the very rapid formation of argillic and calcic horizons.

ACKNOWLEDGEMENTS

I would like to thank L.H. Gile (Las Cruces, New Mexico), C. C. Reeves, Jr. (Geosciences, Texas Tech University) and B. L. Allen (Plant and Soil Science, Texas Tech University) for reviewing a draft of this paper and for sharing their considerable expertise on the geology and soils of the southern High Plains. An anonymous reviewer provided some very helpful comments. Warren Wood (US Geological Survey, Reston, VA) very kindly provided the information on the radiocarbon ages from the Gentry pit, Robert Stuckenrath (Radiocarbon Laboratory, Smithsonian Institution) provided the radiocarbon ages for the Dune Site, and Leland Gile provided Figure 14.9. The Cartography Laboratory of the University of Wisconsin provided aid in preparation of some of the illustrations and Sally Monogue (Geography, The University of Wisconsin) typed the manuscript. And I would like to thank John Boardman for his efforts and patience toward preparation of this paper and, in particular, for organizing the excellent conference that led to this volume.

This paper represents part of the continuing research of the Lubbock Lake Project (The Museum, Texas Tech University; Eileen Johnson, Director), funded by the National Science Foundation (SOC75–14857, BNS7612006, BNS7612006-A01, BNS78-11155), National Geographic Society, Sigma Xi, Texas Historical Commission, City and County of Lubbock, the Center for Field Research (EARTHWATCH), the University of Colorado, Texas Tech University, and the Museum, Texas Tech University.

REFERENCES

Allen, B. L., Harris, B. L., Davis, K. R. and Miller, G. B. (1972). *The Mineralogy and Chemistry of High Plains Playa Lake Soils and Sediments*, Water Resources Center Pub. No. 72-4, Texas Tech University, Lubbock.

Allison, L. E. (1965). 'Organic carbon', in *Methods of Soil Analysis,* Part 2, (Eds C. A. Black *et al.*), pp. 1376–1378, American Society of Agronomy, Madison, Wisconsin.

Bachman, G. O. and Machette, M. N. (1977). Calcic soils and calcretes in the southwestern United States, *US Geological Survey Open File Report*, 77-794.

Birkeland, P. W. (1974). *Pedology, Weathering, and Geomorphological Research*, Oxford University Press, New York.

Black, C. C. (1974). 'History and prehistory of the Lubbock Lake site', *The Museum Journal*, West Texas Museum Association, Lubbock, 15.

Blackstock, D. A. (1979). *Soil Survey of Lubbock County, Texas*, US Department of Agriculture, Soil Conservation Service.

Blair, W. F. (1950). 'The biotic provinces of Texas', *Texas Journal of Science*, **2**, 93–117.

Carr, J. T. (1967). 'The climate and physiography of Texas', *Texas Water Development Board Report*, **53**.

Clanton, J. S. and Reeves, C. C., Jr. (1975). 'Pleistocene drainage on the southern High Plains, west Texas', *Geological Society of America Abstracts with Programs, South-Central Section*, **7**, 152–153.

Day, P. R. (1965). 'Particle fractionation and particle-size analysis', in *Methods of Soil Analysis*, Part 2, (Eds C. A. Black *et al.*), pp. 545–567, American Society of Agronomy, Madison, Wisconsin.

Dreimanis, A. (1962). 'Quantitative determination of calcite and dolomite by using the Chittick apparatus', *Journal of Sedimentary Petrology*, **32**, 520–529.

Gile, L. H. (1977). 'Holocene soils and soil-geomorphic relations in a semi-arid region of southern New Mexico', *Quaternary Research*, **7**, 112–132.

Gile, L. H. (1979). 'Holocene soils in eolian sediments of Bailey County, Texas', *Soil Science Society of America Journal*, **43**, 994–1003.

Gile, L. H. (1981). *Soils and Stratigraphy of Dunes Along a Segment of Farm Road 1731, Bailey County, Texas*, ICASALS, Texas Tech University, Publication No. 81–2.

Gile, L. H. (1983). 'Holocene soils exposed in a Sandhills blowout on the southern High Plains', in *Guidebook to the Central Llano Estacado* (Ed. V. T. Holliday), pp. 135–165, Friends of the Pleistocene, South-Central cell field trip, ICASALS and The Museum, Texas Tech University.

Gile, L. H., Peterson, F. F. and Grossman, R. B. (1966). 'Morphological and genetic sequences of carbonate accumulation in desert soils'. *Soil Science*, **101**, 347–360.

Harris, B. L., Davis, K. R., Miller, G. B. and Allen, B. L. (1972). 'Mineralogical and selected chemical properties of High Plains playa soils and sediments', in *Playa Lake Symposium* (Ed. C. C. Reeves, Jr.), pp. 287–300, ICASALS Publication 4, Texas Tech University.

Hawley, J. W., Bachman, G. O. and Manley, K. (1976). 'Quaternary stratigraphy in the Basin and Range and Great Plains provences, New Mexico and western Texas', in *Quaternary Stratigraphy of North America* (Ed. W. C. Mahaney), pp. 235–275, Dowden, Hutchinson, and Ross, Stroudsburg, Pennsylvania.

Haynes, C. V. (1975). 'Pleistocene and recent stratigraphy', in *Late Pleistocene Environments of the Southern High Plains*, (Eds F. Wendorf and J. J. Hester), pp. 59–96. Fort Burgwin Research Center, Publication 9, Taos.

Haynes, C. V. and Agogino, G. A. (1966). 'Prehistoric springs and geochronology of Blackwater No. 1 locality, New Mexico', *American Antiquity*, **31**, 812–821.

Hefley, H. M. and Sidwell, R. (1945). 'Geological and Ecological Observations of some High Plains dunes', *American Journal of Science*, **243**, 361–376.

Holliday, V. T. (1982). *Morphological and Chemical Trends in Holocene Soils at the Lubbock Lake Archeological Site, Texas*, Unpublished PhD thesis, University of Colorado.

Holliday, V. T. (Ed.) (1983). *Guidebook to the Central Llano Estacado*, Friends of the Pleistocene, South-Central cell field trip, ICASALS and The Museum, Texas Technical University.

Holliday, V. T. (in press a). 'Historical perspective', in *Lubbock Lake: Late Quaternary Studies on the Southern High Plains* (Ed. E. Johnson), Center for the Study of Early Man, Orono.

Holliday, V. T. (in press b). 'Early Holocene soils at the Lubbock Lake archeological site, Texas'. *Catena*.

Holliday, V. T. (in press c). 'Morphology of late Holocene soils at the Lubbock Lake archeological site, Lubbock County, Texas'. *Soil Science Society of America Journal*.

Holliday, V. T. (in press d). 'New data on the stratigraphy and pedology of the Clovis and Plainview sites, southern High Plains'. *Quaternary Research*.

Holliday, V. T., Johnson, E., Haas, H. and Stuckenrath, R. (1983). 'Radiocarbon ages from the Lubbock Lake site, 1950–1980: framework for cultural and ecological change on the southern High Plains', *Plains Anthropologist*, **28**, 165–182.

Holliday, V. T., Johnson, E., Haas, H. and Stuckenrath, R. (in prep.). 'Radiocarbon ages from the Lubbock Lake site: 1981–1983', manuscript submitted for publication.

Hunt, C. B. (1974). *Natural Regions of the United States and Canada*, W. H. Freeman, San Francisco.

Jenny, H. (1941). *The Factors of Soil Formation*, McGraw-Hill, New York.

Jenny, H. (1980). *The Soil Resource*, Springer-Verlag, New York.

Johnson, C. A. (1974). 'Geologic investigations at the Lubbock Lake site', in 'History and prehistory of the Lubbock Lake site', *The Museum Journal* (Ed. C. C. Black), pp. 79–106, West Texas Museum Association, Lubbock, Texas, 15.

Laprade, K. E. (1957). 'Dust-storm sediments of Lubbock area, Texas', *Bulletin of the American Association of Petroleum Geologists*, **41**, 709–726.

Lotspeich, F. B. and Everhart, M. E. (1962). 'Climate and vegetation as soil forming factors on the Llano Estacado', *Journal of Range Management*, **15**, 134–141.

Machette, M. N. (1975a). *The Quaternary Geology of the Lafayette Quadrangle, Colorado*, Unpublished MS Thesis, University of Colorado.

Machette, M. N. (1975b). 'Geologic map of the Lafayette Quadrangle, Adams, Boulder, and Jefferson counties, Colorado, 1:24,000', *US Geological Survey Miscellaneous Field Studies*, MF-656.

Melton, F. A. (1940). 'A tentative classification of sand dunes — its application to dune history in the southern High Plains', *The Journal of Geology*, **48**, 113–174.

NOAA (1982). 'The Climate of Texas', *Climatography of the United States*, **60**, National Climatic Research Center, Asheville, North Carolina.

Reeves, C. C., Jr. (1965). 'Chronology of west Texas lake dunes', *The Journal of Geology*, **73**, 504–508.

Reeves, C. C., Jr. (1966). 'Pluvial lake basins of west Texas', *The Journal of Geology*, **74**, 269–291.

Reeves, C. C., Jr. (1970). *Some Geomorphic Stratigraphic, and Structural Aspects of the Pliocene and Pleistocene Sediments of the Southern High Plains*, Unpublished PhD Dissertation, Texas Tech University.

Reeves, C. C., Jr. (1972). 'Tertiary-Quaternary stratigraphy and geomorphology of west Texas and southeastern New Mexico', in *Guidebook of East-Central New Mexico* (Eds V. C. Kelley and F. D. Trauger), pp. 103–117, New Mexico Geological Society.

Reeves, C. C., Jr. (1976). 'Quaternary stratigraphy and geologic history of the southern High Plains, Texas and New Mexico', in *Quaternary Stratigraphy of North America*, (Ed. W. C. Mahaney), pp. 213–223, Dowden, Hutchinson and Ross, Stroudsburg, Pennsylvania.

Reeves, C. C., Jr. and Parry, W. T. (1969). 'Age and morphology of small lake basins, southern High Plains, Texas and eastern New Mexico', *Texas Journal of Science*, **20**, 349–354.

Soil Survey Staff (1951). *Soil Survey Manual,* US Department of Agriculture Soil Conservation Service, Agriculture Handbook 18.

Soil Survey Staff (1975). *Soil Taxonomy,* US Department of Agriculture, Soil Conservation Service, Agriculture Handbook 436.

Stafford, T. W. (1977). 'Late Quaternary alluvial stratigraphy of Yellowhouse Draw, Lubbock, Texas', in *Cultural Adaptation to Ecological Change on the Llano Estacado,* Preliminary report of the 1976 field season of the Lubbock Lake Project, submitted to the National Science Foundation (Eds E. Johnson and T. W. Stafford), pp. 72–176.

Stafford, T. W. (1978). 'Late Quaternary alluvial stratigraphy of Yellowhouse and Blackwater Draws, Llano Estacado, Texas', in *Cultural Adaptation to Ecological Change on the Llano Estacado,* Preliminary report of the 1977 field season of the Lubbock Lake Project, submitted to the National Science Foundation, (Ed. Eileen Johnson), pp. 99–128.

Stafford, T. W. (1981). 'Alluvial geology and archeological potential of the Texas southern High Plains', *American Antiquity,* **46,** 543–565.

Strahler, A. N. and Strahler, A. H. (1983). *Modern Physical Geography,* Wiley, New York.

Warn, G. F. and Cox, W. H. (1951). 'A sedimentary study of dust storms in the vicinity of Lubbock, Texas', *American Journal of Science,* **249,** 553–568.

Wendorf, F. (1961). 'A general introduction to the ecology of the Llano Estacado', in *Paleoecology of the Llano Estacado* (Ed. F. Wendorf), pp. 12–21, Fort Burgwin Research Center, Publication I, Taos.

Woodruff, C. M., Jr., Gustavson, T. C. and Finley, R. J. (1979). 'Playas and draws on the Llano Estacado-tentative findings based on geomorphic mapping of a test area in Texas', *Texas Journal of Science,* **31,** 213–223.

Soils and Quaternary Landscape Evolution
Edited by J. Boardman
© 1985 John Wiley & Sons Ltd.

15

The Belgian Loess Belt in the Last 20 000 Years: Evolution of Soils and Relief in the Zonien Forest

R. Langohr and J. Sanders

ABSTRACT

In forested areas of the Belgian loess belt above the spring line two main geomorphological units are distinguished, (a) plateaux and smooth slopes, and (b) valleys. In the valleys three lower valley forms are observed, (a) concave bottoms, (b) V-shaped gullies, and (c) meandering flat-bottomed gullies. Six soil types, closely related to the various relief positions, are described. Both geomorphological and soil characteristics indicate that the whole area has been remarkably stable in the last 20 000 years. There is evidence of one erosion cycle towards the end of the Weichsel glaciation. The erosion was only active in the valley bottoms creating V-shaped and meandering flat-bottomed gullies.

INTRODUCTION

In the Belgian loess belt erosion and sedimentation processes are widespread under agriculture. Splash and sheet erosion are most common, but rill and occasionally gully erosion are also observed (Bolinne, 1976; De Ploey, 1979). The average soil loss reaches as much as 1 mm year^{-1} (Bolinne, 1977). Most sediment from the slopes has been deposited in the nearby valley bottoms which now have several metres of material called colluvium by the Belgian National Soil Survey (Louis, 1959). In the last 15 years, land degradation associated with intensive land use has been investigated thoroughly by field and laboratory experiments (e.g. Bolinne, 1977; De Ploey, 1977).

When the Romans invaded Belgium, some 2000 years ago, agriculture already existed in part of the loess belt. As the population increased, more forest was cleared. Sixteenth century maps show that the loess belt was already predominantly agricultural. Clearance continued until about 1840 when only a small percentage was still forest. Compared to the agricultural area, erosion and sedimentation processes under forest cover in the Belgian loess belt have attracted little attention. The suggestion that these processes are less active under the protective vegetation and litter cover is confirmed by the Belgian soil maps which indicate that valleys with colluvial soils are much less common under forest than agriculture (Dudal, 1959; Louis, 1959). V-shaped gullies, up to several metres deep, in the bottom of concave depressions are attributed to Holocene erosion which is still active (Arnould-De Bontridder and Paulis, 1966; Gullentops and Arnould-De Bontridder, 1966). These authors also give evidence of creep on some of the steepest slopes under forest.

This chapter presents information on erosion and sedimentation processes under forest cover, obtained by the study of both soils and relief, in an environment which has experienced minimum disturbance by man—the Zonien Forest.

GENERAL SETTING OF THE AREA

The Zonien Forest, south-east of Brussels in the middle of the 8000 km² of the Belgian loess belt is the largest forest (43 km²) on loess. From the eleventh century to 1840, the forest was owned by the Dukes of Brabant and used for hunting, wood and charcoal production. Since the eleventh century there has been no cultivation or other farming activities except grazing. Although for the last two centuries all trees have been planted (mainly beech—*Fagus sylvatica*), human influence on the soil profile and the relief is minimal. The forest represents 43 km² of soil almost identical to the loess soils that man started to clear more than 6000 years ago.

Of great importance in this investigation is the fact that the forest covers the flat interfluve between two important rivers of mid Belgium, the Zenne and Dyle (Figure 15.1). Most other smaller forests of the loess belt are on steep valley slopes. In the Zonien Forest it is possible to investigate complete toposequences from the watershed to the valley bottom, crossing the plateaux and steeper slopes. Wide plateaux with slopes of less than 5° are dissected by a dendritic system of valleys. Valley bottoms are narrow, side slopes relatively short and usually not exceeding 16°. The lowest point in the forest lies around 65 m and the highest at 130 m. The plateaux are at 100–130 m. All the valleys are dry except for those below 70 m (Figure 15.2).

The present climate is humid temperate with mean annual air temperature of 9.4 °C, mean annual precipitation of 840 mm and evapotranspiration of 640 mm (according to Thornthwaite, 1948). From May to September the maximum potential moisture deficit is about 110 mm.

Beech (*Fagus sylvatica*) highstand composes 80 per cent of the forest, largely without arboreal undergrowth and in large areas without any ground vegetation.

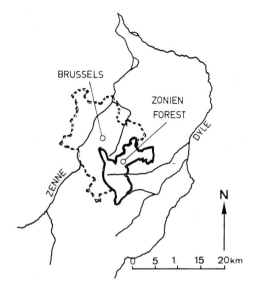

Figure 15.1   Location of the Zonien Forest

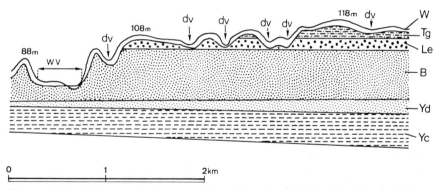

Figure 15.2   Geology of the area (after Mourlon, 1911). W, Weichsel loess; Tg, Tongerian clayey sands; Le, Ledian sands; B, Brusselian sands; Yd, Yperian sands and sandy clays; Yc, Yperian clay (impermeable); 88 m, altitude above sea level; wv, wet valley, with stream; dv, dry valley

There are also stands of oak (*Quercus* sp., 10 per cent) and coniferous trees (10 per cent).

The geology of the Zonien Forest is shown in Figure 15.2. 90 per cent is covered with several metres of loess. The upper part corresponds to the Brabantian loess deposited at the end of the Pleniglacial B ($\pm$ 20 000 years BP, Haesaerts and Bastin, 1977) of the Weichsel glaciation. Beneath the loess occur Tertiary marine deposits, mainly Eocene sands (Brusselian and Ledian) and

Oligocene clayey sands (Tongerian). The latter occur on the highest landscape positions above 100 m altitude.

Soils developed in the loess of this area have been classified as Glossudalfs (Tavernier and Louis, 1971). Yet recent research has shown that nearly all these soils have a fragipan horizon (Langohr and Van Vliet, 1981) or at least a pedogenetically consolidated horizon (Langohr, 1983a). Thus the Fraglossudalf category (Soil Survey Staff, 1975) is better for these soils in which roots are mainly restricted to the upper 35 cm. Nearly all soils in the forest are well drained. Even the valley-bottom soils do not have a ground-water table except for very small areas below the spring line.

METHODS

Following a survey of the geomorphological units, 60 profile pits were studied along selected toposequences. Field study of the soils was very detailed, using a methodology developed by Langohr, combining procedures from pedology, archaeology and geology. Special attention was paid to the combination of both vertical and horizontal sections of the soil profile and to the relative position to each other of the soil characteristics. Samples were collected for physiochemical and mineralogical analysis (Van Ranst *et al.*, 1982; Sanders *et al.*, 1983), for the mesoscopic study of undisturbed soil fragments under stereoscopic microscope (Langohr and Pajares, 1983), and for the preparation of thin sections (Van Vliet and Langohr, 1983).

RESULTS

Geomorphology and topographic position

The major landforms and slope units are shown in Figure 15.3. The area can be subdivided in two major landforms. First, smoothly undulating plateaux with slopes of less than 5°. No further subdivision has been made in this landform which is indicated as one single slope unit (symbol Pl) on all three transects in Figure 15.3. Secondly, valleys, all of which have a convex upper slope (Cv) and a straight mid slope (St) usually of less than 30°. The lower slopes are of three different types. First, concave valley bottoms (transect AB). Two slope units are distinguished here, the concave footslope (Cf) and the concave base of the valley (Cb). Secondly, V-shaped gullies which are eroded in the pre-existing concave valley bottom (transect CD) and have straight gully walls (Gs) usually of 30–40°, and a gully bottom (Gb). Thirdly, meandering flat-bottomed gullies (transect EF) which have very steeply sloping (35–45°) concave banks to the meanders (Fs) and flat valley bottoms (Fb). The convex banks are very weakly developed and are not studied here. The three valley-bottom types, concave, V-shaped and flat-bottomed gully are successively encountered downstream.

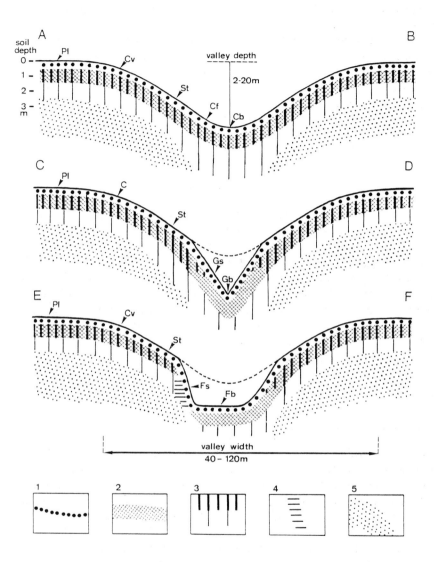

Figure 15.3   Soil toposequence through the three valley bottom types (schematic). AB, transect through a valley with concave bottom. CD, transect through a valley with a V-shaped gully eroded in an initial concave valley bottom. EF, transect through a valley with a meandering flat-bottomed gully, eroded in an initial concave valley bottom. Slope units: Pl, plateau; Cv, convex slope; St, straight slope; Cf, concave footslope; Cb, concave bottom; Gs, gully slope; Gb, gully bottom; Fb, flat valley bottom; Fs, steep concave bank. Soil characteristics: 1, colour and structure (B) horizon with good root penetration; 2, consolidated horizon(s); 3, clay illuviation (upper part = strong, lower part = weak); 4, clay illuviation in bands; 5, calcareous loess (C horizon)

The position of each of these valley types (Figures 15.3 and 15.5) shows that originally, above the spring line, the concave valley bottom type was present everywhere, or at least was dominant. During a later erosion cycle, V-shaped gullies formed where concave valley bottoms had relatively steep longitudinal slopes (more than a few degrees). Where the channel slope became smoother, the streams started meandering, forming the flat-bottomed gullies.

All higher valley bottoms are dry today and from historical data we can deduce that no surface runoff has occurred in the last 600 years.

Soil toposequences

Figure 15.4 shows the six main soil types (SK1-6) observed along the three toposequences and the slope unit positions on Figure 15.3.

SK1, 2 and 3 are found along transect AB (plateau and concave depression). The soils of the SK1 type are found on plateaux (Pl), convex (Cv) and straight (St) slopes of the valleys. Just below the thin humus surface (A1) they have a colour and structure (B) horizon with abundant roots. At 30–40 cm there is a pedogenetically consolidated horizon described as fragipan by Van Vliet and Langohr (1981) through which very few roots pass. As the exact definition of this horizon is rather vague (Smalley and Davin, 1982; Langohr, 1983b), the name 'consolidated horizon' (horizon symbol 'c') is preferred (Langohr, 1983a). The genesis of this consolidated horizon, which occurs in many forest soils of western Europe, is thought to be related to permafrost conditions which occurred at the end of the Weichsel glaciation (Van Vliet and Langohr, 1981; Langohr, 1983a). Within the consolidated horizon of SK1 soils there is a bleached eluvial horizon (A2gc) and a strongly developed brown horizon enriched in illuvial clay (B2tgc). In the latter, most clay coatings are incorporated into the matrix rather than lining the pores (Langohr and Pajares, 1983). The cause of this disturbance is thought to be the freeze-thaw cycles which occurred in these soils during the last cold period(s) of the Weichsel glaciation (Van Vliet and Langohr, 1983).

Both the A2gc and the B2tgc horizons have separate bleached, and Fe-enriched and Mn-enriched hydromorphic mottles characteristic of soils with a seasonal ground-water table. These features are relict as none of the studied soils have a ground-water table today. The bleached A2gc has tongues that penetrate the underlying B2tgc. In a section parallel to the soil surface these tongues form a pattern of streaks delimiting vertical prisms (Van Vliet and Langohr, 1981; Langohr, 1983a). The B2tgc overlies an unconsolidated and unmottled transition horizon (B3t) at 100–120 cm which extends to the calcareous C horizon at 250–350 cm. The B3t is lighter brown in colour than the B2tgc, exhibits weak clay accumulation and has a slightly higher clay content than the C horizon. Undisturbed clay coatings are found along larger pores and the few fissures delimiting coarse prismatic peds.

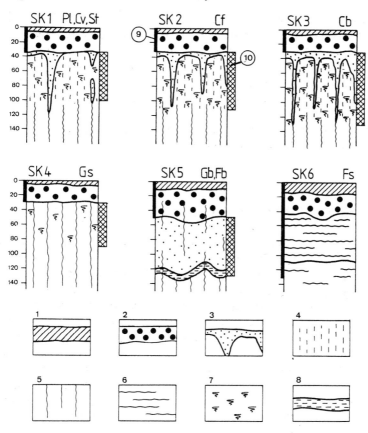

Figure 15.4   Main soil types observed along the toposequences (Figure 15.3). SK: soil type. Slope units: (Figure 15.3): Pl, plateau; Cv, convex slope; St, straight slope; Cf, concave footslope; Cb, concave bottom; Gs, gully slope; Gb, gully bottom; Fb, flat valley bottom; Fs, steep concave bank. Soil characteristics: 1, humiferous surface horizon; 2, colour and structure (B) horizon; 3, bleached A2 horizon; 4, strong clay illuviation with disrupted clay coatings; 5, weak clay illuviation with undisturbed clay coatings; 6, clay illuviation in bands; 7, relict hydromorphic mottles and Fe-Mn concretions; 8, Strong Fe and Mn accumulation; 9, very porous soil with very good root penetration; 10, consolidated soil. Profile depths in cm

The soils of the SK2 and SK3 type, found respectively on the footslope (Cf) and in the bottom (Cb) of concave valleys, while basically similar to SK1 soils, differ in two significant ways. In B2tgc horizons of SK2 and SK3 soils there is a gradual decrease in size of the prisms delimited by the A2gc streaks. This is associated with an increase in the quantity and contrast of relict hydromorphic mottles in the A2gc and the B2tgc horizons.

SK4 and SK5 soils, associated with the gullies (transect CD) have several characteristics similar to the soils described along the concave valley (transect AB). A colour and structure (B) horizon with abundant roots occurs just below the humus horizon and overlies a consolidated horizon with relict hydromorphic mottles. In these soils however the consolidation does not overlap a bleached A2gc or brown B2tgc. In SK4, situated on the gully slopes (slope unit Gs of transect CD), the consolidation coincides with a light brown horizon which has a small percentage higher clay content than the C horizon, clay cutans similar to B3t horizon described for SK1, 2 and 3, and hydromorphic mottles. The consolidation and the mottling end at about 1 m, above a horizon identical to the B3t of SK1, 2 and 3. In SK5, situated in the gully bottoms (slope unit Gb of transect CD) and the flat valley bottoms (relief position Fb of transect EF), the consolidation coincides with a thick, bleached, eluvial A2 horizon with hydromorphic mottles and Fe-Mn concretions (A2gc). The lower limit of the consolidation is formed by a continuous Fe-Mn-enriched horizon which is rusty brown in the upper half and brown to black in the lower half. This horizon varies from a few millimetres to more than 10 cm thick and is smooth or wavy. That these soils possess no ground-water table implies that the hydromorphic properties are also relict. Beneath the A2gc is a B3t horizon as described in SK1, 2 and 3.

SK6 soils are of very limited extent. They are on the very steep concave banks of the relict meanders along the flat-bottomed valleys. Profiles have a colour and structure (B) horizon over banded B2t and B3t horizons. The clay accumulation bands are parallel to the soil surface and up to 15 cm thick. Clay coatings line pores and peds and locally are disturbed by biological activity (mainly roots) which is very active up to the lower limit of the B3t. In these soils the calcareous C horizon (unweathered loess containing 10–15 per cent free calcium carbonate) is within 180 cm depth. No consolidation and no hydromorphic mottles are observed.

DISCUSSION

Both the hydromorphic mottles and the consolidation observed in SK1, 2, 3, 4 and 5, are relict. Today, all these soils are well drained and we know of no soil process active in moist temperate forest environments that can develop consolidation such that roots can only pass through a few fissures. Furthermore this compaction occurs in aeolian deposits from which 10–15 per cent $CaCO_3$ has been leached since deposition. In such material an extremely porous soil would be expected. The hypothesis of a periglacial origin for these relict soil characteristics has been discussed previously (e.g. Van Vliet and Langohr, 1981; Langohr, 1983a).

In SK1, 2 and 3, the upper limit of the B2tgc horizon is situated at a nearly constant depth (30–40 cm) along toposequence AB (Figure 15.3), including the deepest point of the concave valley bottom. As deposition of illuvial clay in this horizon developed before the last, or one of the last, permafrost periods of the Weichsel glaciation (Langohr and Pajares, 1983; Langohr, 1983a), no erosional processes have been active along the slopes of the concave valleys since then, otherwise the B2tgc horizon of the valley bottom would be buried under more recent sediments (slope unit Cb).

The geomorphological setting of the valley types (Figure 15.3) shows that the V-shaped gullies and the meandering flat-bottomed gullies correspond to more recent erosion active after the development of the concave valleys. This is confirmed by the position of the strongly developed B2tgc horizon which is present on the plateaux and the concave valleys (transect AB) but absent along the slopes and in the bottom of the gullies (slope units Gs and Gb of transect CD and slope unit Fb in transect EF). The presence in the soils of the gully slopes (SK4 and slope unit Gs) of a weakly developed illuvial clay-enriched horizon, very similar to the B3t of the soils situated on higher landscape positions supports this hypothesis. These data show that the most recent erosion cycle, (a) post-dates the strong clay-enriched B2tgc horizon of SK1, 2 and 3, and (b) is contemporaneous with, or even occurred before all, or at least part of, the consolidation and the associated hydromorphic mottling. As these two relict soil characteristics are linked to a periglacial environment we conclude that the erosion cycle which generated the gullies is of Weichsel age. However, the erosion occurred after the genesis of the strong illuvial clay-enriched horizon which is only observed along the plateaux and the concave depressions (transect AB). This horizon developed in the loess deposited some 20 000 years BP and its genesis probably took several thousand years, or at least the time necessary to decalcify the 10–15 per cent rich loess up to a depth of 100–150 cm and to reach a degree of clay illuviation characterized by a migration of about one-third of the clay from the eluvial horizon (total clay content of 10, 20 and 15 per cent in respectively the A2gc, the B2tgc and the C horizon). From these data we conclude that this particular erosion cycle must have occurred just before or in the latest cold period of the Weichsel, possibly in the Younger Dryas. Previously it was thought to be of Holocene age (Arnould-De Bontridder and Paulis, 1966), possibly related to the more humid climate of the Atlantic period (Gullentops, 1957).

As the soils of the V-shaped and flat-bottomed gullies are similar, we deduce that they are of the same age. The reason for a difference in form must be sought in other environmental conditions such as the valley slope.

The soils on the very steep slopes (SK6—slope unit Fs) are different from all the others. This is the only soil type without a consolidated horizon. As this horizon developed under a periglacial climate, one could suppose that this soil type is younger. However, the presence of the calcareous C horizon relatively close to the soil surface suggests another interpretation. The hypothesis is put forward (Langohr, 1983a) that under a periglacial climate, compaction developed here in a still calcareous material. During further pedogenesis the calcium carbonate was leached and consequently the consolidation disappeared. The shallowness of the calcareous material on these steep slopes is the result of erosion that occurred along the meandering rivers which exposed calcareous loess (Figure 15.5).

The soils in the V-shaped and flat-bottomed gullies have a particular sequence of pedogenetic horizons (SK5, Figure 15.4). Some of the soil characteristics, such as consolidation, hydromorphic mottles and the Fe-Mn accumulation horizon are relict and can be explained by a periglacial environment with freeze–thaw cycles and permafrost in the subsoil. Since the end of the erosion cycle during which the gullies were cut, no further erosion or aggradation has occurred. This is in contradiction to the soil map of Belgium where the soils in the valley bottoms under forest are shown to be developed in recent colluvial deposits, similar to those observed in the areas under agriculture.

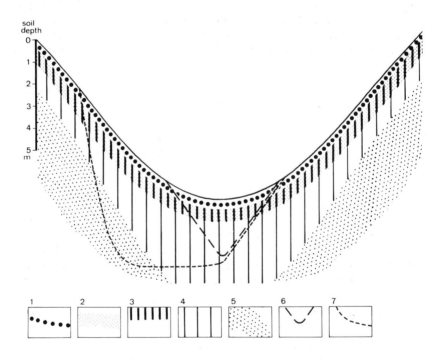

Figure 15.5   Relative position of the eroded gullies in relation to the pre-existing concave valley bottom and its soils (slope exaggeration ×2). Soil characteristics: 1, colour and structure (B) horizon; 2, consolidated horizon; 3, strong clay illuviation (B2t horizon); 4, weak clay illuviation (B3t horizon); 5, calcareous loess (C horizon); 6, V-shaped gully eroded in the concave valley bottom; 7, flat-bottomed gully eroded in the concave valley bottom

A periglacial climate, rather than a moist temperate (Holocene) environment, is indicated by hydromorphic mottles and consolidation. Such a climate can also explain erosion and incision above present-day spring level of the gullies in pre-existing valleys. Soil toposequences (Figure 15.4 and 15.5) show clearly that erosion only occurred in valley bottoms and not on slopes. Gullies with meandering streams were generated by massive quantities of water reaching the valley bottoms without eroding the valley slopes. As all these valleys are now dry (loess cover on thick sandy deposits), this can best be imagined in a landscape where during a short period in spring, water from melting snow cannot percolate sufficiently fast through the soil. This can be caused by the relatively low hydraulic conductivity of the consolidated horizon or by a still frozen soil under the snow cover.

CONCLUSIONS

In the Zonien Forest, a very close relationship exists between geomorphological position and soil. Study of the Quaternary geology, geomorphology and soils suggests the following sequence of processes for the areas situated above the spring line (plateaux, slopes and dry valleys) which were never under intensive agriculture.

1 About 20 000 years BP, calcareous loess was deposited, partially levelling the pre-existing valleys. At the end of loess deposition mainly concave valley bottoms existed.
2 Progressive decalcification of the loess to 100–150 cm with the concurrent development of a strong illuvial clay-enriched horizon (B2t). No erosion or aggradation occurred in this period.
3 Towards the end of the Weichsel glaciation, probably before or in the Younger Dryas, there was a period when in spring concentrated runoff occurred in the lowest landscape positions. It is not yet known if the water came from a water table perched on the consolidated horizon or on a still frozen soil. Where the water flow became important, localized erosion occurred. This erosion created both V-shaped and the flat-bottomed gullies, formed by the lateral erosion of the temporary streams. The hydromorphic mottles in the soils date, at least partially, from this period. The soil consolidation developed during one or more of the coldest periods, when continuous permafrost was present.
4 During the Holocene no erosional processes were active. Soil genesis was limited to horizons above the consolidated layer — the upper 30–40 cm of the soils. Clay migration was only active in soils on very steep slopes in loess deposits eroded by the meandering streams during the previous erosion cycle.

From this sequence it can be concluded that in the last 20 000 years only one relatively short fluviatile erosion cycle has been active in areas of the loess belt protected from man's influence. This erosion was limited to part of the valley bottoms, all the rest of the landscape was remarkably stable. In most soils only the upper 30–40 cm of the profile reflects Holocene pedogenesis, the deeper horizons, up to 100–150 cm are associated with the periglacial environment that existed between 20 000 and 10 000 years BP.

Holocene colluvial sediments do not occur in the forested dry valley bottoms. Several characteristics of valley bottom soils are also relict from a particularly cold period towards the end of the Weichsel glaciation.

REFERENCES

Arnould-De Bontridder, O. and Paulis, L. (1966). 'Étude du ravinement Holocène en Forêt de Soignes', *Acta Geographica Lovaniensia*, **4**, 182–191.
Bolinne, A. (1976). 'L'évolution du relief à l'Holocène. Les processus actuels', in *Géomorphologie de la Belgique, hommage au Prof. P. Macar* (Ed. A. Pissart), pp. 159–168, Université de Liège.
Bolinne, A. (1977). 'La vitesse de l'érosion sous culture en région limoneuse', *Pédologie*, **27**, 191–206.
De Ploey, J. (1977). 'Some experimental data on slopewash and wind action with reference to Quaternary morphogenesis', *Earth Surface Processes*, **2**, 101–115.
De Ploey, J. (1979). 'Aktieve hellingserosie in België', *Bull. Belg. Ver. Geol.* **88**, 137–142.
Dudal, R. (1959). 'Bodemkaart van België. Kaart en verklarende tekst, Tervueren 102 E', *Bodemkaart van België*, IWONL-CVB, Ghent.
Gullentops, F. (1957). 'Quelques phénomènes géomorphologiques depuis le Pléni-Wurm', *Bull. Soc. Belg. de Géol.* **66**, 86–95.
Gullentops, F. and Arnould-De Bontridder, O. (1966). 'Compte rendu de l'excursion du lundi 13 juin 1966. Leuven-Haacht-Wavre-Tervueren-Leuven', in *Evolution des versants, cartographie, géomorphologie et dynamique fluviale*, **1**, 365–375, Un. Geogr. Int., Liège.
Haesaerts, P. and Bastin, B. (1977). 'Chronostratigraphie de la fin de la dernière glaciation, à la lumière des résultats de l'étude lithostratigraphique et palynologique du site de Maisière Canal (Belgique)', *Geobios*, **10**, 123–127.
Langohr, R. (1983a). 'The extension of permafrost in Western Europe in the period between 18,000 and 10,000 years B.P. (Tardiglacial): information from soil studies', in *Permafrost: Fourth international Conference, Proceedings*, pp. 683–688, National Academy Press, Washington, DC.
Langohr, R. (1983b). 'Définition et caractérisation de l'horizon consolidé d'origine périglaciaire. Discussion sur base de descriptions très détaillées de sols', in *Sols lessivés glossiques et fragipans, Nord de la France et Belgique*, pp. 58–62. Comm. pour l'étude des phénomènes périglaciaires, Paris.
Langohr, R. and Pajares, G. (1983). 'The chronosequence of pedogenetic processes in Fraglossudalfs of the Belgian loess belt', in *Soil Micromorphology* (Eds P. Bullock and C. P. Murphy), pp. 503–510, AB Academic Publishers, Berkhamsted.
Langohr, R. and Van Vliet, B. (1981). 'Properties and distribution of Vistulian permafrost traces in today's surface soils of Belgium, with special reference to the data provided by the soil survey', *Biuletyn Peryglacjalny*, **28**, 137–148.
Louis, A. (1959). 'Bodemkaart van België. Kaart en verklarende tekst, Uccle 102 W', *Bodemkaart van België*, IWONL-CVB, Ghent.

Mourlon, M. (1911). 'Texte explicatif du levé géologique de la planchette de Tervueren', Service Géologique de Belgique, Brussels.

Sanders, J., Langohr, R., Pajares, G. and De Corte, M. (1983). 'Topo-hydroséquence de sols dégradés à horizon consolidé sur limons récents dans la Forêt de Soignes', in *Sols lessivés glossiques et fragipan. Nord de la France et Belgique*, pp. 46–57, Comm. pour l'étude des phénomènes périglaciaires, Paris.

Smalley, I. J. and Davin, J. E. (1982). 'Fragipan horizons in soils: a bibliographic study and review of some of the hard layers in loess and other materials', *New Zealand Soil Bureau Bibliographic Report*, 30, Department of Scientific and Industrial Research New Zealand.

Soil Survey Staff (1975). *Soil Taxonomy* US Dept. Agr. Handbook 436.

Tavernier, R. and Louis, A. (1971). 'La dégradation des sols limoneux sous monoculture de hêtres de la forêt de Soignes (Belgique)', *An. Int. St. Cerc. Pédol.* 38, 165–191.

Thornthwaite, C. W. (1948). 'An approach toward a rational classification of climate', *Geogr. Rev.* 38, 55–94.

Van Ranst, E., De Coninck, F., Tavernier, R. and Langohr, R. (1982). 'Mineralogy in silty to loamy soils of Central and High Belgium in respect to autocthonous and allocthonous materials', *Bull. van de Belg. Ver. voor Geologie*, 91, 27–44.

Van Vliet, B. and Langohr, R. (1981). 'Correlation between fragipan and permafrost — with special reference to Weichsel silty deposits in Belgium and northern France', *Catena*, 8, 137–154.

Van Vliet, B. and Langohr, R. (1983). 'Evidence of disturbance by frost of pore ferri-argillans in silty soils of Belgium and northern France', in *Soil Micromorphology* (Eds P. Bullock and C. P. Murphy), pp. 511–518, AB Academic Publishers, Berkhamsted.

Soils and Quaternary Landscape Evolution
Edited by J. Boardman
© 1985 John Wiley & Sons Ltd.

16

Quaternary Paleosols as Indicators of the Changing Landscape on the Northern and Southern Piedmonts of the Alps

ARMELLE BILLARD

ABSTRACT

The paleosols and weathering characteristics in the upper part of several series of glacifluvial terraces on the northern and southern piedmonts of the Alps have been studied by means of field observations, mineralogical, micromorphological, and chemical analyses. The two regions display very different sequence of weathering indicative of a contrast in the paleoclimate and the phytogeographical assemblage under interglacial conditions. On the northern side of the Alps, the weathering remains in the bisiallitization range. The soil sequence begins with Lower Pleistocene paleosols with thick, sandy BC horizons. The Middle and Upper Pleistocene is characterized by reddish and brown lessivé soils. This is interpreted as a change from a hyper-oceanic environment comparable with the present coastal forest of Washington State, USA, to a drier continental climate. To the south of the Alps the lessivé paleosols of Lower Pleistocene age, with their thick, red Bt horizons, show weathering in the monosiallitization range. During the Middle and Upper Pleistocene, intermediate type red lessivé soils developed, followed by brown lessivé soils with weathering in the bisiallitization range. This indicates a change from subtropical conditions, comparable to the present environment in the Batum area of Georgia, USSR, to a drier climate with a moderate Mediterranean influence. The sharp changes affecting the landscape during the glacial stages produced degradation and frost features at the top of the interglacial soils.

INTRODUCTION

Studies of the paleosols developed on glacifluvial and glacial deposits making up moraines and terraces on both northern and southern piedmonts of the Alps have yielded information on the nature and zonation of climatic and biogeographic paleoenvironments in western Europe and the evolution of these zones from the Lower Pleistocene to the present.

Data reported in this paper are taken from two regions (Figure 16.1). The first is a sequence of terraces forming the Iller-Riss plateau in Swabia, south-west Germany, and the related moraines upstream of them. These moraines are the product of the Rhine ice sheet which advanced to the north-east of Lake Constance and discharged its outwash into the Danube basin. The second region lies on the southern piedmonts of the Alps in northern Italy and includes two areas, (a) the north Turin terraces located along the Stura di Lanzo, a left-bank tributary of the Po River, and (b) the north Milan terraces which can be traced upstream to the moraines left by the glaciers which occupied the Como trough.

Present climatic characteristics are as follows. In the south Ulm area, average annual temperature of 8 °C, and precipitation of 700 mm yr^{-1} in Ulm with a summer maximum (June, 90 mm; July 89 mm; August 84 mm). Annual temperature of 12.5 °C in the Italian areas; precipitation of 1100 mm yr^{-1} at Lanzo, to the north of Turin and 1400 mm yr^{-1} at Lentate-Seveso between Milan and Como, with a spring and autumn maximum (respectively at Lanzo and Lentate-Seveso, 161 and 179 mm in May; 138 and 136 mm in October), with

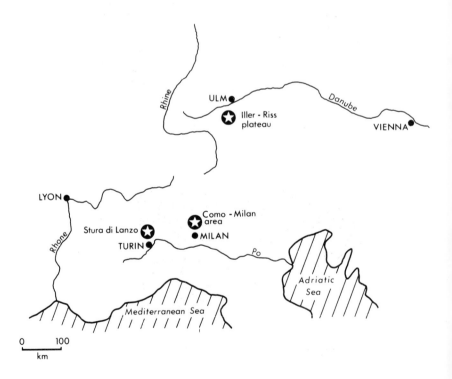

Figure 16.1 Location of the area studied

a secondary summer minimum (99, 81 and 89 mm in June, July and August at Lanzo and 131, 110 and 125 mm at Lentate-Seveso).

In each region paleosols are described from the topographic sequence consisting of very-high, high and low terraces, the results being based mainly on the following types of analysis: field observation (designation of soil horizons after Soil Survey Staff (1975)); micromorphology (in collaboration with N. Fedoroff; clay mineralogy, as determined by X-ray diffractometry and differential thermal analysis of the $<2\ \mu$m fraction of the soil matrix and from disintegrated pebbles of aluminosilicate rocks as well as analysis of isolated minerals sampled from the latter (analysis undertaken by the Laboratoire de Géomorphologie, CNRS, Caen, and the Laboratoire de Géographie Physique, LA 141 in Paris); chemical analysis (in collaboration with J. Dejou).

Two types of weathering are distinguished: 'bisiallitization' is characterized by 2:1 layer lattice clay minerals and 'monosiallitization' by 1:1 layer lattice clay minerals (Pedro *et al.*, 1969).

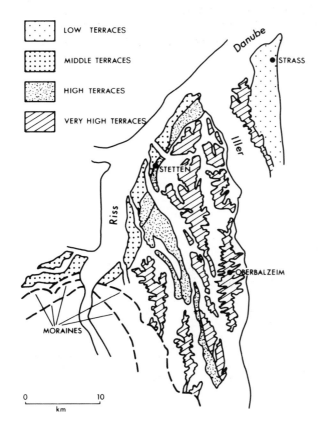

Figure 16.2   Location of sites and terraces in the Iller-Riss region

THE NORTHERN ALPINE PIEDMONT:
THE ILLER-RISS PLATEAU

The glacifluvial gravels of the Iller-Riss plateau in Swabia (Figure 16.2) are composed mainly of limestones, dolomitic limestones, calcareous sandstones, quartz siliceous sandstones and quartzites, schists, gneisses and amphibolites.

The very-high terraces, at least the highest of which (classically attributed to the Donau) are of Lower Pleistocene age, are characterized by thick decalcified paleosols first described by Brunnacker (in Graul, 1962; see also Leger, 1970). A typical example is the Oberbalzeim paleosol developed on the highest Iller-Riss terrace, the profile of which can be summarized (from top downwards) as follows (Figure 16.3).

1   Decalcified loam with scattered pebbles (0.30 m).
2   Colluvium with reworked weathered pebbles (0.50 m). Strong brown sandy, silty clay matrix (7.5 YR 5/6) with small grey, yellow and black patches and yellowish red (5 YR 4/6) coatings.
3.1 Weathered gravel with a Bt or B horizon developed *in situ* (0.20–0.30 m).

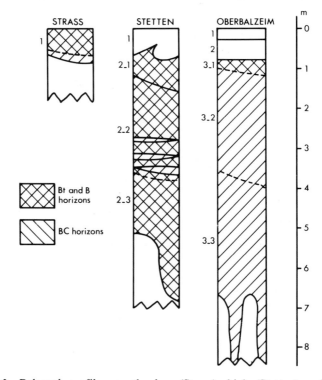

Figure 16.3 Paleosol profiles on the low (Strass), high (Stetten) and very high (Oberbalzeim) terraces on the Iller-Riss plateau in Swabia

Strong brown sandy, silty clay matrix (7.5 YR 5/6) with yellowish red coatings (5 YR 4/8).

3.2 Weathered gravel with a B to BC horizon formed *in situ* (2–3 m). Sandy, silty and slightly clayey, yellowish brown matrix (10 YR 5/8). Occasional thin strong brown coatings (7.5 YR 5/6) near the top.

3.3 Weathered gravel with BC horizon (3 m, reaching 5 m and more in weathering pipes). Brownish yellow, sandy silty matrix (10 YR 6/6).

Below this profile, at the base of the section, an unweathered gravel with a calcareous matrix and limestone pebbles can be observed between the weathering pipes. In the B and BC horizons no limestone or dolomitic limestone pebbles remain, and the calcareous pebbles are decalcified and porous. Even pebbles of gneiss and schist show advanced disintegration. Pebbles of amphibolite and amphibolitic gneiss with a thin, weathered, strong brown, soft cortex (7.5 YR 5/8), corresponding to the initiation of weathering of the amphiboles, are found in the uppermost 1 m of the gravel, the cortex becoming paler (10 YR 8/4) at depths of between 1 and 4 m.

The ratio of very thick, sandy BC horizons and thin argillaceous Bt horizons, which is such a specific feature of the paleosols developed on the highest terraces in Swabia, is reversed in the reddish brown, lessivé, 'ferretto soils' of the succeeding high terraces classically attributed to the Mindel and which belong to the Middle Pleistocene, the Stetten profile being an example.

The profile at Stetten consists of the following sequence, from top to bottom (Figure 16.3).

1 Strongly cryoturbated pale brown (10 YR 6/3) decalcified loam (0.40–0.80 m) with grey and yellow patches.

2.1 Weathered cryoturbated gravel (0.50–1 m) with Bt horizon. Strong brown sandy clay matrix (7.5 YR 5/6) with abundant yellowish red coatings (5 YR 5/6), black patches becoming continuous at the base of the horizon.

2.2 Weathered gravel (2.50 m) with Bt/B horizon. Strong brown sandy clay matrix (7.5 YR 5/6) with yellowish red coatings (5 YR 5/6) less abundant than above. Intercalation in the lower part, of very pale brown layers (10 YR 7/4) with sandy, silty matrix which form a transition to the BC horizon developed below.

2.3 Weathered gravel with BC horizon (1.5–3 m extending deeper along weathering pipes). Light yellowish brown sandy, silty matrix (10 YR 6/4) becoming very pale (2.5 Y 7/3) at the base.

At depth, the gravel is unweathered and contains limestone pebbles and a calcareous matrix.

In this Stetten paleosol, both the matrix and the pebbles of carbonate rocks are decalcified. The schists and gneisses have disintegrated, although not as strongly as at Oberbalzeim, but the amphibolite and amphibolitic gneisses remain essentially unweathered.

The soils of the low terrace series are of lessivé type although total thicknesses decrease progressively with the decreasing age of the glacifluvial deposit, the colour of the Bt horizon changing from reddish hues to brown. The low terraces which pass upstream into the end moraines referred to the last glacial stage (Würm in the Alps) reveal a lessivé soil about 0.70 m thick with a dark yellowish brown (10 YR 4/6) Bt horizon above a very thin decalcified sandy BC horizon. The Strass profile is one example (Figure 16.3). The schistose pebbles in the soil show exfoliation and the gneisses are weakly to moderately disintegrated.

Mineralogical analysis

The clay mineralogy of the paleosol matrix of the low terraces include illite, interstratified illite-vermiculite, Al-intergrade or pedogenic chlorite and a limited amount of kaolinite. As the latter mineral was not found in the disintegrated gneisses, it is concluded that it did not form *in situ* but is probably inherited from limestones and other carbonate-rich sedimentary rocks.

A similar association of minerals is found in the paleosols of the high and very high terraces but montmorillonite is also represented. However, vermiculite replaces the interstratified types and the amount of kaolinite markedly increases, particularly in the very high terraces. The kaolinite appears to have developed *in situ* in the weathered gneisses, the highest amounts occurring at Oberbalzeim (30–44 per cent of the $<2\,\mu$m fraction in the gneisses rich in plagioclase feldspar compared to 10–27 per cent at Stetten). It originates from weathering of the plagioclase while the weathering of the biotite has not passed the vermiculite and Al-intergrade stage.

Throughout the terrace sequence the weathering of the primary minerals seems to be differential depending on their original character, and more or less incomplete depending on the age of the glacifluvial sediments. On the whole, the weathering in the Iller-Riss plateau accords with the state called bisiallitization and occurs in both types of paleosol found in Swabia, namely the lessivé soils (from the brown lessivé soils on the low terraces to the reddish lessivé 'ferretto' soils of the high terraces) with marked Bt horizons; and the thick soils of the very high terraces characterized by deep, sandy, decalcified BC horizons and thin argillaceous Bt horizons.

THE SOUTHERN ALPINE PIEDMONTS: THE STURA DI LANZO AND COMO-MILAN AREA

The two areas studied in the southern piedmonts of the Alps show important differences in the petrographic composition of the gravel. In the Stura di Lanzo terraces (Figure 16.4), limestones and calcareous sandstones are absent but there are large amounts of basic and ultrabasic rocks such as serpentines, amphibolites

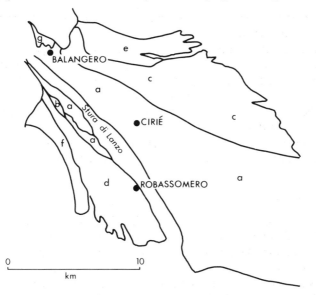

Figure 16.4   Location of sites and terraces in the Stura di Lanzo area. a, low terrace; b, middle terrace; c and d, high terraces (GL5 at Robassomero); e, f and g, very high terraces (GL8 at e, Vaud Grande and GL ⩾9 at g, Balangero). GL = glacial stage

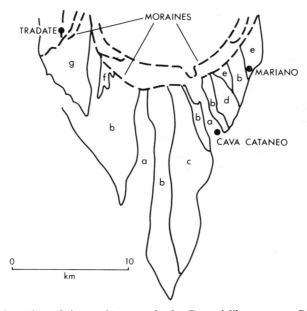

Figure 16.5   Location of sites and terraces in the Como-Milan area. a, Cava Cataneo (GL1); b, Copreno (GL2); c, Barlassina (GL3); d, Meda (GL4): e, Mariano (GL5); f, Lurago (GL6); g, Tradate (GL ⩾7). GL = glacial stage

and pyroxenites mixed with schists, gneisses, aplites and leptynites. In the Como-Milan area (Figure 16.5), limestones, calcareous sandstones and some dolomitic limestones occur along with quartzite, aluminosilicate rocks (gneisses, granites, diorites) and amphibolites. This is very similar to the petrographic composition described in the Iller-Riss plateau.

The Balangero Paleosol

This paleosol occurs in the Stura di Lanzo area on a very high terrace related to at least the ninth glacial stage BP (Billard, in press). The profile consists of the following units (Figure 16.6).

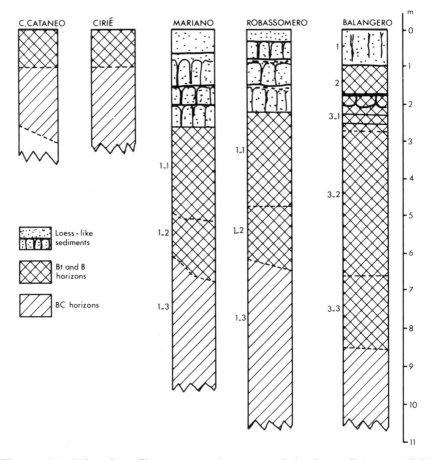

Figure 16.6 Paleosol profiles on stepped terraces of the Stura di Lanzo (Cirié, Robassomero, Balangero: see Figure 16.4) and the Como-Milan area (Cava Cataneo and Mariano: see Figure 16.5)

1   Reddish weathered loess like sediment (0.70–1 m).
2   Colluvially reworked weathered gravel with red matrix, 0.80 m in thickness (2.5 YR 5/8).
3.1 Weathered gravel (0.90 m) with Bt horizon developed *in situ* and subsequently degraded. Strongly argillaceous red matrix (10 R 5/8) becoming brownish yellow (10 YR 6/8) along dense sub-horizontal fissures filled with white clay and silt (10 YR 8/1). This material is deformed at the top by frost features. There are abundant black patches, mainly located around the pebbles.
3.2 Weathered gravel with Bt horizon (3.80 m). Argillaceous red matrix (2.5 YR 5/8) becoming reddish yellow (5 YR 6/8) towards the base. Very abundant red coatings (10 R 5/6 becoming 2.5 YR 5/8 at the base). Concentration of dark red coatings (2.5 YR 3/6) and black patches around the pebbles.
3.3 Weathered gravel with B horizon (2 m). Reddish yellow matrix (5 YR 6/8) becoming yellower (7.5 YR 6/6) towards the base.

Below this sequence the gravel is still strongly weathered with a moderately argillaceous very pale brown matrix (10 YR 7/4) down to a depth of more than 15 m. The main analytical results are as follows.

In the matrix, the ratio SiO_2/Al_2O_3 is about 2 at a depth of 2–5 m. The matrix fraction $<2 \mu$m is composed of kaolinite (30–50 per cent), Al-intergrade, illite and/or sericite down to a depth of 12 m. All pebbles, except those of quartz, are severely weathered and are as soft as the matrix. All chloritic schists and basic and ultrabasic pebbles shown destruction of the ferromagnesian minerals. Release of iron ranges from 40–100 per cent; the rocks have become friable with a marked loss of density and complete loss of the original structure: they have a characteristically bright red colour (10 R 5/8; 2.5 YR 5/8). These transformations affect entire pebbles and cobbles, or a cortex of about 10 cm on larger clasts in stratum 3.2 (and more in stratum 3.1), the thickness of the weathering rinds decreasing with depth in the profile. The aluminosilicate rocks are strongly fragmented and have a much-reduced bulk density to a depth of about 5 m. The fraction finer than 2μm taken from samples of gneiss is made up of kaolinite (30–50 per cent), illite and/or sericite with a mixture of Al-intergrade in the biotite-rich gneisses. Micromorphology and clay mineralogy of the soil shows an evolutionary sequence of primary minerals in alumino-silicate rocks, summarized as follows, from the base to the top of the profile: biotite → illite-vermiculite → aluminous vermiculite → kaolinite; muscovite → kaolinite only at the very top of the profile (remnants of A_2 horizon); plagioclase → kaolinite (completely destroyed in the upper part of the red horizons); sericite (included in plagioclase) → Al-intergrade → kaolinite; K feldspars is fissured in the upper part of the red horizons; quartz is fissured only at the very top of the profile (remnants of A_2 horizon).

It thus appears that most minerals are transformed into kaolinite at different rates, in different amounts, and at different depths in the profile. In contrast to the prevailing bisiallitization at Oberbalzeim in Swabia, there is a general tendency to monosiallitization at Balangero on the southern side of the Alps.

During pedogenesis successive phases of illuviation occurred, the chronology of which can be clearly observed in thin section owing to the absence of pedoturbation (Billard and Fedoroff, 1977). The argillans are of three types as follows:

1   Yellow argillans which reach the base of the profile (15–20 m depth) and are never found in the voids of the weathered minerals.
2   Red argillans reaching the base of the red horizon (about 5 m depth) and partly filling the voids of the weathered minerals in the basic, ultrabasic and aluminosilicate rocks.
3   Grey hydromorphous argillans at the top of the profile in the sub-horizontal fissures.

Two conclusions can be deduced from the study of illuviation. The fact that only red argillans fill the voids of the weathered minerals including those of very resistant rocks shows that these argillans are contemporaneous with the most intense weathering phase. Secondly, as they are not found very deep in the profile, it appears that their deposition, which produced a strongly developed Bt horizon, was the result of heavy rainfall followed by rapid evopotranspiration. It is concluded that the weathering and pedogenesis observed at Balangero developed under warm and wet climatic conditions. More specifically, and considering the latitude, a climate with a warm and wet summer appears to be a necessary condition for their development.

The Tradate Paleosol

A similar assemblage of pedogenetic and weathering features has been observed in the Como–Milan area, at the top of the Tradate till (Figure 16.5) which appears to correlate with at least the seventh glacial stage before the present (Billard, in press). All aluminosilicate rocks are severely weathered and give high percentages of kaolinite mixed with illite and sometimes vermiculite. The microdiorites, amphibolites and calcareous sandstones are weathered to a red colour throughout at the top of the profile and exhibit a red cortex at depths of more than 10 m.

Some of these features are evident at the top of the high terraces in both the Stura di Lanzo and the Como–Milan areas, such as the Robassomero and Mariano terraces respectively, both related to the fifth glacial stage before the present (Billard, in press).

The Robassomero Paleosol

At the top of a high Stura di Lanzo terrace, four loess-like sediments reaching a total thickness of about 2 m, bury the Robassomero paleosol developed on the underlying gravels with a weathering profile as follows (Figure 16.6).

1.1   Weathered gravel (2.60 m) with Bt horizon developed *in situ*. Argillaceous yellowish red matrix (5 YR 5/6) with abundant red coatings (2.5 YR 4/6) in the upper part, becoming yellower (5 YR 5/6) and less abundant below.
1.2   Weathered gravel (1.50 m) with B horizon. Strong brown sandy clay matrix (7.5 YR 5/6) with occasional strong brown coatings (7.5 YR 5/6).

Down to a depth of 5 m the matrix remains slightly argillaceous, with some brown coatings (7.5 YR 5/4) mainly located around the pebbles. The $<2\,\mu m$ fraction extracted from the matrix or the aluminosilicate rocks (mainly gneisses, all of which have disintegrated) is composed of a mixture of illite, chlorite, vermiculite or interstratified illite-vermiculite or aluminous vermiculite and/or weathered chlorite, with kaolinite less abundant than it is at Balangero. The kaolinite originated mainly from the plagioclase: a biotite origin remains to be demonstrated. At the top of the profile, pebbles of basic and ultrabasic rocks and chloritic schists show a soft red cortex (2.5 YR 4/8) about 5 mm thick, in which the chlorite and all the ferromagnesian minerals have been destroyed. However, the centres of these pebbles remain fresh or only slightly weathered.

The Mariano Paleosol

The paleosol, developed at the top of the high Mariano terrace in the Como–Milan area and buried by loess-like silts, is as follows (Figure 16.6).

1.1   Overbank loam (0.70 m) including scattered, small, weathered pebbles, overlying gravel. In both sediments the same Bt horizon is developed to a total thickness of about 2.50 m. The matrix is argillaceous red to reddish yellow in colour (2.5 YR 5/8 to 5 YR 6/6), with red coatings (2.5 YR 5/8) more abundant in the upper part of the horizon, and black patches.
1.2   Weathered gravel (1.40 m) with B horizon. Sandy silty clay, reddish yellow matrix (5 YR 6/6) at the top, becoming less red (7.5 YR 6/6) at the base with occasional red coatings (2.5 YR 5/6).
1.3   Weathered gravel with BC horizon. Sandy silty light yellowish brown matrix (10 YR 6/4).

The gravel is decalcified to a depth of more than 11 m. The pebbles of granite, diorite and gneiss are in a disintegrated state, some throughout and others with a cortex surrounding a more resistant centre. Microdiorite and amphibolite pebbles have a thin (5–10 mm) soft red cortex (2.5 YR 5/8) surrounding a fresh or slightly weathered centre, the colour of which, at a depth of 10 m, changes to yellow. Some decalcified sandstones, weathered and red throughout, occur to a depth of 5 m.

In the Mariano paleosol, the $<2\,\mu$m matrix fraction is composed of illite, vermiculite, gibbsite and kaolinite. A similar composition was found in the $<2\,\mu$m fraction extracted from diorite and granite pebbles: gibbsite appears in the centre of pebbles and decreases or disappears in the more weathered cortex where it is replaced by kaolinite. These two minerals originate mainly from the weathering of plagioclase. Biotite transformation mainly yields vermiculite and limited amounts of kaolinite in the upper part of the profile.

The soils on the low terraces

There is a progressive change from the red lessivé Robassomero and Mariano paleosols to the brown lessivé soils in the middle and low terraces (Billard, in press).

In the Stura di Lanzo area, the Holocene Cirié soil (Figure 16.6) developed on the gravel of the low terrace, exhibits Bt and B horizons 1 m thick with dark brown matrix (10 YR 4/3) and coatings (7.5 YR 4/4). A BC horizon, with a paler sandy matrix (10 YR 6/4), is seen below at a depth of at least 2 m. The basic and ultrabasic rocks remain unweathered or show a thin, pale and still hard cortex resulting from weathering of the feldspars. The schist pebbles are exfoliated. The $<2\,\mu$m fraction extracted from the fragmented gneissic pebbles is composed of illite, vermiculite and/or interstratified illite-vermiculite and small amounts of kaolinite resulting from partial weathering of the plagioclases.

In the Como–Milan area, the Cava Cataneo profile (Figure 16.5), developed in the upper part of a low terrace relating to the last glacial stage, shows a gravel decalcified to depths of 2.5–3 m (Figure 16.6). The matrix is strong brown (7.5 YR 5/6) at the top, becoming paler (10 YR 7/4) towards the base but the amphibolite and microdiorite pebbles are fresh throughout. Weathering has mainly affected the carbonate rocks and the granites and gneisses, most of which have disintegrated. The $<2\,\mu$m fraction, extracted from the matrix and from the aluminosilicate pebbles, shows a combination of illite, vermiculite, interstratified illite-vermiculite, and gibbsite with a small amount of kaolinite, the latter two minerals resulting from partial weathering of the plagioclase.

THE SUPERIMPOSITION OF FROST-DEGRADED SOILS

Frost features are not found in the upper part of Holocene soils developed on the gravels of the low terraces and the Würm loess but, as already indicated, they are seen in the upper part of the older profiles both in southern Germany, at Stetten, and in northern Italy, at Balangero.

Frost-degraded soils with A_2, A and B, and Ct horizons are found superimposed on the argillic Bt horizons of the interglacial lessivé soils, each one corresponding to a glacial-interglacial cycle (Billard and Fedoroff, 1977; Billard, in press). They show the following characteristics.

1 A platy structure owing to the segregation of subhorizontal ice lenses (Fedorova and Yarilova, 1972; Van Vliet-Lanoë, 1976).
2 Strong leaching of iron, clay and silt at the top of the interglacial Bt horizon, giving rise to the formation of the pale grey bleached glossic horizon.
3 Thick accumulation of coarse clay and fine silt in the Ct horizon which is never found integrated with the matrix.

These processes have been observed operating today in Lapland (N. Fedoroff, pers. comm.). They occur in a boreal type of pedogenesis which requires abundant snowfall followed by rapid melting during spring or summer.

Following boreal pedogenesis, there is evidence for two additional processes which appear to indicate a more severe periglacial environment.

1 Ice-wedge casts at the top of the lessivé soil subsequently degraded and probably truncated, developed on gravels or gravels and colluvium such as at Balangero, and on loess-like sediments.
2 Solifluction of reworked gravels, or loess sedimentation burying the lessivé frost-degraded soil. This is assumed to have taken place during the latter part of the glacial stages.

DISCUSSION AND CONCLUSIONS

In spite of great differences in the petrographic composition of the gravels, a similar sequence of interglacial pedogenesis and weathering is observed in the upper part of the glaciofluvial and glacial sediments in the Stura di Lanzo and Como–Milan areas, yet both are quite different from the Iller-Riss plateau sequence. It should be emphasized that the red or reddish lessivé, so-called 'ferretto' soils do not represent the same stratigraphic unit on either side of the Alps. In northern Italy they are developed on the highest terraces and on old moraines of Lower Pleistocene and/or lower Middle Pleistocene age, with thinner intergrade red soils on the Middle Pleistocene high terraces. In Swabia they are found only on the Middle Pleistocene high terraces.

The red lessivé 'ferretto' soils of northern Italy have specific characteristics which clearly differentiate them from the soils referred to as 'ferretto' in the piedmonts of France (Bornand, 1978; Billard, unpublished) and Germany (see above). In summary, these characteristics, as already described from Balangero, are as follows.

1   Softness and significant loss of density in most of the pebbles including those of very resistant rocks.
2   A red coloured matrix and red weathering of the basic and ultrabasic pebbles in which the ferromagnesian minerals have been destroyed, resulting in very severe loss of iron.
3   Formation of kaolinite from numerous minerals other than plagioclase, particularly biotite (also sericite and muscovite in the highest terraces).

These weathering features are also found at the top of the high terraces with a reduced profile thickness and with only a thin red cortex on the basic pebbles. They developed in only a single interglacial period at Robassomero and Mariano before fossilization of the soils occurred under superposed loess-like sediments, on each of which developed a brown lessivé soil which was subsequently degraded by frost action. However, the weathering found in the Robassomero and Mariano buried soils is significantly more advanced than that at the top of the unburied paleosols developed on the very high Swabian terraces. This indicates, in the case of Swabia, that a long period of soil evolution did not compensate for the more limited effects of less aggressive weathering.

The Quaternary paleosols developed on the glacifluvial terrace sequences in the alpine piedmonts appear to be representative of a climatic paleoenvironment of interglacial type. They reflect the paleoclimatic and hence the phytogeographical contrast between the northern and southern piedmonts. This agrees with the palynological data from the Lower Pleistocene obtained at Leffe in northern Italy (Lona, 1950, 1963; Billard *et al.*, 1983) and at Uhlenberg in southern Germany (Filzer and Scheuenpflug, 1970; Frenzel, 1973; Brunnacker *et al.*, 1982).

In addition, the differences noted in successive paleosols on both sides of the Alps indicate progressive modification of the climate and paleoenvironment from the Lower Pleistocene to the Holocene. Profile descriptions, chemical and mineralogical analyses of red lessivé 'ferretto' paleosols of northern Italy show striking similarities to the soil described in the upper part of the terrace (dated at between 4000 and 5000 YR BP) in the Batum area of Georgia, USSR (Cernjakovskij, 1968). This region has high precipitation, 2400 mm yr^{-1} at Batumi, with summer maximum (200–300 mm in August). It is this type of regime which appears to be required to explain the characteristics of the Balangero paleosol, i.e. a wet sub-tropical climate.

There appears to have been a progressive change towards mild, drier climates in the Middle Pleistocene with a decrease in summer precipitation owing to a moderate Mediterranean influence.

In contrast to Balangero, the Oberbalzeim paleosol shows only moderate weathering of all rocks and minerals and a thin Bt horizon. A similar type of pedogenesis (Billard, unpublished) can be observed in the coastal forest of Washington State, USA (Eyre, 1977) in Holocene soils developed on fluvial gravels of late Wisconsin age (Easterbrook and Rahm, 1970) in an area of hyper-oceanic climate with a relatively cool summer. In such environmental conditions, weathering remains moderate and the evapotranspiration does not compensate for the high percolation resulting from the heavy rainfall which prevents the formation of Bt horizons. The present temperate and rather drier continental climate in southern Germany seems to have become established in the Middle Pleistocene. A tentative explanation offered for the formation of the Bt horizon from the high to the low terraces is a rainfall regime with a summer maximum which enhanced the evopotranspiration process.

In both piedmonts of the Alps the changes in climate and vegetation occurred simultaneously with geomorphological changes such as the formation of the terrace sequence and the emplacement of the morainic arcs. The progressive change in the climatic and vegetal environment was sharply disrupted by the effects of the glacial stages. The periglacial conditions which developed as successive ice sheets invaded the piedmonts are recorded in the paleosols developed on the gravels and on the overlying loess-like sediments. These cyclic, abrupt changes in both climate and landscape are an integral part of the general framework of long-term variation which characterized western Europe during the Quaternary period.

REFERENCES

Billard, A. (in press). 'Quaternary chronologies around the Alps', in *Correlation of Quaternary chronologies* (Ed. W. Mahaney), Geo Books, Norwich.

Billard, A. and Fedoroff, N. (1977). 'Interglacial-periglacial pedogenetic cycle in Northern Italy and France', *X INQUA Congress Birmingham, Abs.* 35.

Billard, A., Bucha, V., Horacek, J. and Orombelli, G. (1983). 'Preliminary paleomagnetic investigations on the Pleistocene sequences in Lombardy, Northern Italy', *Riv. Ital. Paleont. Strat.* 83, 2, 295–317.

Bornand, M. (1978). *Altération des matériaux fluvioglaciaires, genèse et évolution des sols sur terrasses quaternaires dans la Moyenne vallée du Rhone.* Thèse Doct. Etat USTL Montpellier, Publ. SES, Montpellier.

Brunnacker, K., Lösher, M., Tillmanns, W. and Urban, B. (1982). 'Correlation of the Quaternary terrace sequence in the Lower Rhine valley and the northern Alpine foothills of Central Europe', *Quaternary Research*, 18, 152–173.

Cernjakovskij, A. G. (1968). 'Extension et âge de l'horizon d'altération en Transcaucasie occidentale', *Dokl. Akad. Nauk S.S.S.R., Ser. Geol.* 182, 1, 171–174.

Easterbrook, D. and Rahm, D. A. (1970). *Landforms of Washington. The Geologic Environment*, Union Printing Company, Bellingham, Washington.

Eyre, S. R. (1977). *Vegetation and Soils, a World Picture*, Edward Arnold, London.

Fedorova, N. N. and Yarilova, E. A. (1972). 'Morphology and genesis of prolonged seasonally frozen soils of Western Siberia', *Geoderma*, 7, 1–13.

Filzer, P. and Scheuenpflug, L. (1970). 'Ein frühpleistozänes Pollenprofil aus dem nördlichen Alpenvorland', *Eiszeitalter u. Gegenwart*, 21, 22–32.

Frenzel, B. (1973). 'Some remarks on the Pleistocene vegetation', *Eiszeitalter u. Gegenwart*, 23/24, 281–292.

Graul, H. (1962). 'Eine Revision des pleistozänen Stratigraphie des schwäbischen Alpenvorlandes (Mit einem bodenkundlichen Beitrag von K. Brunnacker)', *Pet. Mitt.* **106**, 253–271.

Leger, M. (1970). 'Paleosols quaternaires de l'avant-pays au Nord des Alpes', *Bull. A.F.E.Q.* **2/3**, 167–178.

Lona, F. (1950). 'Contributi alla storia della vegetazione e del clima nella Val Padana. Analisi pollinica del giacimento villafranchiano di Leffe (Bergamo)', *Atti Soc. It. Sc. Nat.* **89**, 123–178.

Lona, F. (1963). 'Floristic and glaciologic sequence (from Donau to Mindel) in a complete diagram of the Leffe deposit', *Ber. Geobot. Inst. E.T.H. Stiftg Rubel*, **34**, 64–66.

Pedro, C., Jamagne, M. and Begon, J. C. (1969). 'Mineral interactions and transformations in relation to pedogenesis during the Quaternary', *Soil Sci.* **107**, 6, 462–469.

Soil Survey Staff (1975). *Soil Taxonomy*, US Department of Agriculture, Soil Conservation Service, Agriculture Handbook 436.

Van Vliet-Lanoë, B. (1976). 'Traces de ségrégation de glace en lentilles associées aux sols et phénomènes periglaciaires fossiles', *Biul. Peryglacjalny*, **26**, 41–54.

Index

Subjects indexed are restricted to major references. Locations and authors are not included.